METHODS IN MOLECULAR BIOLOGY

Series Editor
John M. Walker
School of Life Sciences
University of Hertfordshire
Hatfield, Hertfordshire, AL10 9AB, UK

For further volumes:
http://www.springer.com/series/7651

Cardiac Tissue Engineering

Methods and Protocols

Edited by

Milica Radisic

Institute of Biomaterials and Biomedical Engineering, Department of Chemical Engineering and Applied Chemistry, University of Toronto, Toronto, ON, Canada; Toronto General Research Institute, University Health Network, Toronto, ON, Canada

Lauren D. Black III

Department of Biomedical Engineering, Tufts University, Medford, MA, USA; Cellular, Molecular and Developmental Biology Program, Sackler School of Graduate Biomedical Sciences, Tufts University School of Medicine, Boston, MA, USA

 Humana Press

Editors
Milica Radisic
Institute of Biomaterials
  and Biomedical Engineering
Department of Chemical Engineering
  and Applied Chemistry
University of Toronto
Toronto, ON, Canada

Toronto General Research Institute
University Health Network
Toronto, ON, Canada

Lauren D. Black III
Department of Biomedical Engineering
Tufts University
Medford, MA, USA

Cellular, Molecular and Developmental
  Biology Program, Sackler School
  of Graduate Biomedical Sciences
Tufts University School of Medicine
Boston, MA, USA

Videos to this book can be accessed at http://www.springerimages.com/videos/978-1-4939-1046-5

ISSN 1064-3745   ISSN 1940-6029 (electronic)
ISBN 978-1-4939-1046-5   ISBN 978-1-4939-1047-2 (eBook)
DOI 10.1007/978-1-4939-1047-2
Springer New York Heidelberg Dordrecht London

Library of Congress Control Number: 2014940960

Printed on acid-free paper

Humana Press is a brand of Springer
Springer is part of Springer Science+Business Media (www.springer.com)

Preface

Human hearts have a limited regenerative potential, motivating the development of the alternative treatment options for the conditions that result in the loss of beating cardiomyocytes. An example is myocardial infarction that results in a death of tens of millions of ventricular cardiomyocytes that cannot be replaced by the body. It is estimated that five to seven million patients live with myocardial infarction in North America alone. A majority of these patients do not need a surgical intervention, and medical management provides satisfactory results. However, over a period of 5 years, one-half of the patients who experience a myocardial infarction will develop heart failure, ultimately requiring heart transplantation.

The long-term goal of cardiac tissue engineering is to provide a living, beating, ideally autologous, and non-immunogenic myocardial patch that can restore the contractile function of the failing heart. The engineered tissues could also be used for preclinical drug testing to discover new targets for cardiac therapy and eliminate drugs, cardiac and noncardiac, with serious side effects. It generally involves a combination of suitable cell types, human or nonhuman cardiomyocytes and supporting cells, with an appropriate biomaterial made out of either synthetic or natural components and cultivation in an environment that reproduces some of the complexity of the native cardiac environment (e.g., electrical, mechanical stimulation, passive tension, or topographical cues).

This field is still young. The term cardiac tissue engineering usually refers to engineering of myocardial wall in vitro using living and beating cardiomyocytes. The pioneering papers appeared in the late 1990s, and they all utilized either neonatal rat cardiomyocytes or embryonic chick cardiomyocytes as a cell source. Since then, the field has matured significantly to include a range of approaches that all give cardiac tissues in vitro that are capable of developing contractile force and propagating electrical impulses. Advances in human embryonic stem cell research and induced pluripotent stem cell technology now provide the possibility of generating millions of bona fide human cardiomyocytes. When research in cardiac tissue engineering started some 25 years ago, the issue of a human cell source appeared insurmountable; however the researchers continued to make way, and there are many reports now on the use of human pluripotent stem cells as a source of cardiomyocytes for cardiac tissue engineering. Although early researchers thought that having purified cardiomyocytes in three-dimensional structures would be beneficial, based on analogies with monolayer studies where fibroblasts overgrow cardiomyocytes, there is a consensus in the field now that a mixed cell population is optimal for maintenance of cardiac phenotype and survival of cardiomyocytes in engineered tissues both in vitro and in vivo. The mixed population usually contains cardiomyocytes, endothelial cells, and a stromal cell type such as fibroblasts or mesenchymal stem cells. Also, there is a consensus that a form of physical stimulation, either mechanical or electrical, or passive tension is required for cardiomyocytes to achieve and maintain a differentiated phenotype and in vivo-like functional properties during in vitro cultivation.

This book gathers for the first time a collection of protocols on cardiac tissue engineering from pioneering and leading researchers around the globe. Protocols related to cell preparation, biomaterial preparation, cell seeding, and cultivation in various systems are provided.

Our goal is to enable adoption of these protocols in laboratories that are interested in entering the field as well as enable transfer of knowledge between laboratories that are already in this field. We hope that these efforts will lead to standardization, definition of best practices in cardiac tissue cultivation, and direct comparison of various production protocols using controlled in vivo studies that would ultimately lead to translational efforts. Although biomaterial patches alone and hydrogels have been investigated in clinical studies focused on myocardial regeneration, a cardiac patch based on living, beating human cardiomyocytes has not yet been tested in humans. Only patches based on non-cardiomyocytes have been tested in humans with mixed results. Bringing a new therapy to the clinic is an overwhelming task, one that we must approach in a collaborative rather than competitive spirit. We hope that sharing of the best protocols in cardiac tissue engineering will enable this goal.

Toronto, ON, Canada   *Milica Radisic*
Medford, MA, USA   *Lauren D. Black III, Ph.D.*

Contents

Contributors

CHRISTINA M. AMBROSI, PH.D. • *Department of Biomedical Engineering, Institute for Molecular Cardiology, Stony Brook University, Stony Brook, NY, USA*

PETER H. BACKX • *Department of Physiology and Medicine, University of Toronto, Toronto, ON, Canada; The Heart and Stroke/Richard Lewar Centre of Excellence, Toronto, ON, Canada; Division of Cardiology, University Health Network, Toronto, ON, Canada*

NIMA BADIE • *Department of Biomedical Engineering, Duke University, Durham, NC, USA*

CELINE L. BAUWENS • *Centre for Commercialization of Regenerative Medicine, Toronto, ON, Canada*

ALEXANDRA BERDICHEVSKI • *Faculty of Biomedical Engineering, Technion—Israel Institute of Technology, Haifa, Israel*

LAUREN D. BLACK III, PH.D. • *Department of Biomedical Engineering, Tufts University, Medford, MA, USA; Cellular, Molecular and Developmental Biology Program, Sackler School of Graduate Biomedical Sciences, Tufts University School of Medicine, Boston, MA, USA*

REBECCA L. BRADEN, M.S. • *Department of Bioengineering, University of California San Diego, La Jolla, CA, USA; Sanford Consortium for Regenerative Medicine, La Jolla, CA, USA*

NENAD BURSAC, PH.D. • *Department of Biomedical Engineering, Duke University, Durham, NC, USA*

AARON CHEN • *Department of Chemical Engineering and Materials Science, University of California, Irvine, CA, USA*

KAREN L. CHRISTMAN, PH.D. • *Department of Bioengineering, University of California San Diego, La Jolla, CA, USA; Sanford Consortium for Regenerative Medicine, La Jolla, CA, USA*

SMADAR COHEN, PH.D. • *Avram and Stella Goldstein-Goren Department of Biotechnology Engineering, The Center for Regenerative Medicine and Stem Cell (RMSC) Research, Ben-Gurion University of the Negev, Beer-Sheva, Israel; The Ilse Katz Institute for Nanoscale Science and Technology, Ben-Gurion University of the Negev, Beer-Sheva, Israel*

SUZANNE CROWE • *Division of Cardiac Surgery, University of Ottawa Heart Institute, Ottawa, ON, Canada*

MICHAEL E. DAVIS, PH.D. • *Wallace H. Coulter Department of Biomedical Engineering, Emory University and Georgia Institute of Technology, Atlanta, GA, USA*

AMY L. DE JONGH CURRY • *Department of Biomedical Engineering, University of Memphis, Memphis, TN, USA*

SANJIV DHINGRA • *Regenerative Medicine Program, Institute of Cardiovascular Sciences, St. Boniface Research Centre, University of Manitoba, Winnipeg, MB, Canada*

SEBASTIAN DIECKE • *Lorry I. Lokey Stem Cell Research Building, Stanford University School of Medicine, Stanford, CA, USA*

YI DUAN-ARNOLD • *Department of Biomedical Engineering, Columbia University, New York, NY, USA*

ALEXANDRA EDER • *Department of Experimental Pharmacology and Toxicology, University Medical Center Hamburg-Eppendorf (UKE), Hamburg, Germany; DZHK (German Centre for Cardiovascular Research), Hamburg, Germany*

EMILIA ENTCHEVA, PH.D. • *Department of Biomedical Engineering, Institute for Molecular Cardiology, Stony Brook University, Stony Brook, NY, USA*

THOMAS ESCHENHAGEN • *Department of Experimental Pharmacology and Toxicology, University Medical CenterHamburg-Eppendorf (UKE), Hamburg, Germany; DZHK (German Centre for Cardiovascular Research), Hamburg, Germany*

DOMINIC FILICE • *Department of Bioengineering, Institute for Stem Cells and Regenerative Medicine, University of Washington, Seattle, WA, USA*

KRISTIN M. FRENCH • *Wallace H. Coulter Department of Biomedical Engineering, Emory University and Georgia Institute of Technology, Atlanta, GA, USA*

DONALD O. FREYTES • *New York Stem Cell Foundation, New York, NY, USA*

LIOR GEPSTEIN • *The Sohnis Family Research Laboratory for Cardiac Electrophysiology and Regenerative medicine, the Bruce Rappaport Faculty of Medicine, Technion—Israel Institute of Technology, Haifa, Israel*

ARNE HANSEN • *Department of Experimental Pharmacology and Toxicology, University Medical CenterHamburg-Eppendorf (UKE), Hamburg, Germany; DZHK (German Centre for Cardiovascular Research), Hamburg, Germany*

YUJI HARAGUCHI • *Institute of Advanced Biomedical Engineering and Science, TWIns, Tokyo Women's Medical University, Shinjuku-ku, Tokyo, Japan*

TODD D. JOHNSON, M.S. • *Department of Bioengineering, University of California San Diego, La Jolla, CA, USA; Sanford Consortium for Regenerative Medicine, La Jolla, CA, USA*

MAKOTO KANEKO • *Department of Mechanical Engineering, Graduate School of Engineering, Osaka University, Yamadaoka, Suita, Japan*

MICHELLE KHINE, PH.D. • *Department of Biomedical Engineering, University of California, Irvine, CA, USA; Department of Chemical Engineering, University of California, Irvine, CA, USA*

NIGEL G. KOOREMAN • *Lorry I. Lokey Stem Cell Research Building, Stanford University School of Medicine, Stanford, CA, USA*

MICHAEL A. LAFLAMME, M.D., PH.D. • *Department of Pathology, Institute for Stem Cells and Regenerative Medicine, University of Washington, Seattle, WA, USA*

EUGENE LEE • *Department of Biomedical Engineering, University of California, Irvine, CA, USA*

AYELET LESMAN • *Department of Chemistry and Chemical Engineering, California Institute of Technology, Pasadena, CA, USA; Department of Biomedical Engineering, Technion—Israel Institute of Technology, Haifa, Israel*

SHULAMIT LEVENBERG • *Department of Biomedical Engineering, Technion—Israel Institute of Technology, Haifa, Israel*

REN-KE LI, M.D., PH.D. • *Division of Cardiovascular Surgery, Toronto General Research Institute, University Health Network, Toronto, ON, Canada; Division of Cardiac Surgery, Department of Surgery, University of Toronto, Toronto, ON, Canada*

JUN LIAO, PH.D. • *Tissue Bioengineering Laboratory, Department of Biological Engineering, Mississippi State University, Mississippi State, MS, USA*

LESZEK LISOWSKI • *Gene Transfer, Targeting and Therapeutics Facility, Salk Institute for Biological Studies, San Diego, CA, USA*

JIE LIU • *Department of Physiology and Medicine, University of Toronto, Toronto, ON, Canada; Materials Science, University of California, Irvine, CA, USA*

KATSUHISA MATSUURA • *Institute of Advanced Biomedical Engineering and Science, TWIns, Tokyo Women's Medical University, Shinjuku-ku, Tokyo, Japan*

BRIAN MCNEILL • *Division of Cardiac Surgery, University of Ottawa Heart Institute, Ottawa, ON, Canada*

LUKE MCSPADDEN • *Department of Biomedical Engineering, Duke University, Durham, NC, USA*

NICOLE MENDOZA • *Department of Biomedical Engineering, University of California, Irvine, CA, USA*

TIM MEYER • *Institute of Pharmacology, Heart Research Center Göttingen (HRCG), University Medical Center Göttingen, Göttingen, Germany; DZHK (German Center for Cardiovascular Research), partner site Göttingen, Göttingen, Germany*

JASON W. MIKLAS • *Institute of Biomaterials and Biomedical Engineering, University of Toronto, Toronto, ON, Canada*

IRIS MIRONI-HARPAZ • *Faculty of Biomedical Engineering, Technion—Israel Institute of Technology, Haifa, Israel*

KATHY Y. MORGAN • *Department of Biomedical Engineering, Tufts University, Medford, MA, USA*

HUNG NGUYEN • *Department of Biomedical Engineering, Duke University, Durham, NC, USA*

SARA S. NUNES • *Institute of Biomaterials and Biomedical Engineering, University of Toronto, Toronto, ON, Canada; Toronto General Research Institute, University Health Network, Toronto, ON, Canada*

TERUO OKANO • *Institute of Advanced Biomedical Engineering and Science, TWIns, Tokyo Women's Medical University, Shinjuku-ku, Tokyo, Japan*

JOHN D. O'NEILL • *Department of Biomedical Engineering, Columbia University, New York, NY, USA*

ALEKSANDRA OSTOJIC • *Division of Cardiac Surgery, University of Ottawa Heart Institute, Ottawa, ON, Canada*

NATHAN J. PALPANT • *Department of Pathology, Institute for Stem Cells and Regenerative Medicine, University of Washington, Seattle, WA, USA*

DAWN PEDROTTY • *Department of Biomedical Engineering, Duke University, Durham, NC, USA*

BORIS POLYAK, PH.D. • *Department of Surgery, Drexel University College of Medicine, Philadelphia, PA, USA; Department of Pharmacology and Physiology, Drexel University College of Medicine, Philadelphia, PA, USA*

MILICA RADISIC • *Institute of Biomaterials and Biomedical Engineering, Department of Chemical Engineering and Applied Chemistry, University of Toronto, Toronto, ON, Canada; Toronto General Research Institute, University Health Network, Toronto, ON, Canada*

MARC RUEL • *Division of Cardiac Surgery, University of Ottawa Heart Institute, Ottawa, ON, Canada*

EMIL RUVINOV, PH.D. • *Avram and Stella Goldstein-Goren Department of Biotechnology Engineering, Ben-Gurion University of the Negev, Beer-Sheva, Israel*

YULIA SAPIR, M.SC. • *Avram and Stella Goldstein-Goren Department of Biotechnology Engineering, Ben-Gurion University of the Negev, Beer-Sheva, Israel*

SEBASTIAN SCHAAF • *Department of Experimental Pharmacology and Toxicology, University Medical CenterHamburg-Eppendorf (UKE), Hamburg, Germany; DZHK (German Centre for Cardiovascular Research), Hamburg, Germany*

HIDEKAZU SEKINE • *Institute of Advanced Biomedical Engineering and Science, TWIns, Tokyo Women's Medical University, Shinjuku-ku, Tokyo, Japan*

DROR SELIKTAR • *Faculty of Biomedical Engineering, Technion—Israel Institute of Technology, Haifa, Israel; Nanoscience and Nanotechnology Initiative, National University of Singapore, Singapore, Singapore*

TATSUYA SHIMIZU • *Institute of Advanced Biomedical Engineering and Science, TWIns, Tokyo Women's Medical University, Shinjuku-ku, Tokyo, Japan*

POH LOONG SOONG • *Institute of Pharmacology, Heart Research Center Göttingen (HRCG), University Medical Center Göttingen, Göttingen, Germany; DZHK (German Center for Cardiovascular Research), partner site Göttingen, Göttingen, Germany*

ANDREA STÖHR • *Department of Experimental Pharmacology and Toxicology, University Medical Center Hamburg-Eppendorf (UKE), Hamburg, Germany; DZHK (German Centre for Cardiovascular Research), Hamburg, Germany*

ERIK J. SUURONEN • *Division of Cardiac Surgery, University of Ottawa Heart Institute, Ottawa, ON, Canada*

KENJIRO TADAKUMA • *Department of Mechanical Engineering, Graduate School of Engineering, Osaka University, Yamadaoka, Suita, Japan*

NOBUYUKI TANAKA • *Institute of Advanced Biomedical Engineering and Science, TWIns, Tokyo Women's Medical University, Shinjuku-ku, Tokyo, Japan*

MALTE TIBURCY • *Institute of Pharmacology, Heart Research Center Göttingen (HRCG), University Medical Center Göttingen, Göttingen, Germany; DZHK (German Center for Cardiovascular Research), partner site Göttingen, Göttingen, Germany*

ROGER TU • *Department of Biological Sciences, University of California, Irvine, CA, USA*

MARK D. UNGRIN • *Department of Comparative Biology and Experimental Medicine, Faculty of Veterinary Medicine, University of Calgary, Calgary, AB, Canada*

INGRA VOLLERT • *Department of Experimental Pharmacology and Toxicology, University Medical Center Hamburg-Eppendorf (UKE), Hamburg, Germany; DZHK (German Centre for Cardiovascular Research), Hamburg, Germany*

GORDANA VUNJAK-NOVAKOVIC • *Department of Biomedical Engineering, Columbia University, New York, NY, USA*

BO WANG • *Tissue Bioengineering Laboratory, Department of Biological Engineering, Mississippi State University, Mississippi State, MS, USA; Department of Surgery, Comprehensive Transplant Center, Northwestern University Feinberg School of Medicine, Northwestern University, Chicago, IL, USA*

RICHARD D. WEISEL • *Division of Cardiovascular Surgery, Toronto General Research Institute, University Health Network, Toronto, ON, Canada; Department of Surgery, Division of Cardiac Surgery, University of Toronto, Toronto, ON, Canada*

LAKIESHA N. WILLIAMS • *Tissue Bioengineering Laboratory, Department of Biological Engineering, Mississippi State University, Mississippi State, MS, USA*

EMILY A. WRONA • *New York Stem Cell Foundation, New York, NY, USA*

JOSEPH C. WU, M.D., PH.D. • *Lorry I. Lokey Stem Cell Research Building, Stanford University School of Medicine, Stanford, CA, USA*

MASAYUKI YAMATO • *Institute of Advanced Biomedical Engineering and Science, TWIns, Tokyo Women's Medical University, Shinjuku-ku, Tokyo, Japan*

BOYANG ZHANG • *Department of Chemical Engineering and Applied Chemistry, University of Toronto, Toronto, ON, Canada; Institute of Biomaterials and Biomedical Engineering, University of Toronto, Toronto, ON, Canada*

WEI-ZHONG ZHU • *Department of Pathology, Institute for Stem Cells and Regenerative Medicine, University of Washington, Seattle, WA, USA*

WOLFRAM-HUBERTUS ZIMMERMANN, M.D. • *Institute of Pharmacology, Heart Research Center Göttingen (HRCG), University Medical Center Göttingen, Göttingen, Germany; DZHK (German Center for Cardiovascular Research), partner site Göttingen, Göttingen, Germany*

Chapter 1

Second Generation Codon Optimized Minicircle (CoMiC) for Nonviral Reprogramming of Human Adult Fibroblasts

Sebastian Diecke, Leszek Lisowski, Nigel G. Kooreman, and Joseph C. Wu

Abstract

The ability to induce pluripotency in somatic cells is one of the most important scientific achievements in the fields of stem cell research and regenerative medicine. This technique allows researchers to obtain pluripotent stem cells without the controversial use of embryos, providing a novel and powerful tool for disease modeling and drug screening approaches. However, using viruses for the delivery of reprogramming genes and transcription factors may result in integration into the host genome and cause random mutations within the target cell, thus limiting the use of these cells for downstream applications. To overcome this limitation, various non-integrating techniques, including Sendai virus, mRNA, minicircle, and plasmid-based methods, have recently been developed. Utilizing a newly developed codon optimized 4-in-1 minicircle (CoMiC), we were able to reprogram human adult fibroblasts using chemically defined media and without the need for feeder cells.

Key words Human induced pluripotent stem cells (hiPSC), Minicircle, Reprogramming, Disease modeling, Chemically defined, Differentiation

1 Introduction

Somatic cells, such as human adult fibroblasts, can be reverted to an embryonic stem cell (ESC)-like state by inducing the expression of specific genes and transcription factors that regulate the maintenance of pluripotency in embryonic stem cells [1]. These so-called core factors of pluripotency, namely OCT4, SOX2, NANOG, and two oncogenes (c-MYC and LIN28), initiate the resetting of the epigenetic profile of a terminal differentiated somatic cell and activate the molecular circuitry of pluripotency [2–4]. The resultant induced pluripotent stem cells (iPSCs) have similar morphology, growth characteristics, and gene expression profiles as hESCs [5]. Furthermore, the capacity of iPSCs to differentiate into all other cell types of the human body has opened up new clinical opportunities for developing successful personalized stem cell-based therapies

Milica Radisic and Lauren D. Black III (eds.), *Cardiac Tissue Engineering: Methods and Protocols*, Methods in Molecular Biology, vol. 1181, DOI 10.1007/978-1-4939-1047-2_1, © Springer Science+Business Media New York 2014

and drug discovery approaches. Initially iPSCs were produced using retroviral and lentiviral vectors that are known to integrate into the chromosomes of the target cells, potentially causing disruption of gene transcription and subsequent tumor formation [6]. Therefore, various laboratories have developed additional non-integrating reprogramming techniques such as Sendai virus, mRNA, and protein- or DNA-based methods [7–11].

In order to induce the transformation of a nullipotent somatic cell into a pluripotent state, it is necessary to express the right stoichiometry of the reprogramming factors within the same transfected cell [12]. One elegant approach to achieve this requirement uses polycistronic constructs in which the four reprogramming factors are controlled by one promoter separated by a different self-cleaving 2a peptide sequence [13]. The disadvantage of such polycistronic plasmids is their large size, which might decrease the transfection efficiency in the adult human fibroblast. To overcome this problem, we cloned the polycistronic expression cassette into a minicircle plasmid backbone. Minicircles are special episomal DNA vectors devoid of any bacterial plasmid backbone (Fig. 1a). Therefore minicircles are significantly smaller than standard plasmids,

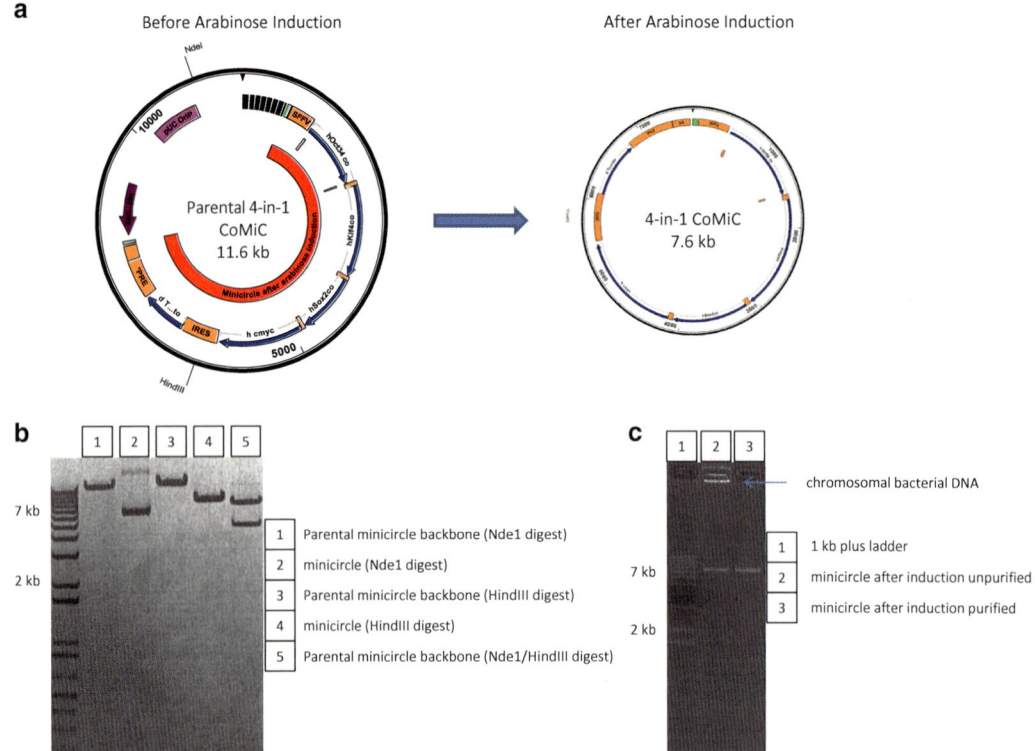

Fig. 1 (**a**) Map of parental 4-in-1 reprogramming minicircle before and after arabinose induction. (**b**) Control digest to verify quality of the minicircle production. (**c**) Gel showing bands representing the minicircle after induction, both before and after purification steps

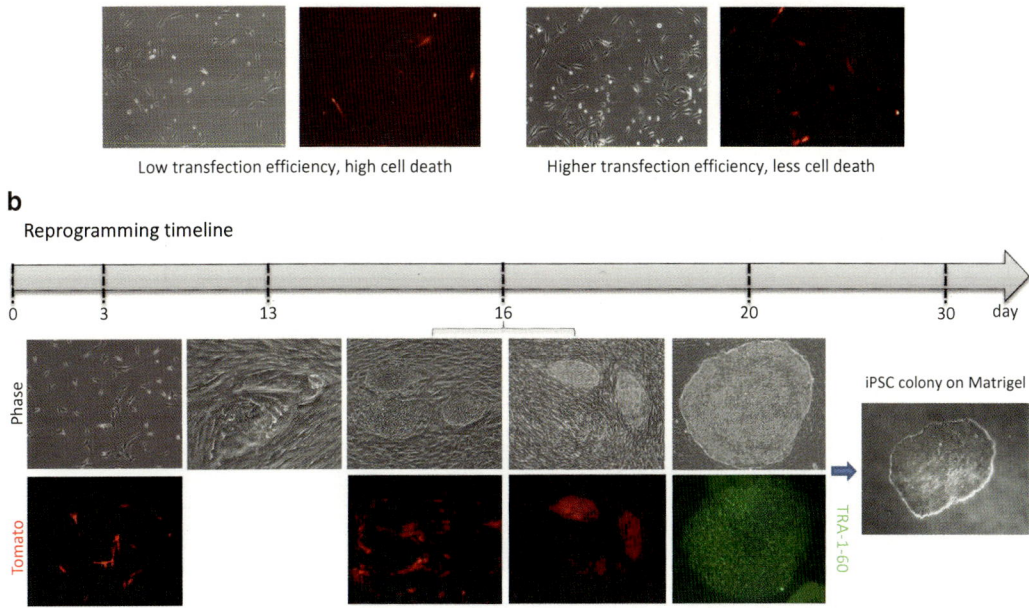

Fig. 2 (**a**) Representative bright field and fluorescent (Tomato) images demonstrating the transfection efficiency and cell survival of unpurified and purified minicircle. (**b**) Timeline showing the expected phase contrast and fluorescent images at various stages of the reprogramming process. At day 20, pluripotency was confirmed by TRA-1-60 immunofluorescent staining

which in turn enhances their transfection efficiency and the survival rate of the target cells [14]. Furthermore, it has been shown that minicircle constructs sustain a higher expression rate of their harboring transgenes over a longer period of time. Although the minicircle theoretically should not integrate into the genome of the target cell, there is still a relatively small chance of integration. Therefore, it is important to screen the derived iPSCs for random integration using specific PCR primers or Southern blot experiments [15].

The following protocol details the entire process of reprogramming human adult fibroblasts into human iPSCs: from dissociating fibroblasts from patient skin biopsy samples, to the reprogramming and characterization of the newly derived iPSC line. The time from induction of pluripotency to expansion of the subsequent iPSC line to passage 3 is approximately 1 month (Fig. 2b). The workflow of the reprogramming process is summarized in Fig. 3. The final part of the protocol explains the pluripotency of the iPSCs is confirmed using immunostaining (Fig. 4a). However, it will also be necessary to perform further experiments, including directed differentiation into specific germ layers and teratoma formation assay, to verify pluripotency (Fig. 4b, c) [16]. It should also be mentioned that the minicircle reprogramming

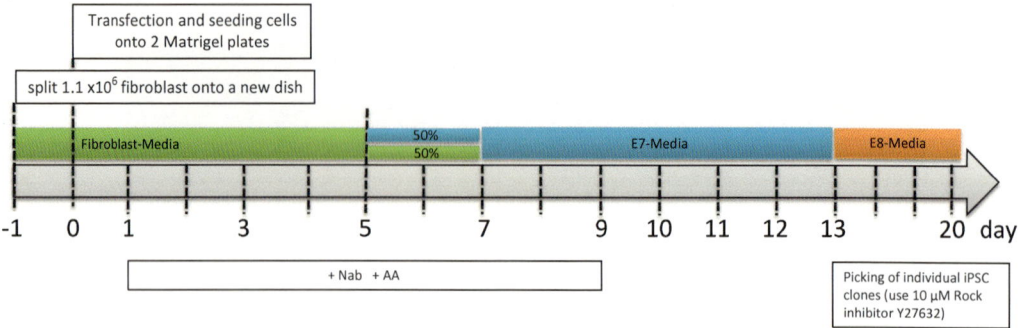

Fig. 3 Timeline for reprogramming adult human fibroblasts into induced pluripotent stem cells using CoMiC

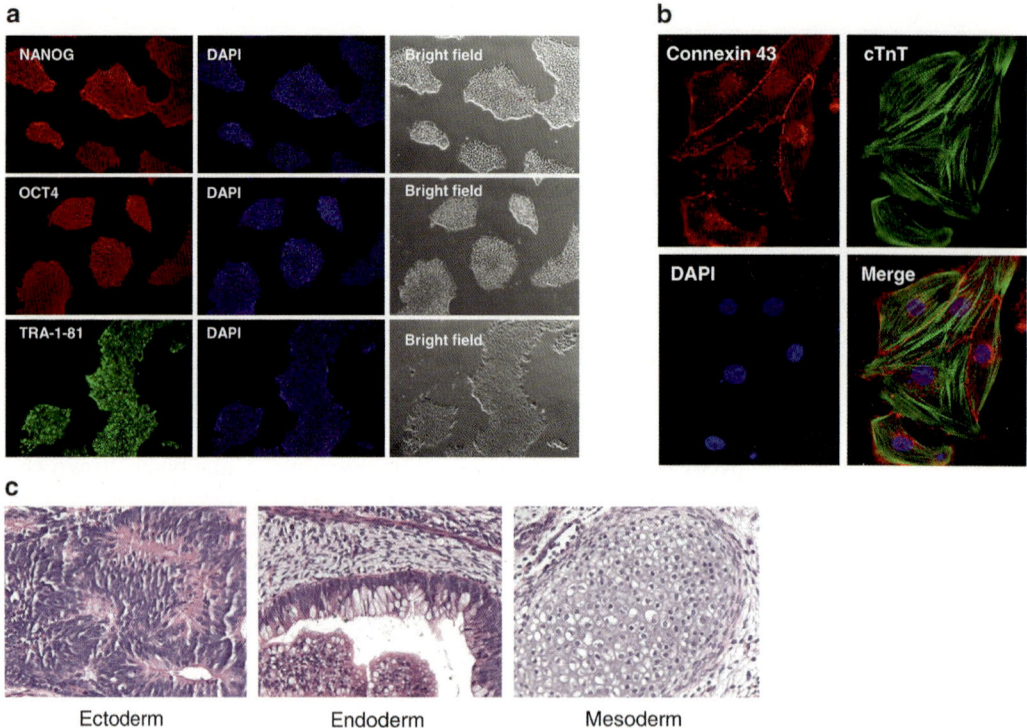

Ectoderm   Endoderm   Mesoderm

Fig. 4 Immunofluorescence staining demonstrating (**a**) pluripotency markers NANOG, OCT4 and TRA-1-81, DAPI, and bright field images. (**b**) Immunofluorescent staining of iPSC-derived cardiomyocytes demonstrating positive staining of cardiac markers connexin 43 and cardiac troponin as well as DAPI. (**c**) Representative H&E staining from teratomas to confirm the pluripotency status of the generated iPSCs

technique still has a very low efficiency and therefore should be only used by researchers with iPSC experience as an alternative method for obtaining integration-free induced pluripotent stem cells, or if it is necessary to compare different reprogramming techniques.

2 Materials

2.1 Media

1. *Fibroblast media*: Low glucose Dulbecco's Modified Eagle Media (DMEM low glucose) with 20 % fetal bovine serum (FBS).
2. *iPSC media*: Essential 8 (combine supplement with base media) and Essential 7 (Essential 6 combined with base media and supplemented with 100 ng/mL FGF2).

2.2 Reagents

1. Phosphate buffered saline (PBS).
2. Collagenase IV.
3. Penicillin/streptomycin.
4. Lennox L Broth (LB) base.
5. Sodium hydroxide (NaOH) solution, 1 M.
6. L-(+)-Arabinose ≥99 %, distilled water (dH$_2$O).
7. Kanamycin sulfate, 100×.
8. Matrigel—Growth factor reduced, hESC-qualified.
9. DMEM/F12.
10. TrypLE Express, store at room temperature (RT), there is no need to warm to 37 °C before use.
11. 0.25 % Trypsin/Ethylenediaminetetraacetic acid (EDTA).
12. Accutase.
13. Y27632 (ROCK inhibitor): 10 mM stock in Ultrapure water, store at −20 °C.
14. Ultrapure water.
15. Sodium butyrate (NaB).
16. Antibodies (*see* Table 1).
17. 4 % paraformaldehyde (PFA) solution.
18. Triton-X 100.
19. Bovine serum albumin (BSA).
20. Goat Serum.
21. Tween-20.
22. Vectashield mounting medium with DAPI.
23. Qiagen Plasmid Plus Maxi Kit.
24. DNAase (2,000 U/ml).
25. Gelatin.
26. Neon Buffer R.
27. Qiagen Plasmid Plus Maxi kit.
28. 4 % Paraformaldehyde (PFA).
29. New England biolabs enzymes HindIII and Nde1.

Table 1
List of antibodies and dilutions used for characterization of human induced pluripotent stem cells and differentiated cardiomyocytes. Staining protocols were optimized with the antibodies from the listed suppliers. Optimization may be required if different antibodies are used

Antigen	Supplier	Product no.	Dilution	Raised in	Secondary (1:400)	Supplier	Product no.
Connexin-43	SIGMA	C6219	1:400	Rabbit	Alexa Fluor® 488 Goat Anti-Rabbit IgG (H+L)	Life Technologies	A-11037
c Troponin-T	Thermo Scientific	MS-295-P	1:400	Mouse	Alexa Fluor® 488 Goat Anti-Mouse IgG (H+L)	Life Technologies	A-11001
NANOG	Santa Cruz	SC-33759	1:100	Rabbit	Alexa Fluor® 488 Goat Anti-Rabbit IgG (H+L)	Life Technologies	A-11037
OCT4	Santa Cruz	SC-9081	1:100	Rabbit	Alexa Fluor® 488 Goat Anti-Rabbit IgG (H+L)	Life Technologies	A-11037
Tra-1-81	Millipore	mab4381	1:250	Mouse	Alexa Fluor® 488 Goat Anti-Mouse IgG (H+L)	Life Technologies	A-11001

2.3 Equipment

1. 6-well cell culture plates (surface area = 9.8 cm^2) coated with 2 mL Matrigel.

2. 12-well cell culture plates (surface area = 3.8 cm^2) coated with 1 mL Matrigel.

3. 10 cm tissue culture plates (surface area = 58.95 cm^2) coated with 12 mL Matrigel.

4. 90 mm petri dishes.

5. 15 and 50 mL polypropylene conical tubes.

6. 2 mL plastic aspiration pipettes.

7. 5, 10, 25, and 50 mL plastic pipettes.

8. 250 and 500 mL polyethersulfone (PES) media filters.

9. 8-chamber slides.

10. T225 flasks.

11. Coverslips.

12. Parafilm.

13. Countess Cell Counter, slides, and trypan blue.

14. Tissue culture incubator capable of 37 °C, 5 % CO_2, and 85 % relative humidity (such as New Brunswick Galaxy 170R).

15. Dual gas tissue culture incubator capable of 37 °C, 5 % CO_2, 5 % O_2, and 85 % relative humidity with split inner door.

16. Centrifuge.

17. Inverted tissue culture microscope (such as Nikon Ti) with heated stage.

18. Autoclave.

19. Orbital shaker.

20. Spectrophotometer.

21. Waterbath.

22. Neon Transfection device Life Technologies.

23. Nalgene centrifuge bottle.

3 Methods

3.1 Minicircle Production

1. Prepare 1,200 mL of LB Broth media by adding 24 g of LB Broth Base to 1,200 mL of distilled water (hH_2O) and divide this over the three culture flasks.

2. Autoclave the flasks using the "liquid cycle."

3. After cooling down of the flasks to room temperature, add 4 mL of Kanamycin 100× to each one of the culture flasks.

4. In the morning, pick one clone from a parental plasmid transformed bacterial plate in 5 mL LB media containing kanamycin

(50 µg/mL) and grow this pre-culture until the evening at 37 °C with shaking at 250 rpm.

5. In the evening, grow an overnight culture by combining 100 µL of the pre-culture with each of the flasks containing media with kanamycin and culture overnight (16–18 h) at 37 °C with shaking at 250 rpm (*see* **Note 1**).

6. The optical density (OD, absorbance) of the overnight culture should be between 3.75 and 4.25.

7. Prepare a minicircle induction mix comprising 1 volume of fresh LB Broth (1,200 mL), 4 % 1 N NaOH (48 mL), and 1 % L-Arabinose (12 g) and mix well.

8. Add 400 mL of the induction mix to each of the three flasks with the overnight culture and mix well.

9. Incubate the combined mix at 32 °C with shaking at 250 rpm for 6–8 h.

10. Harvest all the culture flasks, transfer them to centrifuge bottle, and pellet the bacteria ($6,000 \times g$).

11. Isolate minicircles according to the Qiagen Plasmid Plus Maxi kit instructions with certain modifications (*see* **Notes 2** and **3**).

12. Confirm the successful minicircle induction by digestion using HindIII and Nde1. Nde1 linearizes only the parental backbone but not the minicircle. Therefore, you should expect one band around 7.7 kb when digesting the minicircle with both enzymes (Fig. 1b).

3.2 Fibroblast Isolation from a Patient Skin Punch Biopsy

1. All of these steps should be performed in a sterile tissue culture hood. Prepare a fibroblast digestion solution by dissolving 1 mg/mL Collagenase IV in DMEM/F12 and sterile filter it through a 0.2 µm pore size filter.

2. Transfer the skin punch biopsy into a 15 mL conical tube containing 3 mL PBS + penicillin/streptomycin (1:100), repeat this step three times.

3. Transfer the tissue into a sterile 10 cm plate and add 1 mL Collagenase IV. Use two scalpels to cross cut the tissue into small pieces while keeping the tissue immersed in Collagenase IV.

4. Additionally, add 2 mL Collagenase IV solution to the minced tissue and pipette it up and down. Transfer the Collagenase IV/tissue suspension into a 15 mL conical tube and add 25 µL of 0.5 M EDTA and 20 µL of DNase. Afterwards, incubate the mixture for 3 h in a tissue culture incubator at 37 °C.

5. Following the 3-h incubation time, pipette the tissue pieces up and down several times to obtain single-cell fibroblasts.

6. Add 12 mL fibroblast media containing 1× penicillin/streptomycin to stop the Collagenase IV digestion.

7. Centrifuge for 4 min at $300 \times g$.

8. Resuspend the pellet in 2 mL fibroblast media containing 1×
   penicillin/streptomycin solution, and plate it in one well of a
   0.1 % gelatin-coated 6-well plate. Depending on the size of
   your tissue sample, you may need to plate the cells in more
   than one well of a 6-well plate. To prepare gelatin-coated
   plates, gelatin stock (10 %) is dissolved in PBS, 1 ml of this
   solution is pipetted into a well of a 6-well plate and incubated
   for 1 h at 37 °C.

9. Incubate the cells for at least 4 days without changing the
   media, until some of the fibroblasts start to attach and propa-
   gate. After this, change the media every other day.

10. Once the cells become confluent, split using 0.25 % Trypsin–
    EDTA for 10 min. Plate the cells onto a 0.1 % gelatin-coated
    T225 flask to further expand and freeze multiple vials of fibro-
    blast after the flask becomes confluent.

3.3 Prepare Matrigel Plates

1. Thaw stock bottle of Matrigel at 4 °C overnight.

2. Keep Matrigel on ice, and add 300 µL Matrigel to 50 mL of
   cold DMEM/F12 media. Add 2 mL to each well of a 6-well
   plate. (Alternatively use one Matrigel aliquot per 20 mL cold
   DMEM/F12 media and cover four 10 cm plates using 5 mL.)

3. Place the plates in the fridge and allow the Matrigel to polymerize overnight.

3.4 Reprogramming of Human Fibroblasts Using Neon Transfection and CoMiC

1. The target fibroblast cell line should be thawed and cultured
   for at least 3 days before starting the reprogramming experiment. Additionally, it is beneficial to further purify the minicircle DNA using the Zymoclean Gel DNA Recovery Kit to
   eliminate residual chromosomal bacterial DNA of the minicircle preparation (Fig. 1c). This step will increase the cell survival
   after the electroporation (Fig. 2a).

2. Plate 1.1×10^6 human fibroblast in a 10 cm tissue culture plate
   1 day before starting the reprogramming experiment.

3. On the following day (day 0), trypsinize the cells using Trypsin
   0.25 % EDTA for 10 min to obtain a single cell suspension.

4. Wash off the trypsin using 10 mL of DMEM plus 20 % FBS
   and centrifuge for 3 min at $300 \times g$.

5. Aspirate the media, and dissolve the cell pellet in 100–105 µL
   Neon Buffer R (the volume depends on the amount of plasmid
   you have to use to get 10–12 µg DNA). The final volume of
   buffer, DNA, and cells should be around 110 µL.

6. Perform the transfection using a 100 µL Neon tip and the
   following settings: 1,600 V, 10 ms, and 3 pulses (transfection
   efficiency should be more than 50 %).

7. Plate the cells equally distributed onto two Matrigel-coated plates (5.5×10^5 cells) in fibroblast media.

8. On the following morning (day 1), change the media to fibroblast media and add sodium butyrate (0.2 mM) plus 50 µg/mL ascorbic acid.

9. On day 3, change the media to fibroblast media and add sodium butyrate (0.2 mM) plus 50 µg/mL ascorbic acid.

10. On day 5, change the media to 50:50 fibroblast media:Essential 7 media and add sodium butyrate (0.2 mM) plus 50 µg/mL ascorbic acid.

11. On days 7–13, change the media every other day until the first pre-iPSC colonies appear.

    Change the media to Essential 7 and add sodium butyrate (0.2 mM) plus 50 µg/mL ascorbic acid.

12. Between days 14 and 20, after you recognize the first colonies, switch to Essential 8 media and hypoxic culture conditions. Around day 20, the iPSC colonies should be big enough for manual picking under the microscope. Pick six individual iPSC clones using the Vitrolife stem cell cutting tool with a 10 µL tip and transfer them into six different wells of a Matrigel-coated plate with Essential 8 media in presence of 10 µM ROCK Inhibitor (Y-27632). If it is difficult to detach the iPSC colonies from the surrounding fibroblasts, you also can scrape those fibroblasts away to make room for the outgrowing iPSC colony. After 2 days, the iPSC clone should be large enough to pick.

3.5 Colony Purification and Expansion

1. After 5–7 days, the individual colonies become big enough to dissociate into single cells using 0.5 mL TrypLE for 8 min. Detach the remaining colonies by pipetting them up and down with a 1 mL pipette.

2. Wash the cells once with 5 mL Essential 8 media.

3. Centrifuge for 3 min at $300 \times g$.

4. Passage the cells into one well of a Matrigel-coated 6-well plate containing 2 mL Essential 8 media.

5. After 5–7 days, colonies will have grown out and become dense enough to split onto a Matrigel-coated 10 cm plate. Therefore incubate the cells with 1 mL Accutase for 8 min, wash using 9 mL Essential 8 media, and centrifuge for 3 min at $300 \times g$.

6. Seed the cells onto a Matrigel-coated 10 cm plate in Essential 8 media in the presence of 10 µM Y27632.

7. Once the 10 cm plate become confluent, freeze down 5 vials (passage 3 iPS cells) and continue to grow at a 1:6 split ratio, freezing vials every 10 passages.

8. Grow cells to at least passage 5 to test for the expression of the pluripotency marker genes and to passage 15 before testing for efficient differentiation.

3.6 Immunofluorescence Staining Protocol

1. Coat 8-chamber slides with Matrigel (200 μL per chamber) as described above in Subheading 3.3.

2. Seed iPSCs (1×10^4 cells/chamber) or differentiated cardiomyocytes (1.75×10^4 cells/chamber) as a negative control, on 8-chamber slides in a final volume of 200 μL of the appropriate medium.

3. Place the chamber slides in 90 mm Petri dishes and store in a 37 °C incubator.

4. Allow cells to attach and grow for ~24–72 h and fix as follows (*see* **Note 4**).

5. Aspirate medium and wash cells 3× with PBS.

6. Add to each chamber 100 μL 4 % Paraformaldehyde (PFA) in PBS for 20 min at RT.

7. Aspirate PFA and wash 3× with 200 μL PBS.

8. Aspirate PBS and replace with 500 μL of fresh PBS.

9. Seal slides with parafilm and place at 4 °C until needed for staining (keep for up to 7 days).

10. Incubate with 100 μL of permeabilization solution (0.2 % Triton X-100 in PBS), for 15 min at room temperature (*see* **Note 5**).

11. Wash twice with PBS, for 5 min at room temperature.

12. Incubate with blocking solution (2 % BSA or 4 % Goat Serum or 2 % FBS in PBS), for 1 h at room temperature or overnight at 4 °C.

13. Wash once in PBS, for 5 min at room temperature.

14. Incubate overnight at 4 °C with primary antibody (Table 1) diluted appropriately in blocking solution (100 μL per chamber).

15. Wash in 1 % goat serum, 0.1 % Tween-20 in PBS four times, for 10 min.

16. Incubate with the appropriate secondary antibody diluted in blocking solution, for 1 h at RT (100 μL per chamber). Wrap in aluminum foil as soon as secondary antibody is added.

17. Wash in 1 % Goat Serum, 0.1 % Tween-20 in PBS four times, for 10 min.

18. Remove the chambers from the slide.

19. Place a small amount of Vectashield mounting medium with DAPI over the area where each chamber used to be.

20. Place a clean coverslip over the slide and gently blot off any liquid remaining after the staining protocol.

21. Seal with nail varnish.

22. Store slides at 4 °C in the dark.

23. Image at least 3 h later.

4 Notes

1. Do not start off with too much parental plasmid, as an overgrown culture will limit the induction efficiency.

2. To ensure a high yield, make sure that the pellet is completely resuspended in Qiagen Maxi Kit P1 buffer and use the same volume of P2 and S3. For example, when using 400 mL spinning tubes, resuspend the pellet in 10 mL P1, P2, and S3 buffer instead of 8 mL.

3. To avoid bacterial genomic DNA contamination, avoid excessive shaking of the resuspended pellet when adding buffer P2.

4. Solutions should be made fresh for each experiment. Keep at 4 °C until day 2.

5. The permeabilization step is only required for the detection of nuclear protein (e.g., OCT4, SOX2, or NANOG); when surface markers are detected, skip the permeabilization step and go straight to the primary antibody incubation.

Acknowledgements

We thank Nicholas Mordwinkin for proof-reading the manuscript. We thank funding support from the National Institutes of Health (NIH) R01 HL113006, NIH U01 HL099776, NIH R24 HL117756, and the American Heart Association Established Investigator Award (JCW). This work was also supported by DFG (German Research Foundation) (to S.D.).

References

1. Takahashi K et al (2007) Induction of pluripotent stem cells from adult human fibroblasts by defined factors. Cell 131(5):861–872

2. Mitalipov S, Wolf D (2009) Totipotency, pluripotency and nuclear reprogramming. Adv Biochem Eng Biotechnol 114:185–199

3. Feng B, Ng JH, Heng JC, Ng HH (2009) Molecules that promote or enhance reprogramming of somatic cells to induced pluripotent stem cells. Cell Stem Cell 4(4):301–312

4. Maherali N et al (2007) Directly reprogrammed fibroblasts show global epigenetic remodeling and widespread tissue contribution. Cell Stem Cell 1(1):55–70

5. Wernig M et al (2007) In vitro reprogramming of fibroblasts into a pluripotent ES-cell-like state. Nature 448(7151):318–324

6. Sommer CA et al (2010) Excision of reprogramming transgenes improves the differentiation potential of iPS cells generated with a single excisable vector. Stem Cells 28(1):64–74

7. Stadtfeld M, Hochedlinger K (2010) Induced pluripotency: history, mechanisms, and applications. Genes Dev 24(20):2239–2263

8. Stadtfeld M, Nagaya M, Utikal J, Weir G, Hochedlinger K (2008) Induced pluripotent stem cells generated without viral integration. Science 322(5903):945–949

9. Warren L et al (2010) Highly efficient reprogramming to pluripotency and directed differentiation of human cells with synthetic modified mRNA. Cell Stem Cell 7(5): 618–630

10. Zhou H et al (2009) Generation of induced pluripotent stem cells using recombinant proteins. Cell Stem Cell 4(5):381–384

11. Okita K et al (2011) A more efficient method to generate integration-free human iPS cells. Nat Methods 8(5):409–412

12. Tiemann U et al (2011) Optimal reprogramming factor stoichiometry increases colony numbers and affects molecular characteristics of murine induced pluripotent stem cells. Cytometry A 79(6):426–435

13. Warlich E et al (2011) Lentiviral vector design and imaging approaches to visualize the early stages of cellular reprogramming. Mol Ther 19(4):782–789

14. Jia F et al (2010) A nonviral minicircle vector for deriving human iPS cells. Nat Methods 7(3):197–199

15. Gonzalez F, Boue S, Izpisua Belmonte JC (2011) Methods for making induced pluripotent stem cells: reprogramming a la carte. Nat Rev Genet 12(4):231–242

16. Lan F et al (2013) Abnormal calcium handling properties underlie familial hypertrophic cardiomyopathy pathology in patient-specific induced pluripotent stem cells. Cell Stem Cell 12(1):101–113

Chapter 2

Scalable Cardiac Differentiation of Human Pluripotent Stem Cells as Microwell-Generated, Size Controlled Three-Dimensional Aggregates

Celine L. Bauwens and Mark D. Ungrin

Abstract

The formation of cells into more physiologically relevant three-dimensional multicellular aggregates is an important technique for the differentiation and manipulation of stem cells and their progeny. As industrial and clinical applications for these cells increase, it will be necessary to execute this procedure in a readily scalable format. We present here a method employing microwells to generate large numbers of human pluripotent stem cell aggregates and control their subsequent differentiation towards a cardiac fate.

Key words Cardiac differentiation, Microwells, Human pluripotent stem cells, Size controlled cell aggregates, Forced cell aggregation

1 Introduction

It has been known for a long time that cellular behavior in two-dimensional adherent culture does not fully recapitulate three-dimensional in vivo systems, and consequently there is a growing trend in many fields of research to employ three-dimensional tissue constructs. These may be divided into scaffolded constructs, where cells are cultured on or in an exogenous supporting matrix [1], and unscaffolded systems that consist primarily or exclusively of cells (which may then proceed to generate their own endogenous matrices) [2, 3]. Here, we will discuss a process for generating large numbers of uniform, unscaffolded aggregates of human pluripotent stem cells for differentiation towards a cardiac fate.

Conventionally, size controlled cellular aggregates have been generated in microcentrifuge tubes [4], as hanging drops [5], by micropatterning human embryonic stem cells (hESCs) colonies [6] or by centrifugation into U- or V-bottom plates [7, 8]; however throughput is limited using these approaches. Microwell-based systems are similar in concept to V-bottom plates; however

Milica Radisic and Lauren D. Black III (eds.), *Cardiac Tissue Engineering: Methods and Protocols*, Methods in Molecular Biology, vol. 1181, DOI 10.1007/978-1-4939-1047-2_2, © Springer Science+Business Media New York 2014

the smaller size of the microwells permits very large numbers of uniform aggregates to be generated from a single culture-plate well in a mechanically simple system [9]. This approach has been used for a variety of applications including differentiation of pluripotent stem cells to ectodermal [10], endodermal [11], mesodermal [12], and extraembryonic [13] fates; chondrogenesis from mesenchymal stem cells [14]; and generation of uniform substrates for toxicological screening [15] and investigations of mechanobiology [16].

Using this approach to control aggregate size, we consequently determine surface area to volume ratio, which in turn modulates extraembryonic endoderm (ExE) commitment. Cardiogenic signals from this cell lineage then specify subsequent cardiomyogenesis [12].

2 Materials

All reagent preparation and cell culture handling should be carried out in a biological safety cabinet under sterile conditions.

2.1 General Equipment and Supplies

1. Biological safety cabinet.
2. Pipette aid.
3. Serological pipettes (5–25 mL).
4. Aspirator.
5. Aspirator or Pasteur pipettes.
6. 15 and 50 mL conical tubes.
7. Fume hood.
8. 0.22 μm syringe filter.
9. 5 % CO_2, 5 % O_2, and humidity controlled cell culture incubator (Subheadings 3.1 and 3.4).
10. 5 % CO_2, 20 % O_2, and humidity controlled cell culture incubator (Subheading 3.5).
11. Low speed centrifuge with a swinging bucket rotor fitted with a plate holder.
12. P2, P20, P200, and P1000 micropipettors with appropriate tips.
13. Inverted microscope with 4×, 10×, and 20× phase objectives.
14. Ultra-Low Attachment (ULA) 24 well plates.
15. 1.5 mL microcentrifuge tubes.
16. Bench-top microcentrifuge.

2.2 Basal Medium

1. L-Ascorbic Acid: Prepare a stock solution of 5 mg/mL in 4 °C, sterile ultrapure distilled water. Leave on ice and vortex

Table 1
Cytokines and buffers used for media preparation

Cytokine	Buffer	Stock Conc. (µg/mL)
Human Bone morphogenetic protein 4 (hBMP-4)	4 mM Hydrochloric acid, 0.1 % BSA	10
Human Fibroblast Growth Factor 2 (basic) (hbFGF)	Phosphate buffered saline (PBS), 0.1 % BSA	10
Human Vascular Endothelial Growth Factor (hVEGF)	Phosphate buffered saline (PBS), 0.1 % BSA	5
Activin A	Phosphate buffered saline (PBS), 0.1 % BSA	10
Human Dickkopf 1 homolog (hDkk-1)	Phosphate buffered saline (PBS), 0.1 % BSA	50

periodically until completely dissolved. Filter-sterilize using a 0.22 µm syringe filter, aliquot at 1 mL volume, and store the Ascorbic Acid aliquots at –20 °C. Use a freshly thawed aliquot each time medium is prepared.

2. Monothioglycerol: MTG should be aliquoted (1 mL) and stored frozen (–20 °C). When aliquots are thawed, they can be used for 3 months and then discarded. Aliquoting of MTG is strongly recommended as it minimizes the amount of oxidation due to repeated opening. On the day of media preparation, dilute 13 µL MTG in 1 mL StemPro-34. Discard unused diluted MTG.

3. Transferrin: Transferrin (stock concentration: 30 mg/mL) should be aliquoted (1 mL aliquots) and stored at –20 °C.

4. All cytokines are stored lyophilized at –20 °C until ready to be aliquoted in the following buffers and at the following concentrations, Table 1. Continue to store unused aliquots at –20 °C.

5. 4 mM Hydrochloric acid (HCl), 0.1 % BSA Buffer: In a fume hood, transfer 30 µL of 6.0 N HCl solution to 50 mL of ultrapure distilled water. Filter-sterilize the solution using a 0.22 µm syringe filter. Add 2 mL of 25 % BSA solution to the 50 mL HCl solution.

6. PBS, 0.1 % BSA Buffer: Add 20 µL of 25 % BSA solution for every mL PBS.

7. Supplemented DMEM/F-12: 1 % Penicillin/Streptomycin (Pen/Strep) and 1 % L-glutamine in DMEM/F-12.

8. hESC Wash Medium: 5 % Knockout Serum Replacement in Supplemented DMEM/F12.

9. StemPro-34: StemPro-34 is sold as a kit with two components. The supplement is kept at −20 °C and the liquid media at 4 °C. When combined, the media is stable for 30 days (*see* **Note 1**). StemPro-34 is always used with the supplement added in this protocol.

10. Modified StemPro-34: StemPro-34 modified with the addition of 1 % Pen/Strep, 1 % L-glutamine, 0.5 % Transferrin, 1 % Ascorbic Acid, 0.3 % MTG. This medium should be prepared on the day of use with freshly thawed Ascorbic Acid.

2.3 hESC Aggregate Formation Components

1. One 6 well plate culture of feeder-free or feeder-depleted (*see* **Note 2**) hESCs adapted to single cell passaging (e.g., enzymatic dissociation with TrypLE).

2. TrypLE Select.

3. hESC Wash Medium (preparation previously described in Subheading 2.2, **item 8**).

4. Hemocytometer.

5. Trypan Blue.

6. Aggrewell™ 400 plates (StemCell Technologies 27845).

7. Aggrewell™ Rinsing Solution (StemCell Technologies 07010).

8. Aggregation Medium: Modified StemPro-34, 0.5 ng/mL BMP4, 10 µM Y-27632 ROCK Inhibitor. 1 mL of medium is required per well of a 24 well Aggrewell™ 400 plate.

9. OPTIONAL: 37 µm strainer.

2.4 Cardiac Induction Components

1. Stage 1 Induction Medium: Modified StemPro-34, 10 ng/mL BMP4, 5 ng/mL bFGF, 6 ng/mL Activin A. 1 mL of medium is required per well of a 24 well Aggrewell™ 400 plate (*see* **Note 3**).

2. Stage 2 Induction Medium: Modified StemPro-34, 10 ng/mL VEGF, 150 ng/mL DKK1. 1 mL of medium is required per well of a 24 well ULA plate (*see* **Note 3**).

3. Stage 3 Induction Medium: Modified StemPro-34, 10 ng/mL VEGF, 150 ng/mL DKK1, 5 ng/mL bFGF. 1 mL of medium is required per well of a 24 well ULA plate (*see* **Note 3**).

4. hESC Wash Medium (prepared as previously described in Subheading 2.2, **item 8**).

2.5 Flow Cytometry

1. 1 mg/mL Collagenase Type II in Hank's Balanced Salt Solution.

2. TrypLE Select.

3. hESC Wash Medium (preparation previously described in Subheading 2.2, **item 8**).

4. DNase: DNase should be diluted in ultrapure distilled water to a concentration of 1 mg/mL, aliquoted, and stored at –20 °C.

5. 96 well assay plate (or 1.5 mL microcentrifuge tubes).

6. IntraPrep fixation and permeabilization kit.

7. Cardiac Troponin T antibody.

8. Allophycocyanin (APC)-conjugated goat-anti-mouse IgG antibody.

9. HF buffer solution: 2 % Fetal Bovine Serum (FBS) in Hank's Balanced Salt Solution.

10. 5 mL round-bottom flow cytometry analysis tubes.

11. Flow cytometer with the appropriate laser and filter to excite (i.e., 633–640 nm) and detect (i.e., 660/20 nm) APC.

3 Methods

A schematic of the process is shown in Fig. 1.

3.1 hESC Aggregate Formation

1. Add 0.5 mL Aggrewell™ Rinsing Solution to each well of the Aggrewell™ plate used. Ensure that the solution contacts the entire surface inside each microwell by centrifuging the plate at $840 \times g$ for 2 min. Incubate the Aggrewell™ plate containing the Rinsing Solution for 30–60 min at room temperature. Aspirate the solution and wash twice with 1 mL PBS per well, centrifuging at $840 \times g$ for 2 min after each PBS addition.

2. Completely dissociate the hESCs to single cells: Aspirate the culture medium from each culture well, rinse each well with 1 mL TrypLE Select, and aspirate the TrypLE Select. Residual TrypLE should be sufficient to dissociate the cells. Incubate the plate at 37 °C for 3 min, then add 1 mL of hESC Wash Medium to each well and mechanically dissociate the cells from the tissue culture surface by pipetting with a P1000 micropipettor. Transfer the cells (in hESC Wash Medium) to a 15 mL conical tube.

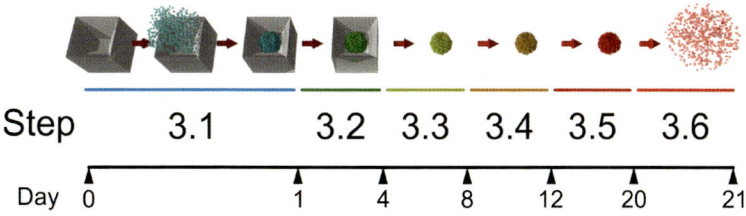

Fig. 1 Schematic of cardiac differentiation protocol. The sequence of steps in Subheadings 3.1–3.6 is diagrammed, along with an indication of the timescale over which they occur

Optionally, if the dissociation appears to be incomplete and cell clumps remain, the suspension may be passed through a strainer at this point.

3. Perform a cell count: Take a 10 μL sample of the cell suspension collected in the previous step. Add 30 μL of Trypan Blue and mix the suspension well by pipetting up and down. Transfer 10 μL of Trypan Blue-stained cells to each chamber of a hemocytometer and visualize under the inverted microscope with the 10× objective. Count the cells.

4. From the cell count results, calculate how many wells can be seeded at 1.2×10^6 cells per well. Resuspend the dissociated hESCs in Aggregation Medium at a density of 1.2×10^6 cells/mL (*see* **Note 4**).

5. Aspirate the PBS wash solution from each well on the Aggrewell™ Plate. Using a P1000 micropipettor, evenly distribute 1 mL of the cell suspension to each well on the Aggrewell™ Plate. If seeding a large numbers of wells, periodically mix the container of cell suspension to prevent settling. Centrifuge the plate at $200 \times g$ for 5 min. Visualize the plate under the microscope to confirm cells have spun to the bottom of each microwell (*see* **Note 5**).

6. Incubate the plate for 24 h at 37 °C in a 5 % CO_2, 5 % O_2 (*hypoxic*) incubator.

3.2 Cardiac Induction Stage 1 (24 ± 2 h Following Aggregation)

1. Prepare the required volume of Stage 1 Induction Medium (for a 24 well plate, volume = 1 mL × number of wells). Place the medium in a 37 °C water bath for at least 15 min.

2. Remove the Aggrewell™ plate from the incubator. Inspect the aggregates under the microscope. Compared to immediately post-centrifuge aggregation, they should appear intact with smooth edges (more round and less square than the previous day).

3. To remove the supernatant but retain the aggregates in their individual microwells: Use a P1000 micropipettor. For each well on the Aggrewell™ plate, place the tip of the micropipettor at the surface of the culture medium in the well and against the edge of the well (keep the plate level horizontally). Slowly remove the medium, being careful not to disturb the aggregates at the bottom of the microwells. Once the medium level has been reduced to about 1–2 mm from the surface of the textured microwell surface, the plate can be slowly tilted to collect medium at one side of the well (once the volume is this low, fluid motion is greatly reduced and it is easier to avoid aggregates getting lifted out of their individual microwells). Slowly pipette out the remaining medium in the well.

4. To add fresh Stage 1 Induction Medium while ensuring the aggregates remain in their individual microwells: Take up 1 mL

of Stage 1 Induction Medium with a P1000 micropipettor. Hold the pipette tip against the inside edge of the well and very slowly dispense the medium against the inside wall of the well. Repeat for the remaining wells.

5. Return the plate to the incubator under *hypoxic* conditions for 4 days.

3.3 Cardiac Induction Stage 2 (Day 4: 96 ± 2 h After Aggregate Formation)

Stage 2 Induction directs mesoderm differentiation. Therefore at this time-point it is very important to remove all residual Activin A from the Stage 1 Induction Medium. Activin A is a very potent signaling molecule and at this stage of differentiation, even trace levels would promote differentiation to the endoderm lineage at the expense of cardiac induction. Therefore, at this stage the aggregates are removed from the Aggrewell™ plates, washed well, and transferred to bulk aggregate cultures in 24 well ULA plates.

1. Place hESC Wash Medium (volume = number of wells × 2 mL) in a 37 °C water bath for 10–15 min.

2. Prepare the necessary volume of Stage 2 Induction medium (volume = number of wells × 1 mL) and place in a 37 °C water bath for 10–15 min.

3. Using a 5 mL serological pipette, harvest the aggregates from each well of the Aggrewell™ plate, and collect the aggregate suspension in a 15 mL conical tube (up to 10 wells per tube).

4. Allow the aggregates to settle for 15 min in hypoxic (5 % O_2) conditions. This step is performed to separate single cells and cellular debris from the intact aggregates. Aspirate the supernatant carefully and resuspend the aggregates in 10 mL hESC Wash Medium to wash out residual inductive cytokines (especially Activin A which is a potent signaling molecule even at very low concentrations).

5. Centrifuge the aggregates at $50 \times g$ for 2 min and aspirate the supernatant.

6. Resuspend the pelleted aggregates in the Induction 2 Medium prepared in **step 2** above.

7. Dispense the aggregate suspension into a 24 well ULA plate at 1 mL per well. Visualize the aggregates under the microscope. They should appear as uniform, tight cell clusters.

8. Incubate under *hypoxic* conditions until day 8.

3.4 Cardiac Induction Stage 3 (Day 8: 192 ± 2 h After Aggregate Formation)

Typically, in a successful cardiac induction, spontaneously contracting aggregates will be observed under the microscope between day 8 and day 12.

1. Prepare the necessary volume of Stage 3 Induction Medium (volume = number of wells × 2 mL) and place in a 37 °C water bath for 10–15 min.

2. Use a 5 mL serological pipette to transfer the aggregates to 15 mL conical tubes, pooling up to 10 mL of aggregates per tube. Allow 10 min for the aggregates to settle. Aspirate the supernatant and resuspend the aggregates in Stage 3 Induction Medium. Using a 5 mL serological pipette, redistribute the aggregates into a 24 well ULA plate at 1 mL per well.

3. Incubate under *hypoxic* conditions until day 12.

3.5 Cardiac Induction Stage 3 (Day 12 to Harvest)

1. Prepare the necessary volume of Stage 3 Induction Medium (volume = number of wells × 1 mL) and place in a 37 °C water bath for 10–15 min.

2. Use a 5 mL serological pipette to transfer the aggregates to 15 mL conical tubes, pooling up to 10 mL of aggregates per tube. Allow 10 min for the aggregates to settle. Aspirate the supernatant and resuspend the aggregates in Stage 3 Induction Medium. Using a 5 mL serological pipette, redistribute the aggregates into a 24 well ULA plate at 1 mL per well.

3. Incubate the cells at *normoxic* oxygen levels for the remainder of the culture period (37 °C, 20 % O_2, 5 % CO_2). *After this time point, cells are no longer cultured under hypoxic conditions.*

4. Repeat this complete medium exchange (**steps 1–3**) at day 16.

5. Cardiac aggregates are ready for harvest on day 20.

3.6 Flow Cytometry Analysis of Cardiac Troponin T Expression Frequency of Aggrewell™ Culture Output

For flow cytometry analysis of cardiac Troponin T (cTnT) expression frequency, it is recommended that 4 wells of day 20 cardiac aggregates be harvested.

1. Using a 5 mL serological pipette, transfer 1 well of aggregates to a 15 mL conical tube. Centrifuge the aggregates at $50 \times g$ for 2 min and carefully aspirate the supernatant.

2. Add 1 mL of freshly dissolved 1 mg/ml collagenase Type II solution to the aggregates, transfer the aggregate suspension to a 1.5 mL microcentrifuge tube, and incubate the aggregates in collagenase Type II overnight at room temperature.

3. The next day, using a P1000 micropipettor, gently pipette the aggregates to dissociate them into a homogenous single cell suspension. If the aggregates do not readily dissociate, settle the aggregates, aspirate the supernatant, and incubate the aggregates in 700 μL TrypLE for 1–2 min at room temperature. Gently pipette the aggregates 1–2 times with a P1000 micropipette to dissociate. Add 700 μL hESC Wash Medium containing 14 μL of 1 mg/mL DNAse stock solution to dilute the TrypLE. Take a 10 μL sample for cell counting, and centrifuge the remaining suspension using a bench-top microcentrifuge for 2 min at $300 \times g$.

4. Stain the 10 μL counting sample with an equal volume of Trypan Blue and count using a hemocytometer.

5. Remove the 1.5 mL microcentrifuge tube containing the remaining cells from the microcentrifuge. Aspirate the supernatant and resuspend the cells in HF at a concentration of 200,000–500,000 cells per 100 µL.

6. Transfer 100 µL per well to 2 wells of a 96 well plate (*see* **Note 6**) per experimental condition or replicate (one well will be for cTnT staining and the other will be the unstained control).

7. Pellet by centrifuging the plate at $300 \times g$ for 2 min in a swinging bucket centrifuge.

8. Fix the cells in 200 µL of Intraprep Reagent 1 per well for 15 min at room temperature.

9. Centrifuge the plate at $300 \times g$ for 2 min. Carefully aspirate the supernatant and dispose of in a paraformaldehyde waste container.

10. Wash the fixed cells twice: Add 200 µL HF to each well. Centrifuge the plate at $300 \times g$ for 2 min, aspirate the supernatant, and repeat the wash step once more. Fixed cells can be stored for up to 1 week in HF at 4 °C.

11. Permeabilize the cells: Centrifuge the plate at $300 \times g$ for 2 min. Add 100 µL of Intraprep Reagent 2 per well and incubate the plate for 5 min at room temperature. Centrifuge the plate at $300 \times g$ for 2 min and aspirate the supernatant.

12. Prepare a staining solution of anti-cTnT (Neomarkers MS-295) in HF at the optimal concentration for the given lot number (determined by titration: typically ranges from 1:500 to 1:2,000 dilution).

13. Add 100 µL staining solution to one well of cells and 100 µL of plain HF to the other well (negative control). Incubate the cells for 30 min at 4 °C.

14. Centrifuge the cells at $300 \times g$ for 2 min. Aspirate the supernatant and add 200 µL HF per well. Repeat the wash step one more time.

15. Prepare a master mix of the secondary antibody: Transfer a volume of HF that corresponds to 100 µL for every well (control and cTnT-stained) being treated. Add 1 µL of goat anti-mouse-APC secondary antibody per 200 µL of HF (1:200 dilution).

16. Stain the samples: Add 100 µL of staining solution to each well (negative and anti-cTnT-stained wells). Incubate cells for 30 min at 4 °C (keep the plate covered or in the dark after adding fluorescent secondary antibody to avoid photobleaching).

17. Centrifuge the cells at $300 \times g$ for 2 min. Aspirate the supernatant and add 200 µL HF per well. Repeat the wash step one more time.

18. Transfer the samples to 5 mL round-bottom flow cytometer analysis tubes and analyze on a flow cytometer using the red laser.

4   Notes

1. If 500 mL of complete StemPro-34 medium (with supplement) will not be used within 30 days, it is recommended that the basal medium and supplement be aliquoted as follows: prepare ten 50 mL aliquots of the basal medium and store at 4 °C, prepare ten 1.3 mL aliquots of the supplement and store at −20 °C. When preparing complete medium from the aliquots, completely thaw 1 aliquot of supplement in a 37 °C water bath and then add the supplement to a 50 mL aliquot of basal medium.

2. To prepare feeder-depleted hESCs, passage the cells onto Growth Factor Reduced Matrigel (BD 354230) coated 6 well plates at a split ratio that will produce $1-2 \times 10^6$ cells per well after 1–2 days. Use the same medium that was being used to maintain the hESCs on feeders.

3. Depending on the volume of Induction Medium to be prepared, the volume of some of the growth factor stock solutions to be added on a given day may be less than 1 μL. To ensure accurate volumes of growth factor stock solutions are being added to the medium preparation, the stock may first be diluted 1:10 in StemPro-34 basal medium, and then the appropriate volume of diluted stock may be added to the medium preparation.

4. At this seeding density, aggregates containing 1,000 cells will be formed. This is based on our observations with cardiac induction from the HES-2 cell line maintained on feeders [12]. The ideal seeding density may vary and should be optimized for the specific cell lines and culture conditions being used.

5. If the cells do not appear to form a tight pellet at the bottom of the microwells, try eliminating air bubbles in the microwells by spinning some medium into the wells prior to adding the cells: Add 0.4 mL of medium per well and centrifuge the plate at $200 \times g$ for 5 min. Then add the appropriate number of cells to each well in 0.6 mL medium per well.

6. If the total number of samples being prepared is small, staining can also be performed in 1.5 mL microcentrifuge tubes. If microcentrifuge tubes are used, supernatant is removed by aspiration and it is recommended that minimum wash volumes of 0.5 mL and centrifuge settings of $900 \times g$ (in a bench-top microcentrifuge) for 3 min be used for the wash steps.

Acknowledgements

We thank Dr. Peter Zandstra, in whose laboratory this protocol was developed, and Drs. Mark Gagliardi and Gordon Keller who provided assistance in establishing the initial methods on which this process was based. Protocol development was supported by an Ontario Graduate Scholarship in Science and Technology to C.B. and a grant from the Heart and Stroke Foundation of Ontario to Peter Zandstra.

References

1. Wintermantel E, Mayer J, Blum J et al (1996) Tissue engineering scaffolds using superstructures. Biomaterials 17:83–91

2. Lazar A, Peshwa MV, Wu FJ et al (1995) Formation of porcine hepatocyte spheroids for use in a bioartificial liver. Cell Transplant 4: 259–268

3. Sachlos E, Auguste DT (2008) Embryoid body morphology influences diffusive transport of inductive biochemicals: a strategy for stem cell differentiation. Biomaterials 29: 4471–4480

4. Johnstone B, Hering TM, Caplan AI et al (1998) In vitro chondrogenesis of bone marrow-derived mesenchymal progenitor cells. Exp Cell Res 238:265–272

5. Steinberg MS (1970) Does differential adhesion govern self-assembly processes in histogenesis? Equilibrium configurations and the emergence of a hierarchy among populations of embryonic cells. J Exp Zool 173:395–433

6. Bauwens CL, Peerani R, Niebruegge S et al (2008) Control of human embryonic stem cell colony and aggregate size heterogeneity influences differentiation trajectories. Stem Cells 26:2300–2310

7. Koike M, Kurosawa H, Amano Y (2005) A round-bottom 96-well polystyrene plate coated with 2-methacryloyloxyethyl phosphorylcholine as an effective tool for embryoid body formation. Cytotechnology 47:3–10

8. Ng ES, Davis RP, Azzola L et al (2005) Forced aggregation of defined numbers of human embryonic stem cells into embryoid bodies fosters robust, reproducible hematopoietic differentiation. Blood 106:1601–1603

9. Ungrin MD, Joshi C, Nica A et al (2008) Reproducible, ultra high-throughput formation of multicellular organization from single cell suspension-derived human embryonic stem cell aggregates. PLoS One 3:e1565

10. Kozhich O, Hamilton R, Mallon B (2013) Standardized generation and differentiation of neural precursor cells from human pluripotent stem cells. Stem Cell Rev Rep 9:531–536

11. Ungrin MD, Clarke G, Yin T et al (2012) Rational bioprocess design for human pluripotent stem cell expansion and endoderm differentiation based on cellular dynamics. Biotechnol Bioeng 109:853–866

12. Bauwens CL, Song H, Thavandiran N et al (2011) Geometric control of cardiomyogenic induction in human pluripotent stem cells. Tissue Eng Part A 17:1901–1909

13. Golos TG, Giakoumopoulos M, Garthwaite MA (2010) Embryonic stem cells as models of trophoblast differentiation: progress, opportunities, and limitations. Reproduction 140:3–9

14. Markway B, Tan G-K, Brooke G et al (2010) Enhanced chondrogenic differentiation of human bone marrow-derived mesenchymal stem cells in low oxygen environment micropellet cultures. Cell Transplant 19(1):29–42

15. Fey SJ, Wrzesinski K (2012) Determination of drug toxicity using 3D spheroids constructed from an immortal human hepatocyte cell line. Toxicol Sci 127(2):403–411

16. Wallace L, Reichelt J (2013) Using 3D culture to investigate the role of mechanical signaling in keratinocyte stem cells. In: Turksen K (ed) Skin stem cells. Humana Press, Totowa, NJ, pp 153–164

Chapter 3

Preparation and Characterization of Circulating Angiogenic Cells for Tissue Engineering Applications

Aleksandra Ostojic, Suzanne Crowe, Brian McNeill, Marc Ruel, and Erik J. Suuronen

Abstract

Circulating angiogenic cells (CACs) are a heterogeneous cell population of bone marrow (BM) origin. These cells are most commonly derived from the peripheral blood, bone marrow, and cord blood, and are one of the leading candidates for promoting vascularization in tissue engineering therapies. CACs can be isolated by culturing peripheral blood mononuclear cells (PBMCs) on fibronectin or by flow cytometry to obtain more specific subpopulations. Here we will describe how to generate a population of CACs, and how to characterize the cells and confirm their phenotype. Also, we will provide select methods that can be used to assess the angiogenic and endothelial cell-like properties of the CACs.

Key words Circulating angiogenic cells (CACs), Angiogenic cell therapy, Neovascularization, Peripheral blood mononuclear cells, Tissue engineering

1 Introduction

Endothelial progenitor cells (EPCs) were initially isolated from peripheral blood in 1997, and identified by their co-expression of CD34 and vascular endothelial growth factor receptor-2 (VEGFR-2) [1]. EPCs have been further shown to originate from bone marrow (BM) and other tissues, including skeletal muscle and adipose, dermis, and vascular tissues [2–4]. A true definition has not yet been established for these cells, since their phenotype and function can vary depending on their source and/or method of selection. Regardless of origin, it is generally accepted that EPCs play a role in contributing to neovascularization in ischemic tissues, directly through participation in the generation of new blood vessels, or indirectly through the production of various pro-angiogenic cytokines [2, 5, 6].

Based on this knowledge, EPCs are actively being tested as cells for therapy (±biomaterials) in multiple diseases such as diabetes,

Milica Radisic and Lauren D. Black III (eds.), *Cardiac Tissue Engineering: Methods and Protocols*, Methods in Molecular Biology, vol. 1181, DOI 10.1007/978-1-4939-1047-2_3, © Springer Science+Business Media New York 2014

wound repair, and cardiovascular disease [7]. The potential for revascularization therapy using EPCs and biomaterials has been demonstrated in animal models [8–11] and the feasibility and safety of cell therapy for cardiovascular regeneration has been demonstrated in clinical studies, albeit with relatively modest structural and functional benefits [12, 13]. Furthermore, EPCs have been used successfully as a strategy to vascularize engineered tissues [14–16]. Together, this evidence demonstrates the promise that EPCs have for promoting vascularization in tissue engineering and regenerative medicine applications.

In 2008, Hirschi et al. assigned the name circulating angiogenic cells (CACs) to a specific type of EPC, due to the fact that they are isolated from the peripheral blood and possess angiogenic properties [17]. The most common method for obtaining CACs from peripheral blood is to isolate the mononuclear cells (MNCs) and culture them on fibronectin-coated plates. After 4–7 days, the non-adherent cells are removed and the remaining adherent cells are considered CACs [18–20]. Starting with ~100 ml of peripheral blood, this isolation procedure can consistently yield $25–60 \times 10^6$ cells, which is plenty for performing multiple experiments using the same donor.

Within the population of CACs, CD34$^+$ cells have been identified as one of the most pro-angiogenic subpopulations [21, 22]. CD34$^+$ cells, or other subpopulations of interest, can be isolated from the CACs using fluorescence-activated cell sorting (FACS). This method labels specific cell-surface antigens to identify and sort specific subpopulations of CACs. Common markers used to identify the endothelial precursor/progenitor population are CD34, CD133, and VEGFR-2 [23–26].

The outlined set of experiments describes the procedure for isolating CACs from human peripheral blood, and how to characterize the phenotype of the population with flow cytometry. Furthermore, methods are provided to evaluate the angiogenic properties of the cells and to confirm that they are CACs (using DiI-acLDL and FITC-lectin staining).

2   Materials

2.1   Isolation of Peripheral Blood Mononuclear Cells

1. Wash buffer: 1 % FBS, 0.11 % EDTA in Phosphate Buffered Saline (PBS).

2. Histopaque 1077 (Sigma).

3. 12×9 ml Ethylenediaminetetraacetic acid (EDTA) vacutainer tubes.

4. 50 ml conical tubes.

5. Human donors to donate peripheral blood.

2.2 Culture of Peripheral Blood Mononuclear Cells	1. Fibronectin: 20 μg/ml fibronectin in PBS. Store at 4 °C.
	2. EBM medium (Lonza): 500 ml EBM supplemented with EGM-2 components (Lonza), containing fetal bovine serum (FBS), recombinant long R insulin-like growth factor-1 in aqueous solution (R^3-IGF-1), endothelial growth factor vascular human recombinant (VEGF), gentamicin sulfate amphotericin-B, and epidermal growth factor human recombinant in a buffered BSA saline solution (*see* **Note 1**). Store at 4 °C.
	3. 4×10 cm culture dishes.

2.3 Lifting Cells

1. PBS. Store at 4 °C.
2. EBM medium + supplements (*see* Subheading 2.2).

2.4 CAC Characterization by Fluorescence-Activated Cell Sorting (FACS)

1. PBS. Store at 4 °C.
2. EBM medium + supplements (*see* Subheading 2.2).
3. Antibodies of interest for flow cytometry.

2.5 Angiogenesis Assay

1. Human umbilical vein endothelial cells (HUVECs).
2. HUVEC Medium: M200 medium supplemented with LSGS kit. Store at 4 °C.
3. ECMatrix™ (EMD Millipore; ECM625): includes 10× diluent and matrix solution.
4. 10 cm plates.
5. PBS. Store at 4 °C.
6. CellTracker™ Orange (Invitrogen).
7. 4′,6-diamidino-2-phenylindol (DAPI): 50 μl stock (1 mg/ml) in 950 μl PBS.
8. 96-well plate.
9. EBM medium (*see* Subheading 2.2).

2.6 DiI and Lectin Staining

1. 1,1′-dioctadecyl-3,3,3′,3′-tetramethylindocarbocyanine (DiI)-labeled acetylated low-density lipoprotein (acLDL, Molecular Probes; L-3484): stock at 1 mg/ml.
2. FITC-labeled Ulex-lectin (Sigma; L-9006): stock at 1 mg/ml.
3. PBS. Store at 4 °C.
4. Cytofix buffer (BD Bioscience).

3 Methods

Carry out all procedures at room temperature unless otherwise specified. All reagents that are stored at 4 °C should first be warmed to room temperature before use, unless otherwise specified.

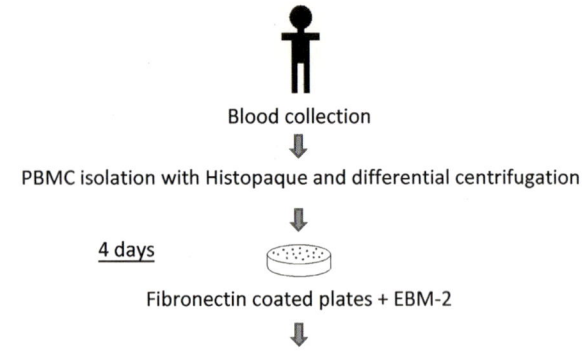

Fig. 1 PBMC isolation and CAC culture protocol

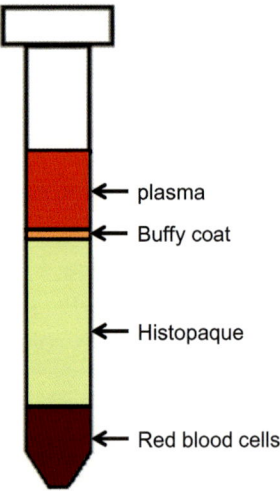

Fig. 2 Following density gradient centrifugation, the peripheral blood is separated into four layers: (1) red blood cell layer; (2) clear Histopaque layer; (3) "Buffy coat" layer consisting of the mononuclear cells; and (4) plasma layer

3.1 Isolation of Mononuclear Cells from Peripheral Blood (Fig. 1)

For ~100 ml blood (12 vials of EDTA vacutainer tubes):

1. Pour 20 ml Histopaque into 4×50 ml tubes (*see* **Note 2**).

2. Collect 12×9 ml blood into EDTA vacutainer tubes. Mix gently for a minimum of 5 min on rocker before processing (*see* **Note 3**).

3. Using a sterile pipette, slowly and carefully layer blood from three tubes on top of the 20 ml Histopaque (*see* **Note 4**). Repeat for each of the 4×50 ml tubes.

4. Centrifuge the tubes for 30 min at $840 \times g$ at room temperature (do not use centrifuge brakes—*see* **Note 5**).

5. Remove the tubes which now contain the separated blood layers (Fig. 2). Aspirate off top plasma layer to within 1 cm of buffy coat (*see* **Note 6**).

6. Using a sterile transfer pipette, remove the top 1 cm of plasma layer and all of buffy coat and place it in clean, sterile 50 ml tubes (should have four separate tubes). Bring final volume in each tube to 40 ml with wash buffer.

7. Centrifuge the tubes at $300 \times g$ for 10 min (do not use centrifuge brakes).

8. Remove and discard the supernatant. Resuspend the cell pellets in a few ml of wash buffer and combine the cell pellet from each of the four tubes into a single tube. Bring the final volume up to 20 ml using wash buffer.

9. Centrifuge the tubes at $300 \times g$ for 10 min (do not use centrifuge brakes).

10. Resuspend the pellet in 1 ml EBM culture medium.

3.2 Plating Fibronectin for Selection of CACs

1. Coat 4×10 cm plate with 3 ml of fibronectin solution (*see* **Note 7**).

2. Incubate at room temperature for 30–60 min.

3. Aspirate the excess liquid and air dry for 15 min.

4. Rinse the plates 2× with PBS: leave PBS on for 5 min, aspirate, and repeat.

5. Add 10 ml EBM medium with supplements, warmed to room temperature.

6. Add ¼ of the cells to each of the four plates and culture at 5 % CO_2 at 37 °C for 4 days without changing the medium.

7. After 4 days, the adherent population is considered to be CACs (Fig. 3).

Fig. 3 Adherent population of CACs generated from the culture of PBMCs on fibronectin-coated plates for 4 days

Table 1
Common markers used to characterize CAC populations. This Table lists some of the markers that have been previously used to characterize the phenotypes of cells in CAC populations. The listed functions are general and not exclusive

Marker	Function	References
CD14	Monocyte marker	[28, 29]
CD31/PECAM-1	Endothelial adhesion molecule	[28–30]
CD34	Progenitor cell marker	[1, 25, 29, 31, 32]
CD45	Leukocyte marker	[28, 29, 33]
CD133	Progenitor cell marker	[25, 29, 31, 32]
CD144/VE-cadherin	Endothelial adhesion molecule	[28, 29, 31]
CD146	Endothelial adhesion molecule	[29, 31, 33]
c-Kit	Progenitor cell marker	[25, 31]
DiI-acLDL uptake	Endothelial/macrophage detection	[28, 31]
VEGFR-2 (KDR, Flk-1)	Endothelial cell receptor	[1, 25, 29, 31]
L-selectin	Leukocyte adhesion molecule	[25, 30]

3.3 Lifting Cells

1. Aspirate the medium.
2. Lift cells using gentle pipetting (*see* **Note 8**).
3. Centrifuge at $300 \times g$ for 5 min (do not use centrifuge brakes).
4. Remove the supernatant and resuspend the CACs in desired volume of EBM medium with supplements, warmed to room temperature.

3.4 CAC Characterization by Fluorescence-Activated Cell Sorting (FACS)

1. Lift and resuspend the cells as previously described.
2. Count cells with an automated system or a hemocytometer.
3. Place 5×10^5 CACs in a 1.5 ml centrifuge tube in 200 μl of PBS.
4. Stain cells for 30 min at 4 °C with pre-conjugated antibodies. Common antibodies used for characterization are CD34, KDR (VEGFR-2), CD133, CD31, and L-selectin, among others (*see* Table 1).
5. Wash with PBS and centrifuge at $300 \times g$ for 5 min.
6. Aspirate medium and resuspend in 500 μl EBM medium with supplements, warmed to room temperature.
7. Place another 5×10^5 CACs in a 1.5 ml centrifuge tube in 200 μl of PBS and stain with appropriate IgG isotype-matched control antibodies (*see* **Note 9**). Repeat **steps 5** and **6**. This can be done concurrently with **steps 3–6**.
8. Analyze and quantify cells with flow cytometry (*see* **Note 10**) [25, 27].

3.5 Angiogenesis Assay

3.5.1 Human Umbilical Vein Endothelial Cell (HUVEC) Culture

1. HUVECs are provided from the manufacturer in a frozen state, and stored this way until needed. Thaw the HUVECs at 37 °C in a water bath.

2. Transfer cells to a 15 ml tube with 4 ml pre-warmed HUVEC medium.

3. Centrifuge at $300 \times g$ for 5 min.

4. Remove supernatant, and resuspend cells in pre-warmed HUVEC medium.

5. Plate cells in 2×10 cm dishes, split 50:50, which is equivalent to $\sim 5 \times 10^5$ cells/dish.

6. Incubate at 37 °C and 5 % CO_2 until 70–80 % confluent (*see* **Note 11**).

7. Change HUVEC medium the following day, and every second day thereafter.

3.5.2 Making the In Vitro Angiogenesis Gels

1. This protocol uses the ECMatrix™ angiogenesis kit (EMD Millipore).

2. Thaw the 10× diluent and the gel solution on ice (*see* **Note 12**).

3. Mix 100 µl diluent with 900 µl of the gel solution. Aliquot 80 µl of gel mixture into individual wells of a 96-well plate (*see* **Note 13**).

4. Allow to set at 37 °C for approximately 1 h.

5. Gels are then ready to use.

3.5.3 Lifting HUVECs

1. Remove the supernatant from the HUVEC cultures.

2. Add 1 ml of 0.25 % trypsin and incubate for 5 min at RT with constant agitation (*see* **Note 14**).

3. Add 5 ml of EBM medium to the plate to inactivate the trypsin. Gently pipette the cells to loosen them from the dish.

4. Rinse the plate with PBS and transfer contents to a 15 ml tube.

5. Centrifuge at $300 \times g$ for 5 min and then remove the supernatant.

6. Stain the cells by resuspending them in CellTracker™ Orange (2 µl aliquot of CellTracker™ Orange + 998 µl PBS) for 30 min at 37 °C (*see* **Note 15**).

7. Add 5 ml of PBS, and then spin the cells as in **step 5**.

8. Resuspend in desired volume of EBM medium with supplements, warmed to room temperature.

3.5.4 Performing the Angiogenesis Assay

1. After completing **step 3** of Subheading 3.3, remove the supernatant.

2. Resuspend the CACs in 4′,6-diamidino-2-phenylindole (DAPI) for 30 min at 37 °C. DAPI solution is prepared by mixing 50 µl of stock (1 mg/ml) with 950 µl PBS.

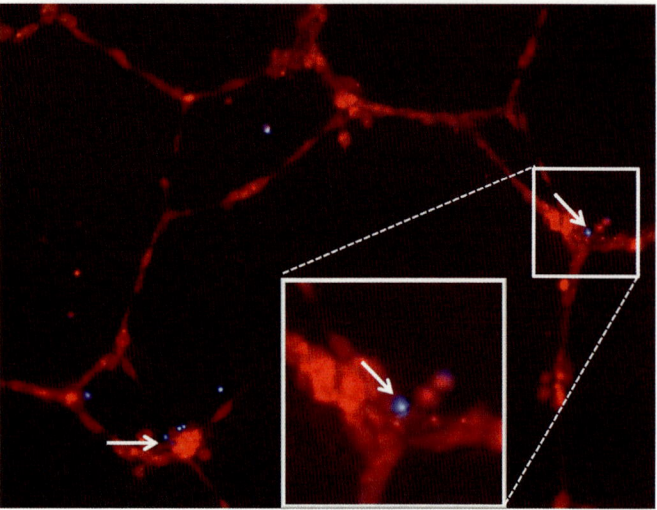

Fig. 4 Angiogenesis assay. HUVECs are stained with CellTracker™ Orange (*red/orange*) and CACs are stained with DAPI (*blue*). *Arrows* indicate CACs that have incorporated into the capillary-like structures

3. Add 5 ml PBS, and centrifuge at $300 \times g$ for 5 min.

4. Remove the supernatant and resuspend CACs in the desired volume of EBM medium with supplements, warmed to room temperature.

5. Count cells using a hemocytometer.

6. Add 10,000 HUVECs and 10,000 CACs together to each well containing ECMatrix™ gel (*see* **Note 16**).

7. Fill the wells with 150 μl EBM medium with supplements and incubate at 37 °C overnight (should be 18 h, and no more than a 24 h incubation) (*see* **Note 17**).

8. Take photographs of the capillary-like structure formations using fluorescence microscopy. Evaluate the number of CACs (blue DAPI$^+$ cells) that have incorporated into the tubule structures (orange HUVECs), as well as the total length of tubule formation (*see* **Note 18**) (Fig. 4).

3.6 DiI and Lectin Staining

1. Prepare (DiI)-labeled acetylated low-density lipoprotein (acLDL) stain (stock at 1 mg/ml). Take 2 μl and add to 998 μl PBS.

2. Aspirate medium from CAC cultures and rinse the plates 2× with PBS.

3. Add 1 ml of DiI-acLDL per dish and incubate at 37 °C for 1 h.

4. Aspirate the DiI-acLDL dye and rinse the cells 2× with PBS.

5. Fix the cells with 1.0 ml of Cytofix buffer for 15 min at 37 °C.

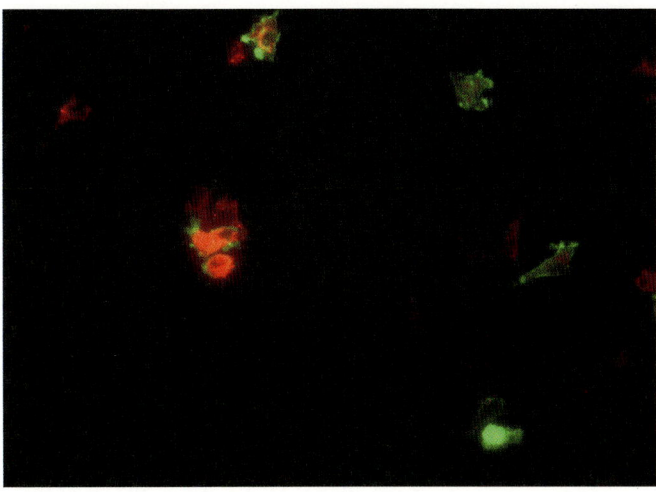

Fig. 5 DiI-acLDL (*red*) and FITC-lectin (*green*) staining of a CAC culture

6. During this incubation step, prepare the lectin stain. Dilute FITC-labeled Ulex-lectin to 10 μg/ml by adding 10 μl of stock solution (1 mg/ml) to 990 μl of PBS.

7. Rinse cells 2× with PBS and add 1 ml FITC-labeled Ulex-lectin per dish and incubate at 37 °C for 1 h.

8. Rinse cells 2× with PBS.

9. Add 1 ml of PBS to each dish and image the cells by fluorescence microscopy (Fig. 5). Dual-staining cells are considered to be CACs (*see* **Note 19**).

---

4 Notes

1. Use a pipette when adding the supplements to minimize volume loss during the transfer.

2. 20 ml of Histopaque is the recommended amount per 30 ml of blood. This amount can be reduced to 10 ml of Histopaque per 30 ml of blood to save on reagents. However, the more Histopaque there is, the easier it is to identify the buffy coat.

3. This step is necessary to mix the blood with the EDTA and prevent clotting. If using the blood right away, mixing can be achieved by inverting the collection tubes multiple times, otherwise leave on the rocker.

4. Cannot have layers mix because it will affect the separation efficacy, and the buffy coat layer will not be as visible.

5. It is very important that braking is not applied during centrifugation, as this may disrupt proper separation of the layers.

6. Make sure not to aspirate any cells that may be stuck to the side of falcon tube. Gently scrape them off with a pipette before proceeding with the transfer.

7. Cover the plate and gently swirl to spread the fibronectin. Make sure all areas are covered; otherwise cells may not uniformly adhere. Fibronectin coating step can be performed immediately prior to CAC isolation. This way the plates will be ready right about the same time that the cells are.

8. PBS can be added and cells can be washed multiple times. If absolutely necessary, a scraper can be used to help lift the cells. Do not allow areas of the plate to dry during this step. Add PBS and look under the microscope to ensure that most of the cells have been removed.

9. If flow cytometry analysis cannot be performed immediately, then the cells can be fixed with 4 % PFA for 20 min, washed with PBS, and spun for 5 min at $300 \times g$. Cells can then be left in the fridge for an extended period of time (weeks) before staining (with primary or IgG control antibodies).

10. Use flow cytometry for analysis using a set number of events (cells). A recommended number of events is 100,000–200,000, but more or less can be used depending on the desired analysis.

11. If cells are over-confluent, they may suffer contact inhibition and lose their growth and proliferative potential, and may also become very difficult to trypsinize. Monitor cell density and make sure to split cells before confluence is reached (>80–90 %).

12. The gel components typically take about 2 h to thaw on ice. When ECMatrix™ is prepared, it is imperative to keep it on ice, otherwise solidification will occur.

13. When pipetting the gel into the wells, do not expel all the way in order to avoid bubbles. As an alternative to 96-well plates, Ibidi microslides can be used, which will reduce reagent consumption.

14. Carefully tap the side of the plate to help lift the cells.

15. Depending on the setup of various microscope systems, other cytoplasmic dyes/fluorochromes may be more suitable to visualize the cells, such as CellTracker™ Green (Invitrogen).

16. Must have 1:1 ratio. Adding too little cells will not allow the HUVECs to form tubule structures and adding too many cells may result in overcrowding or tubules forming too quickly and then regressing.

17. Allowing the angiogenesis assay to proceed for longer than 24 h will lead to regression of the tubule structures. As an alternative to EBM medium, the assay can be performed using M200 medium (this is the medium used for HUVEC culture).

18. It is recommended that a minimum of three, and ideally six images of random fields-of-view (FOV) be taken per well. There are three suggested ways to analyze the results using imaging analysis software (such as ImageJ; http://rsbweb.nih.gov/ij/). (A) Count the number of DAPI$^+$ CACs (blue) that have incorporated into tubule structures per FOV (HUVECs, stained with CellTracker™ Orange). (B) Quantify the total length of tubule formation per field-of-view. This is done by measuring the length of each tubule branch and adding them all together. (C) Quantify the total area of tubule formation. This is done by measuring the area of all the tubule formations and adding them together. The results from all 3–6 FOV are then averaged to obtain the final value per well.

19. DiI-acLDL and FITC-lectin staining is a traditional method originally used to identify adherent endothelial progenitor cells (EPCs) cells from the peripheral blood. However, this means of identifying EPCs is not considered sufficient, and additional characterization is necessary [7]. We suggest this staining simply as a means of confirming the successful generation of an adherent CAC population.

Acknowledgements

This work was supported by the Canadian Institutes of Health Research (to MR and EJS). BM was supported by a University of Ottawa Cardiology Research Endowment Fellowship.

References

1. Asahara T, Murohara T, Sullivan A et al (1997) Isolation of putative progenitor endothelial cells for angiogenesis. Science 275:964–967

2. Asahara T, Takahashi T, Masuda H et al (1999) VEGF contributes to postnatal neovascularization by mobilizing bone marrow-derived endothelial progenitor cells. EMBO J 18:3964–3972

3. Grenier G, Scime A, Le Grand F et al (2007) Resident endothelial precursors in muscle, adipose, and dermis contribute to postnatal vasculogenesis. Stem Cells 25:3101–3110

4. Tamaki T, Akatsuka A, Ando K et al (2002) Identification of myogenic-endothelial progenitor cells in the interstitial spaces of skeletal muscle. J Cell Biol 157:571–577

5. Asahara T, Kawamoto A, Masuda H (2011) Concise review: circulating endothelial progenitor cells for vascular medicine. Stem Cells 29:1650–1655

6. Moreno PR, Sanz J, Fuster V (2009) Promoting mechanisms of vascular health: circulating progenitor cells, angiogenesis, and reverse cholesterol transport. J Am Coll Cardiol 53:2315–2323

7. Fadini GP, Losordo D, Dimmeler S (2012) Critical reevaluation of endothelial progenitor cell phenotypes for therapeutic and diagnostic use. Circ Res 110:624–637

8. Gilbert PM, Blau HM (2011) Engineering a stem cell house into a home. Stem Cell Res Ther 2:3

9. Kuraitis D, Giordano C, Ruel M et al (2012) Exploiting extracellular matrix-stem cell interactions: a review of natural materials for therapeutic muscle regeneration. Biomaterials 33:428–443

10. Segers VF, Lee RT (2011) Biomaterials to enhance stem cell function in the heart. Circ Res 109:910–922

11. Nunes SS, Song H, Chiang CK et al (2011) Stem cell-based cardiac tissue engineering. J Cardiovasc Transl Res 4:592–602

12. Delewi R, Andriessen A, Tijssen JG et al (2013) Impact of intracoronary cell therapy on left ventricular function in the setting of acute myocardial infarction: a meta-analysis of randomised controlled clinical trials. Heart 99:225–232

13. Jeevanantham V, Butler M, Saad A et al (2012) Adult bone marrow cell therapy improves survival and induces long-term improvement in cardiac parameters: a systematic review and meta-analysis. Circulation 126:551–568

14. Amini AR, Laurencin CT, Nukavarapu SP (2012) Differential analysis of peripheral blood- and bone marrow-derived endothelial progenitor cells for enhanced vascularization in bone tissue engineering. J Orthop Res 30:1507–1515

15. Critser PJ, Voytik-Harbin SL, Yoder MC (2011) Isolating and defining cells to engineer human blood vessels. Cell Prolif 44(Suppl 1):15–21

16. Zhang Z, Ito WD, Hopfner U et al (2011) The role of single cell derived vascular resident endothelial progenitor cells in the enhancement of vascularization in scaffold-based skin regeneration. Biomaterials 32:4109–4117

17. Hirschi KK, Ingram DA, Yoder MC (2008) Assessing identity, phenotype, and fate of endothelial progenitor cells. Arterioscler Thromb Vasc Biol 28:1584–1595

18. Favre J, Terborg N, Horrevoets AJ (2013) The diverse identity of angiogenic monocytes. Eur J Clin Invest 43:100–107

19. Kalka C, Masuda H, Takahashi T et al (2000) Transplantation of ex vivo expanded endothelial progenitor cells for therapeutic neovascularization. Proc Natl Acad Sci U S A 97:3422–3427

20. Kuraitis D, Hou C, Zhang Y et al (2011) Ex vivo generation of a highly potent population of circulating angiogenic cells using a collagen matrix. J Mol Cell Cardiol 51:187–197

21. Mackie AR, Losordo DW (2011) CD34-positive stem cells: in the treatment of heart and vascular disease in human beings. Tex Heart Inst J 38:474–485

22. Yang J, Ii M, Kamei N et al (2011) CD34+ cells represent highly functional endothelial progenitor cells in murine bone marrow. PLoS One 6:e20219

23. Garmy-Susini B, Varner JA (2005) Circulating endothelial progenitor cells. Br J Cancer 93:855–858

24. Peichev M, Naiyer AJ, Pereira D et al (2000) Expression of VEGFR-2 and AC133 by circulating human CD34(+) cells identifies a population of functional endothelial precursors. Blood 95:952–958

25. Sofrenovic T, McEwan K, Crowe S et al (2012) Circulating angiogenic cells can be derived from cryopreserved peripheral blood mononuclear cells. PLoS One 7:e48067

26. Werner N, Kosiol S, Schiegl T et al (2005) Circulating endothelial progenitor cells and cardiovascular outcomes. N Engl J Med 353:999–1007

27. Khan SS, Solomon MA, McCoy JP Jr (2005) Detection of circulating endothelial cells and endothelial progenitor cells by flow cytometry. Cytometry B Clin Cytom 64:1–8

28. Shepherd RM, Capoccia BJ, Devine SM et al (2006) Angiogenic cells can be rapidly mobilized and efficiently harvested from the blood following treatment with AMD3100. Blood 108:3662–3667

29. Steinmetz M, Nickenig G, Werner N (2010) Endothelial-regenerating cells: an expanding universe. Hypertension 55:593–599

30. Yip HK, Chang LT, Chang WN et al (2008) Level and value of circulating endothelial progenitor cells in patients after acute ischemic stroke. Stroke 39:69–74

31. Rafii S, Lyden D (2003) Therapeutic stem and progenitor cell transplantation for organ vascularization and regeneration. Nat Med 9:702–712

32. Werner N, Wassmann S, Ahlers P et al (2007) Endothelial progenitor cells correlate with endothelial function in patients with coronary artery disease. Basic Res Cardiol 102:565–571

33. Delorme B, Basire A, Gentile C et al (2005) Presence of endothelial progenitor cells, distinct from mature endothelial cells, within human CD146+ blood cells. Thromb Haemost 94:1270–1279

Chapter 4

Isolation and Expansion of C-Kit-Positive Cardiac Progenitor Cells by Magnetic Cell Sorting

Kristin M. French and Michael E. Davis

Abstract

Cell therapy techniques are a promising option for tissue regeneration; especially in cases such as heart failure where transplantation is limited by donor availability. Multiple cell types have been examined for myocardial regeneration, including mesenchymal stem cells (and other bone marrow-derived cells), induced pluripotent stem cells, embryonic stem cells, cardiosphere-derived cells, and cardiac progenitor cells (CPCs). CPCs are multipotent and clonogenic, can be harvested from mature tissue, and have the distinct advantages of autologous transplant and lack of tumor formation in a clinical setting. Here we focus on the isolation, expansion, and myocardial differentiation of rat CPCs. Brief adaptations of the protocol for isolation from mouse and human tissue are also provided.

Key words Cardiac progenitor cells, C-kit, Cell therapy, MACS, Myocardial infarction

1 Introduction

Cardiac progenitor cells (CPCs), expressing the stem cell marker c-kit, were first identified in the adult heart a decade ago and shown to be self-renewing, clonogenic, and multipotent [1]. In addition to c-kit, CPCs express the stem cell markers multiple drug resistance (MDR) and Sca-1 as well as cardiac markers Nkx2.5 and Gata-4. They are negative for markers of the hematopoietic lineage. CPCs are reported to exist at a density of 1 per 10,000 myocytes in the adult rat heart [1] and as high as 1.4 % of total cardiac cells in humans [2]. The safety and efficacy of CPCs for myocardial repair were demonstrated in the SCIPIO Phase I clinical trial [3]. This trial, in addition to various animal studies, suggests that CPCs can improve cardiac function post-injury through reduction in scar size and myocardial preservation/regeneration [4–6]. CPCs may also be a stem cell candidate for generating cardiac patches for tissue repair or drug screening [7].

In order to isolate CPCs, a tissue section must be excised from the myocardium (atrial tissue is preferred). The cells are then enzymatically

Milica Radisic and Lauren D. Black III (eds.), *Cardiac Tissue Engineering: Methods and Protocols*, Methods in Molecular Biology, vol. 1181, DOI 10.1007/978-1-4939-1047-2_4, © Springer Science+Business Media New York 2014

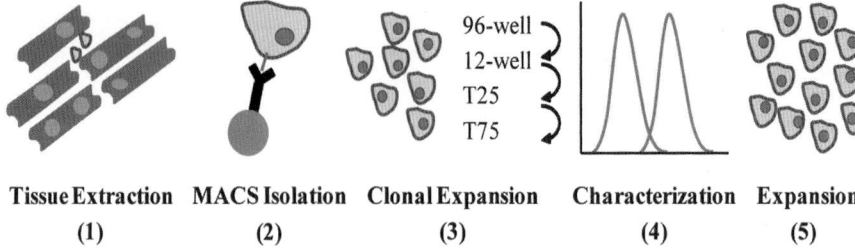

Fig. 1 CPC isolation workflow. Cardiac cells are dissociated from myocardium samples (*1*). Cells are sorted by MACS for c-kit and allowed to expand, 5–6 weeks (*2*). Clonal expansion, if desired, is performed by serial dilution and followed by expansion, 8 weeks (*3*). Clonal candidates are examined by flow cytometry (*4*). Final expansion of c-kit-positive cells allows for cryopreservation and use of CPCs, 2 weeks (*5*)

dissociated from the tissue and collected for magnetic sorting. Immunolabeled beads against the stem cell surface marker c-kit are used to sort CPCs from the total cardiac cells (Subheading 3.2). Cloning individual CPCs (Subheading 3.3) increases the purity of a population, and expansion increases the number of viable cells with which one has to work. The isolation, expansion, and characterization of CPCs are summarized in Fig. 1. Various methods have been reported to induce differentiation of CPCs, with the most popular being dexamethasone (Subheading 3.5) or 5-azacytidine treatment (Subheading 3.6). Variation on this protocol is presented for human (Subheading 3.7) and mouse (Subheading 3.8) samples.

2 Materials

2.1 MACS

1. Tissue.
2. Sterile surgical equipment.
3. Dyna beads (Dynal) (*see* **Notes 1** and **2**).
4. Magnetic sorter (Dynal).
5. 5 mL Falcon tubes, round bottom, with lids.
6. C-kit H-300 (Santa Cruz), sterile.
7. Hanks' balanced salt solution (HBSS), sterile.
8. 0.1 % Bovine serum albumin (BSA)–phosphate buffered saline (PBS), sterile.
9. Collagenase: 1 mg/mL in HBSS, pH 7; sterile.
10. 70 μm Cell strainer.

2.2 Cell Culture

1. 96-Well and 12-well culture plates.
2. T25 and T75 culture flasks.
3. 15 mL Falcon tubes.
4. Treatment media: Ham's F-12, 1× insulin transferrin selenium (ITS), 1× penicillin/streptomycin/glutamine (P/S/G).

5. Culture media: Ham's F-12, 10 % fetal bovine serum (FBS), 10 ng/mL basic fibroblast growth factor (bFGF), 10 ng/mL leukemia inhibitory factor (LIF), 1× P/S/G.

6. Erythropoietin (EPO) (used for human CPC media only).

7. Dexamethasone differentiation media: Ham's F-12, 10 % FBS, and 10 nM dexamethasone.

8. 5-Azacytidine differentiation media: Dulbecco's modified Eagle medium (DMEM), 10 % FBS, and 10 μmol/L 5-azacytidine.

9. Trypsin.

10. Cryoprotective media.

11. Cryo tubes.

12. Freezing container.

13. Pipettes and tips.

14. Serological pipettes.

15. Universal centrifuge.

16. Cell culture incubator at 37 °C and 5 % CO_2.

17. Laminar flow hood.

18. Light microscope.

2.3 Characterization

1. C-kit antibody.

2. MDR antibody.

3. Sca-1 antibody.

4. Gata-4 antibody.

5. Nkx2.5 antibody.

6. Other lineage markers (optional).

7. Appropriate secondary antibodies.

8. 3 % BSA–PBS or Ham's F-12.

9. Flow buffer: 0.1 % BSA, 1× PBS, 5 mM ethylenediamine-tetraacetic acid (EDTA) (*see* **Note 3**).

10. 2 % Paraformaldehyde (PFA) in 1× PBS, ice cold.

11. 90 % Methanol, ice cold.

12. 10 % Isotype appropriate blocking serum.

3 Methods

3.1 Magnetic Bead Preparation

This section should be performed under sterile conditions.

1. Vortex the beads thoroughly.

2. Place 50 μL of beads into 2 mL of 0.1 % BSA–PBS in a 5 mL round-bottom Falcon tube.

3. Concentrate the beads on the magnetic sorter.

4. Aspirate the supernatant away from the beads, and resuspend the beads in 2 mL of 0.1 % BSA–PBS (Subheading 3.1, **steps 3** and **4**, is one wash) (*see* **Note 4**).

5. Add 5 µL of c-kit antibody (*see* **Note 2**).

6. Incubate on a rotator for 2 h at 37 °C.

7. Wash the beads 3× with 2 mL of 0.1 % BSA–PBS.

8. Resuspend the beads in 2 mL of same solution and store at 4 °C.

9. Use the antibody-conjugated beads within 2 weeks.

3.2 Isolation

All animal work should be approved by the appropriate institutional committee. From **step 2**, all work should be performed under sterile conditions.

1. Sacrifice the rats, and isolate the heart following approved procedures.

2. Fill 6 wells of a 12-well plate with cold HBSS.

3. Wash the heart in cold HBSS by serial washing between wells to remove blood.

4. Remove extra tissues (i.e., aorta) from the heart using the lid of the 12-well plate as working surface (*see* **Note 5**).

5. Mince the heart into very small pieces (as fine as possible), and incubate them in 50 mL of fresh, filtered collagenase for 30 min at 37 °C.

6. Shake or vortex the suspension briefly. The tissue will look digested giving a more viscous pasty appearance.

7. Pass the solution through a 70 µm cell strainer into a fresh 50 mL centrifuge tube.

8. Spin down the solution at $1,250 \times g$ for 5 min. The pellet has the required cells.

9. Meanwhile, concentrate the prepared beads, aspirate the supernatant, and resuspend in 2 mL of treatment media.

10. Aspirate the supernatant from the cell pellet. Resuspend the pellet with the c-kit-coated beads in treatment media, and allow it to incubate for 2 h at 37 °C while rotating.

11. After incubation with the beads, concentrate the beads with a magnetic sorter and aspirate the supernatant away (*see* **Note 6**).

12. Wash the beads 4–5 times with sterile Ham's F-12 culture media.

13. Resuspend the beads in the culture media.

14. Seed the resuspended beads in a T25 culture flask with 4 mL of culture media.

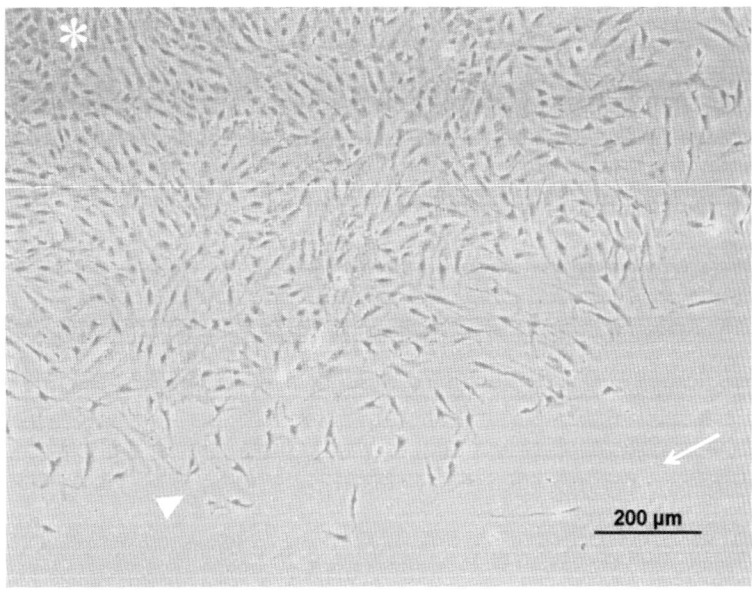

Fig. 2 Expansion of hCPCs. Proliferation of a population of uncloned hCPCs. Newly formed cells (*arrow head*) migrate away from the dense "clone" center (*asterisk*). Magnetic beads from sorting are still visible (*arrows*)

15. Culture the cells at 37 °C and 5 % CO_2. Exchange the culture media every 7 days. Colonies of cells will start appearing in 2–3 weeks. Figure 2 shows an expanding colony.

16. Allow the cells to proliferate until the flask has reached ~80 % confluency; this will take an additional 1–2 weeks.

17. Trypsinize the cells, and transfer them into a T75 culture flask with 10 mL of culture media (*see* **Note 7**).

18. Allow the cells to proliferate until the flask has reached ~80 % confluency. At this point, the cells are ready for characterization and expansion.

3.3 Cloning and Expansion

All steps should be performed under sterile conditions.

1. Trypsinize the cells.

2. Count the cells.

3. Divide the cells in three ways: ~10,000 for clonal expansion (if desired), at least 5×10^5 cells for characterization (1×10^5 cells/ antibody + appropriate controls; *see* Subheading 3.4), and 5×10^5 to 1×10^6 cells for expansion of the uncloned cells.

4. Culture uncloned cells by seeding 5×10^5 to 1×10^6 cells in a T75 with 10 mL of culture media. If additional cells remain, multiple flasks of uncloned cells can be cultured for cryopreservation in order to maintain a stock of cells. *See* Subheading 3.3, **steps 40–44**, for cryopreservation protocol once cells reach 80 % confluency.

	1	2	3	4	5	6	7	8	9	10	11	12
A	*	1000.0	500.0	250.0	125.0	62.5	31.3	15.6	7.8	3.9	2.0	1.0
B	n/a	500.0	250.0	125.0	62.5	31.3	15.6	7.8	3.9	2.0	1.0	0.5
C	n/a	250.0	125.0	62.5	31.3	15.6	7.8	3.9	2.0	1.0	0.5	0.2
D	n/a	125.0	62.5	31.3	15.6	7.8	3.9	2.0	1.0	0.5	0.2	0.1
E	n/a	62.5	31.3	15.6	7.8	3.9	2.0	1.0	0.5	0.2	0.1	0.1
F	n/a	31.3	15.6	7.8	3.9	2.0	1.0	0.5	0.2	0.1	0.1	0.0
G	n/a	15.6	7.8	3.9	2.0	1.0	0.5	0.2	0.1	0.1	0.0	0.0
H	n/a	7.8	3.9	2.0	1.0	0.5	0.2	0.1	0.1	0.0	0.0	0.0

Fig. 3 Cloning by serial dilutions. Theoretical 96-well plate layout for cloning by serial dilutions. The number within each well indicates the theoretical number of cells present after dilutions are performed. Well A1 (*) begins with 2,000 cells. Wells marked "n/a" do not contain cells, nor media. The *grey boxes* represent the theoretical line at which wells containing only a single cell are reached

5. For clonal expansion, dilute cells to 10,000 cells/mL in culture media (*see* **Note 8**).

6. Prepare a 96-well plate by adding 100 μL of culture media to each well, except for A1 (*see* **Note 9**).

7. From the above dilution, take 200 μL (i.e., 2,000 cells) and place it in well A1 of a 96-well plate.

8. Transfer 100 μL from A1 to B1, and mix by pipetting.

9. Transfer 100 μL from B1 to C1, and mix by pipetting. Continue to serially dilute down column 1, until you reach H1. Discard the final 100 μL.

10. Transfer 100 μL from A1 to A2, and mix by pipetting. Continue to serially dilute across row A, until you reach A12. Discard the final 100 μL.

11. Repeat Subheading 3.3, **step 10**, for each row of the 96-well plate.

12. Culture the 96-well plate for 24–48 h.

13. Examine each well of the 96-well plate on a light microscope.

14. Identify wells which contain a single cell. *See* Fig. 3 for theoretical seeding density. Note: On the plate lid, circle identified wells with a colored permanent marker using a different color each time the plate is examined.

15. Culture the 96-well plate for an additional 5 days, exchanging media on day 7.

16. Examine the previously identified wells on the 96-well plate for colony formation.

17. Exclude any wells which appear to have more than one colony growing (i.e., from opposite side of the plate) (*see* **Note 10**).

18. Allow for identified wells to grow to 80 % confluency. This may take 1–2 weeks. Exchange the media every 7 days.

19. Trypsinize the cells, and directly transfer them (without pelleting) from a single well to a single well of a 12-well plate.

20. Repeat Subheading 3.3, **step 19**, for each identified well from the 96-well plate (*see* **Note 11**).

21. Number each clonal candidate for identification purposes.

22. Culture the clonal candidates until they reach 80 % confluency, 1–2 weeks, exchanging media every 7 days.

23. Trypsinize the cells, transfer them from a single well to a T25 culture flask, and culture with 4 mL of culture media. Repeat for all clonal candidates (*see* **Note 12**).

24. Culture the clonal candidates until they reach 80 % confluency, 1–2 weeks, exchanging media every 7 days.

25. Trypsinize the cells, and transfer them to a T75 culture flask with 10 mL of culture media.

26. Culture until the clonal candidates reach 80 % confluency in the T75 culture flasks, 1–2 weeks. Exchange media every 7 days.

27. The cells are now ready for characterization and further expansion of c-kit-positive clones.

28. Trypsinize and pellet the cells.

29. Resuspend the cells in culture media and count.

30. Set aside 1×10^5 cells per desired stain (at least 5×10^5 cells) for characterization.

31. Seed the remaining cells at 5×10^5 to 1×10^6 cells per T75 culture flask (*see* **Note 13**).

32. Complete Subheading 3.4 for characterizing the clonal candidates.

33. Once c-kit-positive clones have been identified, discard culture flasks from any clones which are not >90 % c-kit positive.

34. With the remaining clones, trypsinize and pellet the cells from the flasks.

35. Resuspend the cells in growth media and count.

36. Seed the remaining cells at 5×10^5 to 1×10^6 cells per T75 culture flask. Exchange media every 7 days (*see* **Note 14**).

37. Once the culture flasks are 80 % confluent, about 1 week, trypsinize and pellet the cells.

38. Resuspend the cells in a small volume of culture media (1 mL) and count.

39. Dilute the resuspended cells to 2×10^6 cell/mL in culture media.

40. Prepare the cell-freezing container.

41. For cryopreservation, place 1 mL of the resuspended cells into a labeled cryovial.

42. Add 0.5 mL of cryoprotective media to each cryovial, and gently rock.

43. Quickly place the cryovials into the freezing container and place in −80 °C overnight.

44. Transfer the cryovials to liquid nitrogen.

45. For use, thaw cells by quickly warming the cryovial at 37 °C.

46. Transfer the contents of the cryovial to a T75 culture flask with 10 mL of culture media.

47. Exchange the culture media the next day.

48. Cells should be ~80 % confluent within 1 week and can then be further expanded. Media should be exchanged every 7 days (*see* **Note 15**).

3.4 Characterization

Work in the dark for fluorescently tagged antibodies.

1. Pellet the cells by centrifuging at $1,250 \times g$ for 7 min at 4 °C (*see* **Note 16**). (Skip **steps 1** and **2** if you have already set aside cells for characterization.)

2. Resuspend the cells in 3 % BSA–PBS or culture media.

3. Aliquot the cells in the required number of Falcon tubes (*see* **Notes 17** and **18**).

4. Centrifuge the cells to pellet.

5. Aspirate supernatant, being careful not to disturb the cell pellet.

6. For fixation, add 1 mL of ice-cold 2 % paraformaldehyde to each tube, vortex briefly, and incubate for 30 min on a rocker at 4 °C (*see* **Note 19**).

7. Centrifuge the cells to pellet.

8. For permeabilization, aspirate supernatant, add 1 mL of ice-cold 90 % methanol, vortex briefly, and incubate for 30 min on a rocker at 4 °C (*see* **Notes 20** and **21**).

9. Centrifuge the cells to pellet.

10. Add 300 μL of flow buffer to each tube.

11. Add 300 μL of blocking serum to each tube, vortex briefly, and incubate for 2–4 h on a rocker at 4 °C (*see* **Note 22**).

12. Centrifuge the cells to pellet.

13. Aspirate the supernatant, and add 300 μL of flow buffer.

14. Add 3 μL of the primary antibody, vortex, and incubate on rocker overnight at 4 °C (*see* **Notes 23** and **24**).

15. Centrifuge the cells to pellet, and wash twice with 300 μL flow buffer (*see* **Note 25**). Aspirate the supernatant, and resuspend the cells in 300 μL of flow buffer.

16. Working in the dark, add 1 μL of the appropriate secondary antibody and vortex briefly (*see* **Note 26**).

17. Cover the tubes in foil and incubate on a rocker for 2 h at 37 °C (*see* **Note 27**).

18. Working in the dark, centrifuge the cells to pellet and wash twice with 300 μL flow buffer.

19. Aspirate the supernatant, and resuspend the cells in 300 μL of flow buffer.

20. Transfer to round-bottom Falcon tubes for flow cytometry.

21. Based on flow cytometry results, select clones which are >90 % c-kit positive.

3.5 Differentiation with Dexamethasone [8]

1. Trypsinize cells.

2. Resuspend cells in culture media and count.

3. Seed cells at desired density.

4. Treat the cells with dexamethasone differentiation media for 5 days, exchanging media on day 3.

5. Assess differentiation by flow cytometry (Subheading 3.4) or immunocytochemistry.

3.6 Differentiation with 5-Azacytidine [9]

1. Trypsinize cells.

2. Resuspend cells in culture media and count.

3. Seed cells at a density of 2×10^4 cells/cm^2.

4. Treat the cells with 5-azacytidine differentiation media for 24 h.

5. Following treatment, exchange the media for DMEM with 2 % serum.

6. Culture for 2 weeks, exchanging the media every 3–4 days.

7. Assess differentiation by flow cytometry (Subheading 3.4) or immunocytochemistry.

3.7 Variations for Human Cardiac Progenitor Cells

1. Culture media: Ham's F-12, 10 % FBS, 10 ng/mL bFGF, 1x P/S/G.

2. Exchange media every 3–4 days.

3. Choose appropriate antibodies for characterization.

3.8 Variations for Mouse Cardiac Progenitor Cells

1. Culture media: Ham's F-12, 10 % FBS, 10 ng/mL bFGF, 1× P/S/G.

2. Choose appropriate antibodies for characterization.

4   Notes

1. A bead size of 4.5 μm is better for cell isolation, but 2.8 μm can also be used if the isotype is difficult to match commercially.

2. Take care to buy the appropriate antibody to match the isotype of the beads.

3. EDTA may be removed for sensitive antibodies.

4. After adding to tissue solution, sometimes the beads will fail to concentrate to the side, in which case more concentrated beads have to be prepared (about 100 μl of beads during bead preparation instead of 50 μl of beads).

5. CPCs are thought to be concentrated in the atrial tissue, but the exact locations are unknown. Using the whole heart increases the likelihood of a successful isolation.

6. When target cell population is very less, antibody can be incubated with the digested tissue solution directly after which the incubated solution is incubated with the beads (normal beads; not c-kit-coated beads) overnight at 4 °C.

7. Trypsinize by adding 2 mL of warm trypsin or equivalent to the T25 (or 4 mL to the T75) and incubating for 5 min at 37 °C. Quench the reaction by adding 4 mL of culture media. Spin at $1,250 \times g$ for 5 min at 4 °C. Resuspend the pellet in culture media.

8. Multiple dilution steps may need to be performed to ensure accurate diluting.

9. Preparing two plates in parallel increases the number of clones achieved.

10. Also reexamine wells which have theoretically <1 cell for any growing colonies.

11. One 96-well plate typically yields 5–10 clonal candidates.

12. Continue to maintain the clonal candidate numbering for future identification.

13. Expansion is not critical until after characterization has been performed, but starting to expand at this point maintains the cells at a lower passage.

14. Expand as much as possible at this point.

15. Clones typically maintain c-kit for >15 passages; characterization can be repeated at later passages to confirm that cells are >90 % c-kit positive.

16. The centrifugation time is increased throughout Subheading 3.4 due to the reduced number of cells.

17. Polystyrene centrifuge tubes result in a greater cell yield in this procedure than polypropylene microcentrifuge tubes.

18. You will need one tube for each stain, an unstained control, plus appropriate isotype controls. At the least, cells should be evaluated for c-kit and cardiac lineage markers (i.e., Gata-4 and Nkx2.5). The cells may also be evaluated for smooth muscle and endothelial markers (i.e., smooth muscle alpha-actin and von Willebrand factor) and for the absence of hematopoietic markers (i.e., CD34).

19. Vortex to resuspend the pellet instead of pipetting to reduce cell loss.

20. This time may be reduced if debris is evident during flow cytometry.

21. Permeabilization is only required for staining of cytoplasmic proteins. For cell surface markers and appropriate controls, skip to Subheading 3.4, **step 10**.

22. This step may be performed overnight.

23. To reduce antibody consumption, if staining multiple clones simultaneously, make a 1:100 master mix of the primary antibody and reduce the volume to 100–200 μL.

24. If antibody is fluorescently conjugated skip to **step 19**.

25. Skip this step for the unstained and isotype controls to preserve cells.

26. To reduce antibody consumption, make a 1:300 master mix of the secondary antibody and reduce the volume to 100–200 μL.

27. This step may be performed overnight at 4 °C.

Acknowledgments

This material is based upon work supported by the National Science Foundation Graduate Research Fellowship under Grant No. DGE-1148903 to KMF as well as a grant from the National Institutes of Health HL094527 to MED.

References

1. Beltrami AP, Barlucchi L, Torella D, Baker M, Limana F, Chimenti S et al (2003) Adult cardiac stem cells are multipotent and support myocardial regeneration. Cell 114:763–776

2. Choi S, Jung S, Asahara T, Suh W, Kwon S-M, Baek S (2012) Direct comparison of distinct cardiomyogenic induction methodologies in human cardiac-derived c-kit positive progenitor cells. Tissue Eng Regen Med 9:311–319

3. Bolli R, Chugh AR, D'Amario D, Loughran JH, Stoddard MF, Ikram S et al (2011) Cardiac stem cells in patients with ischaemic cardiomyopathy (SCIPIO): initial results of a randomised phase 1 trial. Lancet 378:1847–1857

4. Makkar RR, Smith RR, Cheng K, Malliaras K, Thomson LE, Berman D et al (2012) Intracoronary cardiosphere-derived cells for heart regeneration after myocardial infarction

(CADUCEUS): a prospective, randomised phase 1 trial. Lancet 379:895–904

5. Linke A, Muller P, Nurzynska D, Casarsa C, Torella D, Nascimbene A et al (2005) Stem cells in the dog heart are self-renewing, clonogenic, and multipotent and regenerate infarcted myocardium, improving cardiac function. Proc Natl Acad Sci U S A 102:8966–8971

6. Urbanek K, Rota M, Cascapera S, Bearzi C, Nascimbene A, De Angelis A et al (2005) Cardiac stem cells possess growth factor-receptor systems that after activation regenerate the infarcted myocardium, improving ventricular function and long-term survival. Circ Res 97:663–673

7. Leor J, Amsalem Y, Cohen S (2005) Cells, scaffolds, and molecules for myocardial tissue engineering. Pharmacol Ther 105:151–163

8. D'Amario D, Fiorini C, Campbell PM, Goichberg P, Sanada F, Zheng H et al (2011) Functionally competent cardiac stem cells can be isolated from endomyocardial biopsies of patients with Advanced cardiomyopathies. Circ Res 108:857–861

9. Itzhaki-Alfia A, Leor J, Raanani E, Sternik L, Spiegelstein D, Netser S et al (2009) Patient characteristics and cell source determine the number of isolated human cardiac progenitor cells. Circulation 120:2559–2566

Chapter 5

Synthesis of Aliphatic Polyester Hydrogel for Cardiac Tissue Engineering

Sanjiv Dhingra, Richard D. Weisel, and Ren-Ke Li

Abstract

Despite clinical advances, ischemic heart disease continues to be a major cause of morbidity and mortality worldwide. Prolonged cardiac ischemia and loss of cardiomyocytes frequently result in progressive pathological remodeling of the myocardium. If the heart is unable to adapt, patients may succumb to terminal heart failure. Cardiac tissue regeneration combining biodegradable biomaterials and stem cells has emerged as a new approach to restore heart function. Biomaterials, including injectable hydrogels and spongy scaffolds, can facilitate stem cell engraftment and survival and prevent adverse ventricular remodeling. Promising early results with injectable, biodegradable hydrogels for cardiac repair have provided new opportunities for designing innovative therapies to treat injured hearts.

Hydrogels can be made from natural or synthetic polymers and have a water content, flexibility, and other physiochemical characteristics similar to those of living tissue, which makes them excellent candidates for tissue repair. In addition, hydrogels can be used as a vehicle to deliver cytokines or cells to the heart and can be employed to encapsulate biological macromolecules or cells and release them into the surrounding tissues during degradation. Hydrogels undergo physicochemical modifications in response to changes in temperature or pH, depending upon their polymer composition, converting from a liquid to a gel. The gel form retains cytokine molecules, allows their prolonged, controlled release, and preserves their bioactivity for extended periods. Polyethylene glycol is a water-soluble, biocompatible polymer that has negligible immunogenicity and can produce efficient conjugation of hydrogels to growth factors. In this chapter, we provide insight into the composition, polymerization, and use of a temperature-sensitive, biodegradable, aliphatic polyester hydrogel that transforms to a gel at physiological temperatures and is a potential candidate for cardiac tissue regeneration.

Key words Cardiac tissue regeneration, Hydrogel, Biomaterial, Ischemia, Stem cells, Growth factors

1 Introduction

1.1 Biomaterials for Heart Repair

Ischemic heart disease continues to be a major source of morbidity and mortality, accounting for more than 12 % of deaths globally. Following a myocardial infarction, the heart is a disturbed microenvironment with the loss of cardiomyocytes and the surviving cells have impaired calcium handling and contractile

Milica Radisic and Lauren D. Black III (eds.), *Cardiac Tissue Engineering: Methods and Protocols*, Methods in Molecular Biology, vol. 1181, DOI 10.1007/978-1-4939-1047-2_5, © Springer Science+Business Media New York 2014

dysfunction. The infarct scar may undergo progressive ventricular thinning and dilatation, resulting in heart failure and death. Injectable hydrogels have been employed to stabilize the infarct and prevent progressive adverse remodeling [1–5]. However, returning ventricular function to near normal may require the combination of injectable biomaterials, the slow release of biologically active cytokines, and regenerative stem cells [3]. Experimental studies and clinical trials suggest that stem cells have the ability to regenerate the heart and prevent pathological remodeling by paracrine mechanisms, including stimulating cardiac-resident stem cells [6–10]. Cardiac tissue engineering techniques apply stem cell biology and engineering principles to the repair and regeneration of cardiac tissue. Three-dimensional polymer scaffolds support stem cell proliferation and stimulate their differentiation and the synthesis of matrix proteins. Successful engraftment of a tissue-engineered construct requires that the scaffold have the optimal biocompatible polymer composition, with close interaction between the cells and the biomaterial. Recently, several biomaterials of various polymer compositions, either alone or conjugated to cytokines, growth factors, or immunosuppressive soluble factors, have been shown to improve the survival and proliferation of the stem cells in the injured tissue [3, 11, 12]. The scaffold material can be of biological origin, derived from various animal tissues such as the urinary bladder, dermis, intestine, or pericardium [11, 13, 14]. However, biologically originated scaffolds may induce host immune responses. Synthetic scaffolds prepared from either natural or synthetic polymers are promising vehicles for cytokine or drug delivery and provide a supportive frame to enhance the engraftment of transplanted cells. So far, hydrogels from polymers such as fibrin, collagen, alginates, peptides, and polyethylene glycol (PEG) have been tested for cardiac tissue engineering [15]. In recent years, hydrogels formed by polymerization of macromolecules with PEG have been extensively investigated as matrices for the controlled release of biologically important molecules and for regenerative applications [16, 17]. PEG is a water-soluble, biocompatible polymer that has negligible immunogenicity and can produce efficient conjugation of growth factors and cells to biomaterials.

1.2 Hydrogels in Cardiac Tissue Regeneration

Temperature-sensitive, biodegradable, PEG hydrogels may be preferred for cardiac repair because they undergo mild solidification under physiological conditions that allows them to maintain cellular bioactivity and viability. In situ gelation provides a huge advantage of hydrogels over solid scaffolds because hydrogels can be injected but solid scaffolds require an extensive surgical procedure to insert them into the injured cardiac region. The injection of cells with hydrogels in the liquid form permits retention of the cells at the site of injection as the hydrogel becomes a gel.

Moreover, hydrogels possess high, tissue-like water content and moldable characteristics [18]. Hydrogels are insoluble, hydrophilic polymer networks produced by cross-linking water-soluble monomers through covalent or hydrogen bonding or by van der Waals interactions between the monomer chains. Hydrogels also have the special ability to swell in liquid solutions, producing an aqueous environment for encapsulated cells [19]. The porous structure of the gel allows the free movement of low-molecular-weight solutes and nutrients and the transport of cellular waste out of the hydrogel, thus mimicking a tissue microenvironment.

Hydrogel-mediated repair and regeneration of the injured myocardium can be accomplished with different combinations of biomaterials, cytokines, and cells [20]. Hydrogels can be carriers for stem cells or biological molecules, which are either physically mixed with or conjugated to the hydrogel. The aqueous hydrogel is loaded with stem cells and biological molecules such as cytokines, matrix proteins, cell adhesive proteins, pro-angiogenic factors, or immunosuppressive factors, and the combination is injected into the damaged myocardium. In situ gelation of the hydrogel provides a niche to nurture engraftment of the injected and recruited stem cells and a microenvironment that is supported by the delayed release of biological molecules. Injected hydrogels are porous and highly permissive, thereby producing a supportive milieu by allowing free diffusion of nutrients and signaling molecules [3, 20–22]. An alternate approach is to separately inject the stem cells and the hydrogel. Under these circumstances, the hydrogel slowly releases the conjugated material and facilitates the survival, proliferation, and differentiation of the stem cells [23].

In this chapter, we provide the method to synthesize a temperature-sensitive, aliphatic polyester hydrogel: PVL–PEG–PVL triblock polymer [poly(d-valerolactone)-block–poly(ethylene glycol)-block–poly(d-valerolactone)]. To explain the conjugation procedure, we describe the procedure for conjugating vascular endothelial growth factor (VEGF) to this hydrogel.

2   Materials

2.1   Chemicals Required

PEG, dichloromethane, calcium hydride, δ-valerolactone, nitrogen gas, trifluoromethanesulfonic acid, $NaHCO_3$, hexane, phosphate-buffered saline (PBS at pH 7.4), N,N'-dicyclohexylcarbodiimide, succinic anhydride, cold diethyl ether, VEGF.

2.2   Equipment Required

Vacuum pump, glass tubes, rubber septum to seal the tube, magnetic stirrer, Bunsen burner, ice bath, Spectra/Por 2 dialysis membrane, SDS electrophoresis system.

To determine the integrity of the conjugate in the polymer, an SDS-PAGE apparatus and consumables are required.

3 Methods

3.1 Synthesis of the PVL–PEG–PVL Polymer (Fig. 1)

1. Dissolve 5 g of PEG in 10 mL of dichloromethane (in a 15 mL glass tube). Add 50 mg of calcium hydride to remove the water, and seal the tube with the rubber septum. Leave at room temperature for 12 h.

2. Add 10 mL of δ-valerolactone and 0.5 g of calcium hydride to the glass tube, and seal the tube with the rubber septum. Leave at room temperature for 12 h. Remove the hydrogen by vacuum, and then fill the tube with high-purity nitrogen gas.

3. To another dry glass tube, add 2 mL of PEG solution (from **step 1**), 2.33 mL of dried valerolactone (from **step 2**), and 5 mL of dichloromethane. Cool this mixture to 0 °C in an ice bath for about 20 min. Add 25 μL of trifluoromethanesulfonic acid dropwise and maintain at 0 °C for 2 h (*see* **Note 1**). Remove from the ice bath, and maintain the reaction for 12–24 h at room temperature.

4. Add 0.2 g of NaHCO$_3$ to neutralize the trifluoromethanesulfonic acid, and stir for 2 h (*see* **Note 2**). Then, precipitate the block copolymer in 50–100 mL of cold hexane, stir for 5–10 min, and leave at room temperature for 2 h. Pour out the hexane, and let the precipitated polymer dry overnight in the fume hood (*see* **Note 3**).

3.2 Gelation of the PVL–PEG–PVL Polymer

To prepare the temperature-sensitive PVL–PEG–PVL gel, employ the polymer prepared in Subheading 3.1.

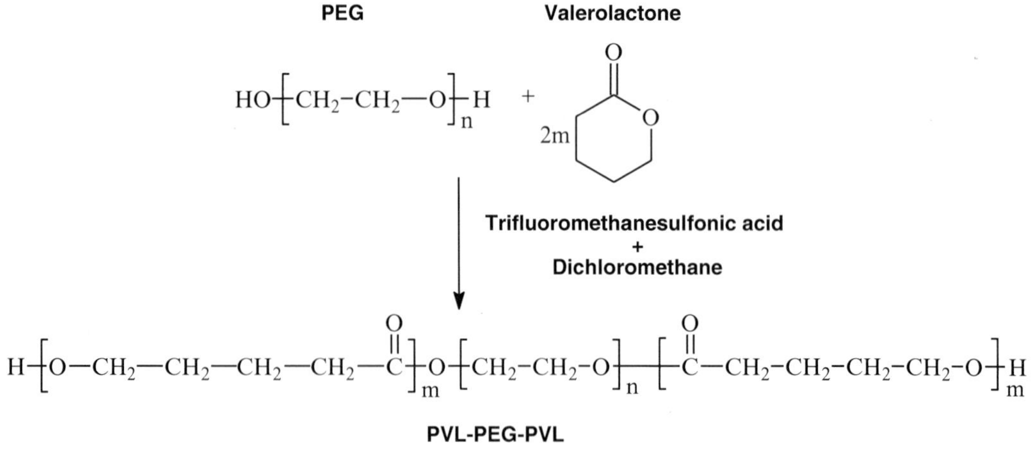

Fig. 1 Synthesis of the PVL–PEG–PVL polymer. Reprinted from Biomaterials 32, Wu J et al., Infarct stabilization and cardiac repair with a VEGF-conjugated, injectable hydrogel, page 580, 2011, with permission from Elsevier

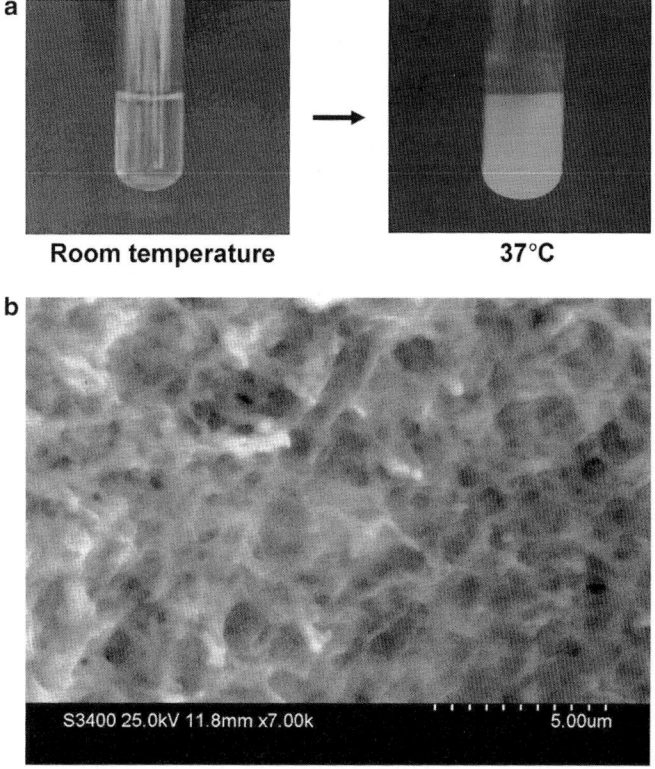

Fig. 2 Gelation and ultrastructure of the hydrogel. (**a**) The PVL–PEG–PVL hydrogel forms a clear liquid at room temperature but solidifies at 37 °C. (**b**) Scanning electron micrograph shows the honeycomb structure of the hydrogel with a pore size of approximately 1 μm. Reprinted from Biomaterials 32, Wu J et al., Infarct stabilization and cardiac repair with a VEGF-conjugated, injectable hydrogel, page 581, 2011, with permission from Elsevier

1. Dissolve 0.4 g of the PVL–PEG–PVL polymer in 1.0 mL of PBS (pH 7.4) in a test tube.

2. Heat the mixture at 60 °C in a water bath and stir until the polymer is completely dissolved. Then cool the mixture to 10 °C by rinsing in cold water. A clear polymer solution will be formed.

The PVL–PEG–PVL triblock polymer is a clear solution when dissolved in water at ambient temperature and forms a hydrogel within 10 min at 37 °C (Fig. 2a). The gelation temperature was determined by maintaining the polymer solution at 37 °C for 10 min (*see* **Note 4**). The gel appears as a honeycomb structure under the scanning electron microscope, with a pore size of approximately 1 μm (Fig. 2b).

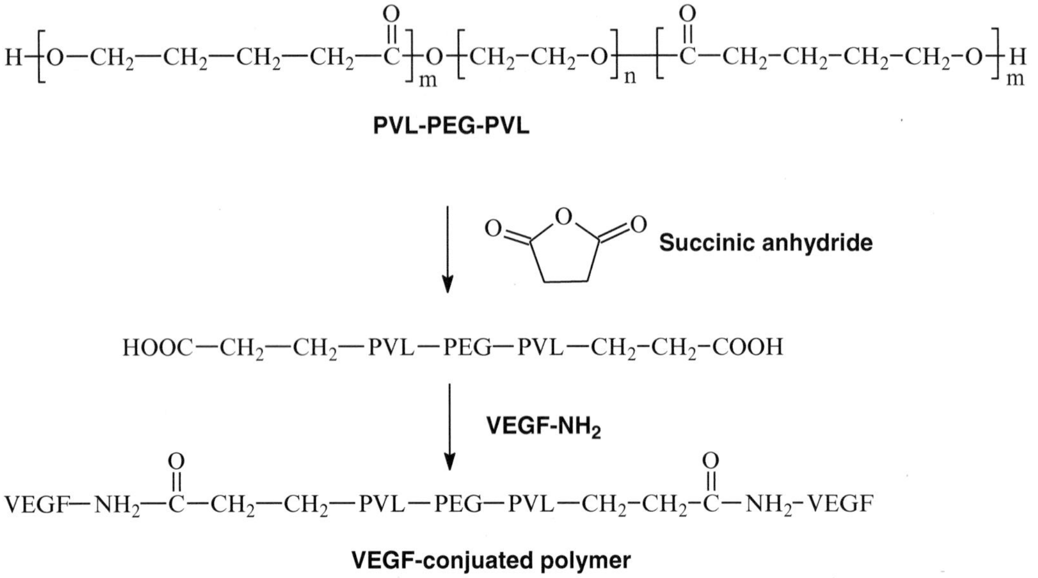

PVL-PEG-PVL

Succinic anhydride

$$HOOC-CH_2-CH_2-PVL-PEG-PVL-CH_2-CH_2-COOH$$

VEGF-NH₂

$$VEGF-NH_2-\overset{O}{\overset{\|}{C}}-CH_2-CH_2-PVL-PEG-PVL-CH_2-CH_2-\overset{O}{\overset{\|}{C}}-NH_2-VEGF$$

VEGF-conjugated polymer

Fig. 3 Conjugation of VEGF to the polymer. Reprinted from Biomaterials 32, Wu J et al., Infarct stabilization and cardiac repair with a VEGF-conjugated, injectable hydrogel, page 580, 2011, with permission from Elsevier

3.3 Modification of the PVL–PEG–PVL Polymer for Conjugation to Biomolecules

For conjugation to a biomolecule, the PVL–PEG–PVL hydrogel obtained above requires modification that can facilitate its conjugation to the biomolecule of the interest. The end hydroxyl groups of the PEG molecules can be easily modified with various functional groups (*see* **Note 5**). We describe here the procedure for conjugation of VEGF to this hydrogel (Fig. 3).

3.4 Synthesis of N-Hydroxysuccinimide-Terminated PEG–PVL–PEG for VEGF Conjugation

The carboxyl-terminated block copolymer is synthesized by reacting the hydroxyl-terminated PVL–PEG–PVL polymer (from Subheading 3.1) with succinic anhydride.

1. Mix 0.5 g of dicarboxyl-terminated block polymer, 0.0396 g of *N,N'*-dicyclohexylcarbodiimide, 0.0221 g of succinic anhydride, and 5 mL of dichloromethane.
2. Maintain the reaction at room temperature for 24 h.
3. Filter the reaction mixture, and add cold diethyl ether to precipitate the NHS–PVL–PEG–PVL–NHS.

3.5 VEGF Conjugation to NHS–PVL–PEG–PVL–NHS

1. To synthesize the VEGF-conjugated hydrogel, take 500 μL of PBS at pH 7.4 (*see* **Note 6**).
2. Add 10 mg of NHS–PVL–PEG–PVL–NHS and 100 ng of VEGF.
3. Maintain the reaction at room temperature for 24 h. VEGF will conjugate to the polymer.

4. Remove the unconjugated VEGF by dialyzing the reaction mixture against water using a Spectra/Por 2 dialysis membrane tubing with a molecular weight cutoff of 12–14 kDa for 48 h.

3.6 Gelation of the VEGF-Conjugated PVL–PEG–PVL

To prepare the temperature-sensitive, VEGF-conjugated gel, follow the steps in Subheading 3.2.

3.7 Integrity of the Conjugate in the Polymer

To verify the conjugation of VEGF to the hydrogel, run an SDS-PAGE and analyze the VEGF conjugated to the PVL–PEG–PVL (VEGF conjugate) by Coomassie brilliant blue staining.

3.8 Hydrogel Stability and Degradation

To check the stability of the hydrogel in vitro, pour different volumes of the hydrogel (e.g., 10, 20, or 30 µL) onto cell culture plates at 37 °C. The gel will solidify and form a soft pellet. The pellets are soaked in culture media and incubated at 37 °C for 35 days. The diameter of each pellet is measured initially and then again at 7-day intervals to assess hydrogel degradation.

To check the stability of the hydrogel in vivo, inject different volumes of the hydrogel (e.g., 10, 20, or 30 µL) subcutaneously into the dorsum of a rat. The implanted gel will absorb water, and nodules will appear underneath the skin. Observe and record the change in nodule size over an extended period of time (from 1 to 42 days).

3.9 Toxicity of the Hydrogel In Vitro

To determine the toxicity of the hydrogel, culture the cells of interest in the presence of the hydrogel for different intervals of time. Measure the level of cytotoxicity (if any) caused by the hydrogel to the cells by recording the amount of lactate dehydrogenase (LDH) release or by using another viability assay. Compare with the untreated control plates.

4 Notes

1. Trifluoromethanesulfonic acid acts as a catalyst in the polymerization and will hasten the reaction. Add it slowly, preferably dropwise, to the reaction tube.

2. NaHCO$_3$ will neutralize the trifluoromethanesulfonic acid and thus terminate the reaction.

3. Cool the hexane to 0 °C in an ice bath for 20 min. The hexane precipitation step is light sensitive. Therefore, wrap the tube with aluminum foil.

4. The gelation temperature is the point where the clear liquid solution converts to a solid gel.

5. The terminal hydroxyl groups of the PEG molecules can be modified with various functional groups, such as carboxyl, thiol, and acrylate, or attached to other molecules or bioactive agents.

6. Like VEGF, other molecules, including cytokines, extracellular matrix proteins, and immunosuppressive factors, can be used for conjugation. The procedure can be modified accordingly.

Acknowledgement

Dr. Ren-Ke Li holds a Canada Research Chair in Cardiac Regeneration.

References

1. Leor J, Tuvia S, Guetta V et al (2009) Intracoronary injection of in situ forming alginate hydrogel reverses left ventricular remodeling after myocardial infarction in swine. J Am Coll Cardiol 54:1014–1023
2. Ifkovits JL, Tous E, Minakawa M et al (2010) Injectable hydrogel properties influence infarct expansion and extent of postinfarction left ventricular remodeling in an ovine model. Proc Natl Acad Sci U S A 107:11507–11512
3. Wu J, Zeng F, Huang XP et al (2011) Infarct stabilization and cardiac repair with a VEGF-conjugated, injectable hydrogel. Biomaterials 32:579–586
4. Singelyn JM, Sundaramurthy P, Johnson TD et al (2012) Catheter-deliverable hydrogel derived from decellularized ventricular extracellular matrix increases endogenous cardiomyocytes and preserves cardiac function post-myocardial infarction. J Am Coll Cardiol 59:751–763. doi:10.1016/j.jacc.2011.10.888
5. Sabbah HN, Wang M, Gupta RC et al (2013) Augmentation of left ventricular wall thickness with alginate hydrogel implants improves left ventricular function and prevents progressive remodeling in dogs with chronic heart failure. J Am Coll Cardiol Heart Failure 1:252–258. doi:10.1016/j.jchf.2013.02.006
6. Balsam LB, Wagers AJ, Christensen JL et al (2004) Haematopoietic stem cells adopt mature haematopoietic fates in ischaemic myocardium. Nature 428:668–673
7. Orlic D, Kajstura J, Chimenti S et al (2001) Bone marrow cells regenerate infarcted myocardium. Nature 410:701–705
8. Tomita S, Li RK, Weisel RD et al (1999) Autologous transplantation of bone marrow cells improves damaged heart function. Circulation 100:II247–II256
9. Assmus B, Honold J, Schachinger V et al (2006) Transcoronary transplantation of progenitor cells after myocardial infarction. N Engl J Med 355:1222–1232
10. Schachinger V, Erbs S, Elsasser A et al (2006) Intracoronary bone marrow-derived progenitor cells in acute myocardial infarction. N Engl J Med 355:1210–1221
11. Christman KL, Lee RJ (2006) Biomaterials for the treatment of myocardial infarction. J Am Coll Cardiol 48:907–913
12. Kang K, Sun L, Xiao Y et al (2012) Aged human cells rejuvenated by cytokine enhancement of biomaterials for surgical ventricular restoration. J Am Coll Cardiol 60:2237–2249. doi:10.1016/j.jacc.2012.08.985
13. Ueno T, Pickett LC, de la Fuente SG, Lawson DC, Pappas TN (2004) Clinical application of porcine small intestinal submucosa in the management of infected or potentially contaminated abdominal defects. J Gastrointest Surg 8:109–112
14. Zhao ZQ, Puskas JD, Xu D et al (2010) Improvement in cardiac function with small intestine extracellular matrix is associated with recruitment of C-kit cells, myofibroblasts, and macrophages after myocardial infarction. J Am Coll Cardiol 55:1250–1261
15. Christman KL, Vardanian AJ, Fang Q et al (2004) Injectable fibrin scaffold improves cell transplant survival, reduces infarct expansion, and induces neovasculature formation in ischemic myocardium. J Am Coll Cardiol 44:654–660
16. Peppas NA, Keys KB, Torres-Lugo M, Lowman AM (1999) Poly(ethylene glycol)-containing hydrogels in drug delivery. J Control Release 62:81–87
17. Yonet-Tanyeri N, Rich MH, Lee M et al (2013) The spatiotemporal control of erosion and molecular release from micropatterned poly(ethylene glycol)-based hydrogel. Biomaterials 34:8416–8423

18. Torres DS, Freyman TM, Yannas IV, Spector M (2000) Tendon cell contraction of collagen-GAG matrices in vitro: effect of cross-linking. Biomaterials 21:1607–1619

19. Brannon-Peppas L (1990) Preparation and characterization of crosslinked hydrophilic networks. In: Brannon-Peppas L, Harland RS (eds) Absorbent polymer technology. Elsevier, Amsterdam, pp 45–66

20. Miyagi Y, Zeng F, Huang XP et al (2010) Surgical ventricular restoration with a cell- and cytokine-seeded biodegradable scaffold. Biomaterials 31:7684–7694

21. Salimath AS, Phelps EA, Boopathy AV et al (2012) Dual delivery of hepatocyte and vascular endothelial growth factors via a protease-degradable hydrogel improves cardiac function in rats. PLoS One 7:e50980. doi:10.1371/journal.pone.0050980

22. Hassan W, Dong Y, Wang W (2013) Encapsulation and 3D culture of human adipose-derived stem cells in an in-situ cross-linked hybrid hydrogel composed of PEG-based hyperbranched copolymer and hyaluronic acid. Stem Cell Res Ther 4:32

23. Dhingra S, Li P, Huang XP et al (2013) Preserving PGE2 level prevents rejection of implanted allogeneic mesenchymal stem cells and restores post-infarction ventricular function. Circulation 128:S1–S10

Chapter 6

Fabrication of PEGylated Fibrinogen: A Versatile Injectable Hydrogel Biomaterial

Iris Mironi-Harpaz, Alexandra Berdichevski, and Dror Seliktar

Abstract

Hydrogels are one of the most versatile biomaterials in use for tissue engineering and regenerative medicine. They are assembled from either natural or synthetic polymers, and their high water content gives these materials practical advantages in numerous biomedical applications. Semisynthetic hydrogels, such as those that combine synthetic and biological building blocks, have the added advantage of controlled bioactivity and material properties. In myocardial regeneration, injectable hydrogels premised on a semisynthetic design are advantageous both as bioactive bulking agents and as a delivery vehicle for controlled release of bioactive factors and/or cardiomyocytes. A new semisynthetic hydrogel based on PEGylated fibrinogen has been developed to address the many requirements of an injectable biomaterial in cardiac restoration. This chapter highlights the fundamental aspects of making this biomimetic hydrogel matrix for cardiac applications.

Key words Hydrogel, Myocardial infarction, Polyethylene glycol

1 Introduction

Cardiac restoration therapies as treatments for myocardial infarction (MI) are intended to augment the fibrotic scar tissue in the infarct with functional cardiac muscle and to prevent the deterioration of the remaining healthy cardiac muscle. Several approaches have been suggested to achieve these goals; generally they have been divided into cellular and acellular approaches [1]. The use of an acellular approach in cardiac restoration is premised on the hypothesis that a biomaterial can prevent the destructive cardiac remodeling process following MI by mechanically stabilizing the myocardial wall that proceeds an acute infarction. A number of different biomaterials were tested for this purpose, including biological polymers such as collagen, alginate, Matrigel, fibrin, self-assembling polypeptides, hyaluronic acid-based hydrogels and naturally derived myocardial matrix, as well as synthetic polymers such as poly(propylene) (Marlex) and polyester [2].

Milica Radisic and Lauren D. Black III (eds.), *Cardiac Tissue Engineering: Methods and Protocols*, Methods in Molecular Biology, vol. 1181, DOI 10.1007/978-1-4939-1047-2_6, © Springer Science+Business Media New York 2014

The simplest method to stabilize the myocardium is to use an inert injectable biomaterial matrix in order to passively preserve cardiac geometry, i.e., cardiac bulking agent. Injecting a hydrogel into the wall of the left ventricle (LV), for example, affects the heart function as described by the law of Laplace, which stipulates that wall stress is proportional to pressure and radius and inversely proportional to the wall thickness. In cardiac mechanics, remodeling following MI promotes the thinning of the LV wall and thereby increases the wall stresses. This results in apoptotic cell death that progressively enhances adverse remodeling toward scarring and dilated cardiomyopathies. Hence, by increasing wall thickness using an injectable biomaterial, the wall stress can be reduced either by attenuating the adverse cardiac remodeling or increasing the bulk properties of the myocardial tissue (i.e., bulking agent). In theory, altering cardiac mechanics vis-à-vis the law of Laplace using a cardiac bulking agent can decrease apoptosis and provide a more effective preservation of cardiac function. However, Rane et al. recently demonstrated that passive structural reinforcement alone is not sufficient to prevent adverse remodeling following MI, implying that cardiac bulking agents must also actively alter cardiac remodeling to attenuate post-MI scar formation. Accordingly, bioactive hydrogels, such as those that can promote cell adhesion, proliferation, or inflammatory biodegradation, are necessary for post-MI cellular remodeling leading to improvements in cardiac function. The role of bioactivity notwithstanding, few studies have examined biomechanical augmentation on cardiac remodeling using injectable bioactive bulking agents of wide-ranging mechanical properties.

Biomaterials that act as both passive and active bulking agents can be prepared by introducing bioactivity through sequestered biomolecules within the matrix (e.g., pro-angiogenic factors, growth factors, and differentiation factors). These bioactive molecules in the matrix can be designed to initiate a healing response by stimulating neovascularization, angiogenesis, and cellular recruitment to the scar tissue [3–8]. The biomaterial may also be designed to control the release of these biomolecules, thereby optimizing the kinetics of the healing response and improving the neo-tissue formation and integration in the host [9–12]. Moreover, an injectable bioactive biomaterial can be administered by a less invasive arthroscopic procedure or catheterization, which has a tremendous clinical advantage over open-chest surgeries. Accordingly, injectable biomaterials should undergo a nontoxic in situ liquid-to-solid transition (polymerization), with the transition being amenable to a suspension of cardiomyocytes dispersed in the liquid precursor. The compliance of the polymer matrix must not obstruct the cellular remodeling in the myocardium [13], nor distort the myocardial geometry [14]. In the case where cells are incorporated in the hydrogel matrix, it is also important to consider the impact that material

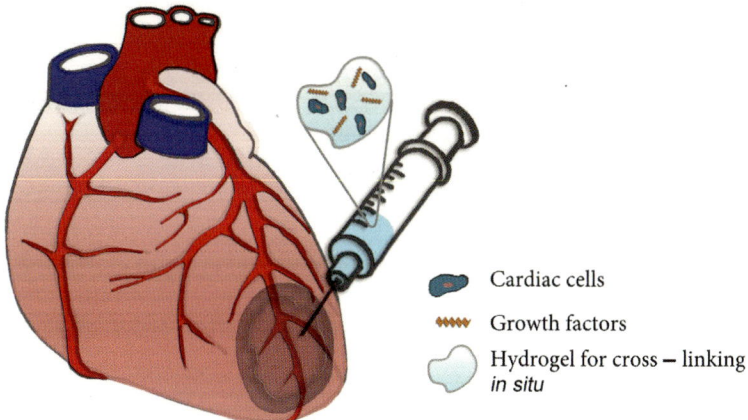

Cardiac cells

Growth factors

Hydrogel for cross – linking *in situ*

Fig. 1 Injectable hydrogel biomaterials for cardiac repair. A liquid precursor solution can be combined with cells and/or bioactive factors and injected into the infarcted myocardium. A nontoxic polymerization reaction causes the liquid-to-solid transition that forms an elastic matrix in situ (image adapted from Williams et al. 2014)

compliance has on cardiomyocyte survival and phenotype [15]. Finally, the biomaterial needs to promote a suitable environment for myocardial cells to migrate and integrate within the host tissue, including the proper vascularization for these cells, all the while supporting their contractility and function.

To address many of the stringent requirements for injectable biomaterial in cardiac restoration, a new biosynthetic material based on PEGylated fibrinogen has been developed [16]. The PEGylated fibrinogen is an injectable polymer matrix with bioactivity similar to native fibrin and the added advantage of controllable physical properties and biodegradation [17, 18]. Several studies have been conducted with this matrix, most notably for treating MI through the transplantation of cardiomyocytes or angiogenic growth factors, or as a bioactive bulking agent alone. The synthetic constituent of the PEGylated fibrinogen hydrogel is poly(ethylene glycol) (PEG)—a widely used polymer in biomedical and pharmacological applications due to its inert characteristics and low immunogenicity [19, 20]. The injectable hydrogel is made by conjugating PEG with denatured fibrinogen in solution to form a PEGylated fibrinogen liquid precursor. The liquid precursor is assembled into a 3-dimensional (3-D) hydrogel scaffold in situ using a nontoxic photopolymerization [21] (Fig. 1).

The fibrinogen molecule contains functional biological sequences that are essential for cell invasion, cell migration, and tissue remodeling. These include proteolytic cleavage sites sensitive to matrix metalloproteinases (MMPs) and plasmin, two common proteases in cardiac remodeling, as well as adhesion sites such as RGD [22]. Moreover, fibrin degradation products

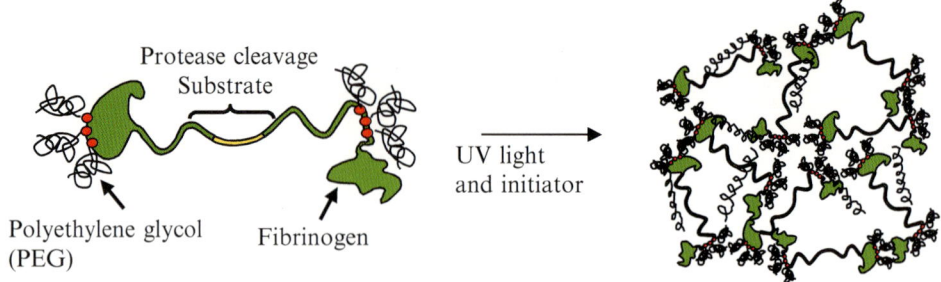

Fig. 2 PEGylated fibrinogen assembles into a solid hydrogel network with UV photopolymerization (adapted from ref. 12)

induce angiogenesis by stimulating the proliferation, migration, and differentiation of endothelial cells [5]. Hence, the fibrinogen-based hydrogels not only provide the cardiac cells with a 3-D environment containing adhesion and proteolytic sites but will also induce an angiogenesis response, which should increase the long-term in vivo survival of a cellular graft. Based on these favorable characteristics, the PEG-conjugated fibrinogen allows one to create a hydrogel with a controlled degradation rate, reduced immune response to the graft, and controlled mechanical properties (e.g., by altering the composition of the matrix) [18].

The PEGylated fibrinogen is a very versatile material that has been used as a bioactive implant alone [23, 24], as a cell carrier [23], or as a growth factor delivery platform for controlled release of bioactive factors [25, 26]. The preparation of the PEGylated fibrinogen is straightforward and involves a Michael-type addition reaction of the denatured fibrinogen to the di-functional PEG-diacrylate (PEG-DA), precipitation and purification steps of the PEGylated fibrinogen, and cross-linking into a hydrogel in the presence of a photoinitiator (Fig. 2). Prior to cross-linking, cardiomyocytes or growth factors such as vascular endothelial growth factor (VEGF) may be added to the PEGylated fibrinogen hydrogel precursor solution. This chapter covers the methods used for the preparation of PEGylated fibrinogen liquid precursor, the formation of a PEGylated fibrinogen hydrogel from the liquid precursor, and the encapsulation of a growth factor into the PEGylated fibrinogen matrix during the polymerization reaction.

2 Materials

1. Fibrinogen (Type I-S from bovine plasma, 65 %, 166 kDa).
2. Linear PEG-DA, 10 kDa.
3. Tris(2-carboxyethyl)phosphine hydrochloride (TCEP).
4. Phosphate buffer solution (PBS).
5. PBS with 8 M urea at pH 7.4.

6. Dialysis bag (MWCO 12,000–14,000).

7. Purified VEGF.

8. Photoinitiator (Ciba Irgacure 2959).

9. Acetone.

10. 100 ml Erlenmeyer flaks.

11. Magnetic stirrer.

12. 1 L Separation funnel.

13. 50 ml Centrifuge tube.

14. Microcentrifuge tubes.

---

3 Methods

3.1 Product Reaction

1. Clean all dishes to be used with acetone to remove the organics in the dish.

2. Weigh 0.25 g fibrinogen (this is the weight of pure fibrinogen) (*see* **Notes 1** and **2**) and place in a 100 ml Erlenmeyer flask.

3. Dissolve fibrinogen in 8 M urea in PBS (pH 7.4) to a final concentration of 7 mg/ml of fibrinogen (add 35.7 ml PBS–urea buffer). Add a magnetic stir bar, and stir slowly (*see* **Note 3**).

4. Add 1.5-fold excess of TCEP (excess to fibrinogen cysteines, 62) (add 18 mg TCEP), and continue to stir.

5. Adjust pH of solution to pH = 8.

6. Weigh fourfold excess PEG-DA ($1.82 \times g$) in a centrifuge tube, dissolve in 8 M urea in PBS to final concentration of 280 mg/ml (6.5 ml), vortex, and centrifuge for 1–2 min at 2,500 RPM.

7. Slowly add PEG-DA solution to flask.

8. Cover the flask in aluminum foil completely, and allow reaction to proceed at room temperature for 3 h.

3.2 Precipitation

1. Double the volume of the reaction solution with 8 M urea in PBS pH 7.4 (42 ml), and continue to stir for 5 min.

2. Transfer the final product to a 1 L separation funnel containing 4 volumes of acetone (328 ml), and mix vigorously.

3. PEGylated fibrinogen should precipitate almost immediately and migrate toward the bottom of the funnel. Allow product to gather at the bottom completely (5–15 min).

4. Transfer precipitate to a 50 ml centrifuge tube.

3.3 Purification

1. Centrifuge for 2 min at 1342 r.c.f. to separate product from acetone. Aspirate acetone supernatant. Repeat this step to collect all the remaining product. The expected product should contain about 5 ml of the precipitate.

2. Weigh the precipitate and dissolve with ~2.2 volume of 8 M urea in PBS pH 7.4 using a homogenizer at low speed.

3. Centrifuge for 1–2 min at 1342 r.c.f to remove air bubbles—the solution should be clear.

4. Transfer the solution to a dialysis bag (*see* **Note 4**), and secure with clips.

5. Dialyze product against 4 L PBS (×1) over 24 h at 4 °C in a dark environment. The PBS should be changed three times over the 24-h period.

6. Remove product solution from the dialysis bag and aliquot into sterile microcentrifuge tubes. Store in a dark environment at –80 °C.

3.4 Characterization

1. Test the ability of solution to form a hydrogel by adding 1 % photoinitiator solution (*see* **Note 5**) and exposing to UV light (*see* **Note 6**) for 5 min for 0.1–1 ml hydrogel precursor.

2. Check the gels under phase contrast-enhanced light microscope to ensure homogeneous polymerization.

3. Perform BCA analysis to obtain protein (fibrinogen) concentration (usually 7–10 mg/ml).

4. Lyophilize overnight an exact volume of product in microcentrifuge tubes to determine total product concentration and PEGylation efficiency (*see* **Notes 7** and **8**).

3.5 VEGF Incorporation

1. Prepare VEGF-containing solution according to the manufacturer's instructions and with desired concentration.

2. Prepare PEG–fibrinogen solution.

3. Gently mix PEG–fibrinogen with VEGF according to the calculations (*see* **Note 9**).

4. Store at 4 °C in a tube wrapped with aluminum foil to allow binding between fibrinogen and VEGF.

5. Add 0.1% w/vol photoinitiator, and use.

4 Notes

1. Warm up fibrinogen to room temperature before opening.

2. In case of 65 % fibrinogen, weigh 0.385 g.

3. It is very important to avoid foaming.

4. Remove as many air bubbles as you can by squeezing them out.

5. To prepare the photoinitiator solution dissolve 10 % of Irgacure 2959 powder in 90 % ethanol. Wrap the tube containing the solution in aluminum foil and keep at room temperature. Use 1 % solution of photoinitiator in PEG–fibrinogen solution.

6. A 365 nm UV lamp is used.

7. To obtain the total product concentration, subtract the PBS salt concentration (9.88 mg/ml) from the dry net product concentration (after lyophilization). To obtain the total PEG concentration (should be 10–20 mg/ml), subtract both PBS salts and protein concentrations.

8. The PEGylation efficiency is the ratio of the actual PEG concentration in the PEG–fibrinogen to the theoretical calculated PEG concentration that can be achieved in this process.

9. Avoid bubble formation.

References

1. Murry CE, Field LJ, Menasche P (2005) Cell-based cardiac repair: reflections at the 10-year point. Circulation 112:3174–3183

2. Christman KL, Lee RJ (2006) Biomaterials for the treatment of myocardial infarction. J Am Coll Cardiol 48:907–913

3. Christman KL, Fok HH, Sievers RE, Fang Q, Lee RJ (2004) Fibrin glue alone and skeletal myoblasts in a fibrin scaffold preserve cardiac function after myocardial infarction. Tissue Eng 10:403–409

4. Christman KL, Vardanian AJ, Fang Q, Sievers RE, Fok HH, Lee RJ (2004) Injectable fibrin scaffold improves cell transplant survival, reduces infarct expansion, and induces neovasculature formation in ischemic myocardium. J Am Coll Cardiol 44:654–660

5. Bootle-Wilbraham CA, Tazzyman S, Thompson WD, Stirk CM, Lewis CE (2001) Fibrin fragment E stimulates the proliferation, migration and differentiation of human microvascular endothelial cells in vitro. Angiogenesis 4:269–275

6. Huang NF, Yu J, Sievers R, Li S, Lee RJ (2005) Injectable biopolymers enhance angiogenesis after myocardial infarction. Tissue Eng 11: 1860–1866

7. Dai W, Wold LE, Dow JS, Kloner RA (2005) Thickening of the infarcted wall by collagen injection improves left ventricular function in rats: a novel approach to preserve cardiac function after myocardial infarction. J Am Coll Cardiol 46:714–719

8. Landa N, Miller L, Feinberg MS, Holbova R, Shachar M, Freeman I et al (2008) Effect of injectable alginate implant on cardiac remodeling and function after recent and old infarcts in rat. Circulation 117:1388–1396

9. Salimath AS, Phelps EA, Boopathy AV, Che PL, Brown M, García AJ (2012) Dual delivery of hepatocyte and vascular endothelial growth factors via a protease-degradable hydrogel improves cardiac function in rats. PLoS One 7(11)

10. Davies NH, Schmidt C, Bezuidenhout D, Zilla P (2012) Sustaining neovascularization of a scaffold through staged release of vascular endothelial growth factor-A and platelet-derived growth factor-BB. Tissue Eng A 18: 26–34

11. Hao X, Silva EA, Månsson-Broberg A, Grinnemo KH, Siddiqui AJ, Dellgren G, Wärdell E, Brodin LA, Mooney DJ, Sylvén C (2007) Angiogenic effects of sequential release of VEGF-A165 and PDGF-BB with alginate hydrogels after myocardial infarction. Cardiovasc Res 75:178–185

12. Moon JJ, Saik JE, Poché RA, Leslie-Barbick JE, Lee SH, Smith AA, Dickinson ME, West JL (2010) Biomimetic hydrogels with pro-angiogenic properties. Biomaterials 31: 3840–3847

13. Davis ME, Motion JP, Narmoneva DA, Takahashi T, Hakuno D, Kamm RD et al (2005) Injectable self-assembling peptide nanofibers create intramyocardial microenvironments for endothelial cells. Circulation 111:442–450

14. Kofidis T, Lebl DR, Martinez EC, Hoyt G, Tanaka M, Robbins RC (2005) Novel injectable bioartificial tissue facilitates targeted, less invasive, large-scale tissue restoration on the beating heart after myocardial injury. Circulation 112:173–177

15. McDevitt TC, Woodhouse KA, Hauschka SD, Murry CE, Stayton PS (2003) Spatially organized layers of cardiomyocytes on biodegradable polyurethane films for myocardial repair. J Biomed Mater Res A 66:586–595

16. Seliktar D (2005) Extracellular stimulation in tissue engineering. Ann N Y Acad Sci 1047: 386–394

17. Almany L, Seliktar D (2005) Biosynthetic hydrogel scaffolds made from fibrinogen and polyethylene glycol for 3D cell cultures. Biomaterials 26:2467–2477

18. Dikovsky D, Bianco-Peled H, Seliktar D (2006) The effect of structural alterations of PEG-fibrinogen hydrogel scaffolds on 3-D cellular morphology and cellular migration. Biomaterials 27:1496–1506

19. Barker TH, Fuller GM, Klinger MM, Feldman DS, Hagood JS (2001) Modification of fibrinogen with poly(ethylene glycol) and its effects on fibrin clot characteristics. J Biomed Mater Res 56:529–535

20. Bailon P, Berthold W (1998) Polyethylene glycol-conjugated pharmaceutical proteins. Pharm Sci Technol Today 1:352–356

21. Williams CG, Malik AN, Kim TK, Manson PN, Elisseeff JH (2005) Variable cytocompatibility of six cell lines with photoinitiators used for polymerizing hydrogels and cell encapsulation. Biomaterials 26:1211–1218

22. Hersel U, Dahmen C, Kessler H (2003) RGD modified polymers: biomaterials for stimulated cell adhesion and beyond. Biomaterials 24: 4385–4415

23. Habib M, Shapira-Schweitzer K, Caspi O, Gepstein A, Arbel G, Aronson D, Seliktar D, Gepstein L (2011) A combined cell therapy and in-situ tissue-engineering approach for myocardial repair. Biomaterials 32:7514–7523

24. Plotkin M, Vaibavi SR, Rufaihah AJ, Nithya V, Wang J, Shachaf Y, Kofidis T, Seliktar D (2014) The effects of matrix stiffness of injectable hydrogels on the preservation of cardiac function after a heart attack. Biomaterials 35(5):1429–1438

25. Rufaihah AJ, Vaibavi SR, Plotkin M, Shen J, Nithya V, Wang J, Seliktar D, Kofidis T (2013) Enhanced infarct stabilization and neovascularization mediated by VEGF-loaded PEGylated fibrinogen hydrogel in a rodent myocardial infarction model. Biomaterials 34: 8195–8202

26. Berdichevski A, Simaan-Yameen H, Dafni H, Neeman M, Seliktar D (2014) Bimodal MRI/Fluorescence imaging identifies host angiogenic response to implant geometry. (In Review)

Chapter 7

Natural Cardiac Extracellular Matrix Hydrogels for Cultivation of Human Stem Cell-Derived Cardiomyocytes

Donald O. Freytes, John D. O'Neill, Yi Duan-Arnold, Emily A. Wrona, and Gordana Vunjak-Novakovic

Abstract

Biomaterial scaffolds made of natural and synthetic materials are designed to serve as a structural and informational template for cell attachment and tissue formation. The use of native extracellular matrix (ECM) is of special interest for the culture of cardiac stem and progenitor cells due to the presence of intrinsic regulatory factors regulating cardiac function. We describe here how to obtain native ECM hydrogels from porcine hearts for the culture of human embryonic, induced pluripotent, and somatic stem cells for cardiac tissue engineering and regenerative medicine applications.

Key words Native, Heart, Cardiac, Extracellular matrix, ECM, Hydrogel, In vitro, Stem cells

1 Introduction

In our body, cells reside in the extracellular matrix (ECM) providing tissue-specific molecular, structural, and mechanical signals that regulate cell behavior. In culture, biomaterial scaffolds provide the functions of the native ECM during the formation and subsequent culture of cells and engineered tissue constructs [1]. Hydrogels are the most frequently used scaffolds for the study of living cells and the delivery of cells to repair sites. Ideally, a hydrogel should mimic the native tissue matrix in order to provide a "natural" environment for the cells. In addition, it is of interest to have a range of hydrogel properties, as the environments encountered by cells can change with time, location within a tissue region, and the state of health or disease.

We have developed a method for generating hydrogels from a number of tissues, including the heart. For research purposes, we routinely derive hydrogels from animal heart tissues (such as porcine or murine) and human heart tissues that we obtain through protocols approved by the Institutional Research Ethics Board (e.g., explanted human hearts, cores taken during the implantation of the pump in left ventricular assist device (LVAD) patients). Using this method, we can also obtain hydrogels consisting of tissue matrix from a specific region (such as the left or the right ventricle) or tissue matrix associated with disease (such as the infarct scar) [2]. The resulting hydrogels have tunable properties and can be blended with other components, such as collagen. The hydrogels are easy to use, as cells are simply mixed with hydrogel solution and culture medium, and polymerize after being incubated briefly in a culture well or being delivered into the heart muscle itself.

We investigated the effects of cardiac ECM on differentiation and maturation of cardiomyocytes derived from human embryonic stem cells (hESCs) [3, 4]. To this end, we prepared a series of hydrogels by blending decellularized ECM with collagen type I at varying ratios. We found that hydrogels with high ECM content increased the expression of cardiac marker troponin T, promoted the maturation of cellular ultrastructure, and improved the contractile function of cardiac cells. The ability of native cardiac ECM hydrogel to induce cardiac differentiation of hESCs without the addition of soluble factors makes it an attractive biomaterial for basic studies of cardiac development and potentially for cell delivery into the heart. We describe here the preparation of hydrogels derived from native heart ECM and the use of these hydrogels for the culture of human cardiac stem and progenitor cells.

2 Materials

Prepare all decellularization solutions using deionized water. Prepare and store all reagents at room temperature (unless indicated otherwise). Diligently follow all waste disposal regulations when disposing hazardous materials (*see* **Note 1**). Most solutions can be sterile-filtered to improve sterility.

2.1 Decellularization Reagents

1. Hypertonic phosphate-buffered saline (PBS): 2× Solution in water.

2. Trypsin: 0.02 % Solution, 1.5 mM ethylenediaminetetraacetic acid (EDTA) in water.

3. Tween-20®: 3 % Solution, 1.5 mM EDTA, in water.

4. Sodium deoxycholate: 102 mM Solution in water (*see* **Note 2**).

5. Peracetic acid: 0.1 % Solution, 4 % ethanol, in water (*see* **Note 3**).

6. PBS: 1× Solution, sterile.

7. Deionized water: Sterile.

8. Penicillin/streptomycin.

2.2 Decellularization Components

1. Tweezers.
2. 50 mL Centrifuge tubes.
3. 50 mL Centrifuge tube drain caps (*see* **Note 4**).
4. Orbital shaker.
5. Four-way tube racks.
6. Sterile filters.
7. Meat slicer.
8. Non-stick wax paper.

2.3 Hydrogel Reagents

1. Cardiac extracellular matrix powder, lyophilized (prepared in Subheading 3.2).
2. Pepsin from porcine gastric mucosa, lyophilized, 3,200–4,500 U/mg protein.
3. Hydrochloric acid: 0.01 N HCl in water, pH 1.8.
4. Sodium hydroxide: 0.1 N NaOH in water, sterile.
5. 1× PBS: Sterile.
6. 10× PBS: Sterile.
7. Storage solution: 1 % Penicillin/streptomycin in 1× PBS, sterile.

2.4 Hydrogel Equipment

1. Tweezers.
2. Scalpel or razor blade.
3. Cutting board.
4. Paper towels.
5. Liquid nitrogen.
6. Pestle and mortar.
7. Kimwipes.
8. Rubber band.
9. FreezeZone lyophilizer.
10. Magnetic stir bar.
11. Magnetic stir plate.
12. Parafilm.
13. Microcentrifuge tubes.

2.5 Cell Culture Reagents

1. StemPro34 Kit (*see* **Note 5**).
2. DMEM/F12.
3. Penicillin/streptomycin.
4. PBS (with Ca^{2+} and Mg^{2+}).
5. PBS (without Ca^{2+} and Mg^{2+}).
6. Glutamine (Invitrogen) (*see* **Note 6**).

7. Knockout serum replacement.

8. Fetal calf serum.

9. DNase.

10. Transferrin: Store transferrin into 2 mL aliquots at 4 °C.

11. Ascorbic acid (*see* **Note 7**).

12. Monothioglycerol (*see* **Note 8**).

13. Bovine serum albumin (BSA): 30 %.

14. Cytokines:
    All cytokines are stored either in the original supplied form or aliquots after reconstitution at −20 to −80 °C. They are stable as the supplied form for 12 months and 3 months after reconstitution unless otherwise indicated. If stored at 2–8 °C under sterile condition, it must be discarded after 1 month of use.

    (a) BMP4: Reconstitute at 10 μg/mL in sterile 4 mM HCl containing 0.1 % BSA.

    (b) VEGF: Reconstitute at 5 μg/mL in sterile PBS containing 0.1 % BSA.

    (c) DKK: Reconstitute at 50 μg/mL in sterile PBS containing 0.1 % BSA.

    (d) bFGF: Reconstitute at 10 μg/mL in sterile PBS containing 0.1 % BSA. Stable for up to 6 months at −20 °C after reconstitution.

    (e) Activin A: Reconstitute at 10 μg/mL in sterile PBS containing 0.1 % BSA.

15. Trypsin–EDTA: Final concentration is 0.25 % (*see* **Note 9**).

16. Collagenase type I: Reconstitute collagenase type I at 2 mg/mL in sterile PBS (with Ca^{2+} and Mg^{2+}) containing 20 % fetal calf serum. Filter and store as 12 mL aliquots at −20 °C (*see* **Note 10**).

2.6 Cell Culture Components

1. Base medium:
   (a) StemPro34 Kit.
   (b) Glutamine (1 %).
   (c) Transferrin (150 μg/mL).
   (d) Ascorbic acid (50 ng/mL).
   (e) Monothioglycerol (0.4 μM).

2. Aggregation medium:
   (a) Base medium.
   (b) BMP4 (0.5 ng/mL).

3. Stage I medium:
   (a) Base medium.

    (b) BMP4 (10 ng/mL).

    (c) bFGF (5 ng/mL).

    (d) Activin A (3 ng/mL).

4. Stage II medium:

    (a) Base medium.

    (b) VEGF (10 ng/mL).

    (c) DKK1 (150 ng/mL).

5. Stage III medium:

    (a) Base medium.

    (b) VEGF (10 ng/mL).

    (c) bFGF (5 ng/mL).

6. Stop medium:

    (a) DMEM/F12.

    (b) Penicillin/streptomycin (0.5 %).

    (c) Glutamine (0.5 %).

    (d) Knockout serum replacement (2.5 %).

    (e) Fetal calf serum (50 %).

    (f) DNase (30 μg/mL): Reconstitute with cold distilled water at 1 mg/mL. Filter and store in 1 mL aliquots at −20 °C. Use once, and discard the rest.

7. Low attachment 6-well culture plate.

8. 24-Well plate.

9. Hypoxic incubator/hypoxia chamber.

2.7 Fluorescence-Activated Cell Sorting Analysis and Immunofluorescence Staining Components

1. Iscove's Modified Dulbecco's Medium (IMDM).

2. Collagen solution for coating the chamber slides: Prepare 3 mg/mL stock solution of collagen in 0.1 % acetic acid. Dilute the stock solution 1:20 in distilled water. Apply this solution to the surface of the slides ensuring even coating. Air-dry at room temperature (under UV light for sterilization).

3. Triton X-100.

4. Goat serum.

5. 4 % Paraformaldehyde: Prepare 8 % stock solution by mixing 8 g paraformaldehyde in 100 mL distilled water. Heat to 60 °C while stirring. Slowly add 1–3 drops of 1 M NaOH until solution is clear. Filter and store at 4 °C. Prior to use, mix stock solution with equal volume of PBS.

6. TO-PRO®-3 Stain (Life Technologies).

7. Vectashield mounting medium.

8. Primary antibodies:

   (a) Cardiac troponin T.

   (b) KDR.

   (c) C-kit.

9. Microscope slides.

10. Cover slips.

11. Tweezer.

12. 12×75 mm Tubes with cell strainer cap (BD Falcon): Wet the tubes first with a small amount of buffer.

13. Syringes.

14. 20 Gauge needles.

3   Methods

Use deionized water for all steps unless indicated otherwise. Following decellularization, all steps should be performed using sterile technique and instruments. To improve sterility, the last steps should be performed inside a laminar flow hood when possible.

3.1   Decellularization of Heart Tissue

1. Procure porcine hearts from pigs (65–70 kg) immediately following euthanasia. Process immediately or place in saline (4 °C) for no more than 24 h.

2. Thoroughly clean the heart tissue by removing excess connective tissue, blood, and debris and rinsing with water at room temperature.

3. Blot the excess water, wrap in non-stick wax paper, and freeze at −80 °C for at least 24 h. Maintain the shape of the heart in order to facilitate the preparation of thin myocardial tissue slices.

4. Slice frozen myocardium (left and right ventricles) into thin (<1 mm) slices using a meat slicer. Blot the excess water, and record the weight of each slice.

5. Place each slice within a 50 mL tube, add 40 mL of water, and load each tube into a four-way tube rack mounted on an orbital shaker (*see* **Note 11**). Wash for 5 min.

6. Drain water and replace with 40 mL of 2× PBS. Wash for 15 min.

7. Calculate the appropriate volume of decellularization solution for each slice according to a ratio of 15 g tissue mass to 100 mL of solution. Drain 2× PBS, and replace with the amount of 0.02 % trypsin solution calculated for each slice based on the mass of tissue. Wash all slices with trypsin for 2 h.

8. Drain the trypsin from each tube, replace with water, and wash for 5 min. Perform an additional wash with 2× PBS for 15 min (40 mL each).

9. Drain 2× PBS, and replace with the amount of 3 % Tween-20® solution calculated for each slice based on the mass of the tissue (15 g tissue mass to 100 mL of solution). Wash for 2 h.

10. Drain Tween-20®, replace with water, and wash all slices for 5 min. Perform an additional wash with 2× PBS for 15 min.

11. Drain 2× PBS, and replace with the amount of 102 mM sodium deoxycholate solution calculated for each slice based on the mass of tissue. Wash all slices with sodium deoxycholate for 2 h.

12. Drain sodium deoxycholate, replace with water, and wash for 5 min. Perform an additional wash with 2× PBS for 15 min.

13. Drain 2× PBS, and replace with the amount of 0.1 % peracetic acid solution calculated for each slice based on the mass of tissue. Wash and sterilize all slices for 1 h.

14. Drain peracetic acid, replace with *sterile* 1× PBS, and wash for 15 min.

15. Replace 1× PBS with *sterile* water, and wash for 5 min. Repeat wash.

16. Drain water, replace with *sterile* 1× PBS, and wash for 15 min.

17. Store tissues in *sterile* 1× PBS with 1 % penicillin/ streptomycin.

18. Decellularization can be confirmed by DNA quantification and histological analysis. To quantify DNA content following decellularization, use a Quant-iT PicoGreen Assay Kit (Invitrogen) according to the manufacturer's instructions. Briefly digest ~10 mg samples in 100 µL proteinase K solution for 24 h at 60 °C. Prepare proteinase K solution by mixing 2.2 mg proteinase K in 2 mL TE buffer, provided with PicoGreen Assay Kit. For histological analysis, such as hematoxylin and eosin staining, fix samples in Accustain (Sigma) for 20 min at room temperature, and then prepare samples for paraffin embedding, sectioning, and staining.

3.2 Preparation of Native ECM Hydrogels

1. Use a scalpel to cut decellularized heart tissue slices into smaller sections. Blot sections with a Kimwipe to remove excess water.

2. Snap-freeze sections by immersing the tissues in a small amount of liquid nitrogen (~5 mL) in a mortar. Pulverize the frozen tissue sections into a fine powder using a mortar and a pestle (*see* **Note 12**).

3. Collect the frozen ECM powder in a tube, cover the uncapped tube with a Kimwipe, secure the Kimwipe with a rubber band, and load into a lyophilizer (*see* **Note 13**). Lyophilize frozen ECM powder overnight.

4. After lyophilization, mix the ECM powder with pepsin in a ratio of 10:1 w/w per 100 mL 0.01 N HCl. Otherwise, for long-term storage at room temperature, wrap the cap of the tube containing the lyophilized cardiac ECM powder with parafilm.

5. Digest the solution for 48 h at room temperature under constant stirring using a magnetic stir bar and plate until the solution becomes viscous with no visibly undigested granules.

6. Aliquot the 10 mg/mL digested ECM solution, and freeze aliquots at –80 °C to terminate pepsin digestion (*see* **Note 14**).

3.3 Reconstitution of Native ECM Hydrogels

1. Thaw frozen aliquot of 10 mg/mL digested ECM solution at room temperature.

2. Place thawed aliquot of 10 mg/mL digested ECM solution and sterile hydrogel reagents (0.1 M NaOH, 10× PBS, 1× PBS) on ice and allow to cool to 4 °C.

3. Generate 1 mL of 6 mg/mL cardiac ECM hydrogel by adding 60 µL 0.1 N NaOH, 67 µL 10× PBS, 275 µL 1× PBS, and 600 µL cardiac ECM digest in a microcentrifuge tube, and mix thoroughly with pipette tip (*see* **Note 15**). Keep cold at all times.

4. Add desired volume of mixture to cell culture plate, and incubate at 37 °C for 45 min.

5. Recover cardiac ECM hydrogel for the in vitro culture of cardiac stem cells or other desired applications.

3.4 Encapsulation of Human Embryonic Bodies in Native ECM Hydrogels

1. Maintain hESCs in 6-well low attachment plates under hypoxic condition for 4 days in aggregation medium (day 0) and then in stage I medium (days 1–4) to form human embryonic bodies (hEBs, aggregates of hESCs), and induce mesoderm differentiation. The seeding density of hESCs is 0.5–1 million cells/well of a 6-well plate. Use 2 mL media for every 0.5–1 million cells.

2. At day 4, hESCs have formed hEBs (*see* **Note 16**).

3. Thaw native ECM hydrogel overnight at 4 °C the night before the encapsulation.

4. Coat 24-well plates with 50 µL hydrogel solution/well and incubate at 37 °C for at least 30 min to allow for gelation before encapsulating the hEBs. This is to prevent hEBs from adhering to the bottom of the culture well because of gravity and to ensure a true 3D culture system throughout the differentiation period.

5. Remove hEBs with a 5 mL pipette, and pool 3 wells of a 6-well plate per 15 mL conical tube.

6. Allow hEBs to settle for 20 min in hypoxic incubator/chamber to separate hEBs from single cells and cell debris.

7. Remove supernatants carefully, and leave hEBs undisturbed at the bottom of the tube.

8. Wash hEBs with 10 mL IMDM medium supplemented with antibiotics (penicillin/streptomycin) to wash out the residual cytokines (*see* **Note 17**).

9. Centrifuge the hEBs at 137 r.c.f. for 5 min, and remove supernatant.

10. Resuspend hEBs in culture medium either with or without growth factors at a concentration of 0.5 million cells/mL. *See* Subheading 3.5 for counting method of hEBs.

11. Add 50 μL of hEB suspension to 1 mL of ECM hydrogel solution (*see* **Note 18**).

12. After gentle mixing, transfer 250 μL of each hEB/hydrogel mixture to a coated well of a 24-well plate and incubate at 37 °C for 30–60 min to gel.

13. Add 2 mL of fresh stage II medium or base medium to each well, and maintain the culture in hypoxia chamber (*see* **Note 19**).

14. At day 8, remove culture plates from hypoxia chamber and change culture medium to stage III medium or base medium (*see* **Note 20**).

15. Place culture plates in hypoxia chamber for an additional 4 days.

16. Repeat **step 12**, and incubate cells in stage III medium or base medium at ambient oxygen levels for the rest of the induction procedure.

17. Stain for cardiac markers at days 14–16 (*see* Subheading 3.6 for staining of cardiac troponin T).

3.5 Disassociation of Human Embryonic Bodies for Cell Counting or FACS Analysis

1. Pool 1–3 wells of hEBs in a 15 mL conical tube for 15–20 min to separate hEBs from cell debris.

2. Aspirate medium, and add 2 mL trypsin–EDTA.

3. Incubate at 37 °C for 5 min, and then stop the reaction with 1 mL stop medium.

4. Prepare a single-cell suspension by mechanically passing hEBs four to six times through a 20 gauge needle.

5. Add 4 mL IMDM supplemented with 10 % fetal calf serum.

6. Centrifuge at 215 r.c.f. for 5 min.

7. Remove supernatant, and resuspend cells in IMDM supplemented with 5 % fetal calf serum (500 μL/well).

8. Take 100 μL of the cell suspension for cell counting and viability determination using trypan blue exclusion. Use the rest of the cells for FACS analysis.

9. For FACS analysis, filter the suspension through round-bottom tubes with a cell strainer cap to remove the remaining cell aggregates and ensure a uniform single-cell suspension.

10. Stain for desired antibodies according to the manufacturer's instructions, and perform FACS analysis. KDR and c-kit can be used to monitor the induction of cardiac mesoderm [4].

3.6 Cardiac Troponin T Staining for hEBs Encapsulated in Native ECM Hydrogel

1. Gently remove culture medium from one well of the 6-well ultralow attachment plate.

2. Add 1 mL collagenase type I (containing 10 μg/mL DNase), and incubate at 37 °C for 1 h.

3. Aspirate hEBs with a 5 mL pipette into a 15 mL conical tube, and add 9 mL of IMDM.

4. Wash hEBs by gently pipetting up and down three times.

5. Centrifuge at 137 r.c.f. for 5 min.

6. Repeat **steps 2–6** from Subheading 3.5.

7. Remove supernatant, and resuspend cells in stage III medium.

8. Seed cells on microscope slides coated with type I collagen and culture for 2 days.

9. Fix cells in 4 % paraformaldehyde for 15 min at room temperature.

10. Wash fixed cells three times in PBS (without Ca^{2+} and Mg^{2+}).

11. Quench cells with 0.5 M NH_4Cl in PBS containing 0.1 % BSA for 15 min.

12. Permeabilize cells using PBS containing 5 % fetal calf serum and 0.1 % Triton X-100 for 10 min at room temperature.

13. Wash cells in PBS for 5 min. Repeat three times.

14. Block cells for 15 min in blocking buffer (10 % goat serum in PBS containing 0.1 % BSA).

15. Stain cells with primary antibody (cardiac troponin T) overnight at 4 °C in blocking buffer (*see* **Note 21**).

16. Wash cells with PBS containing 0.1 % BSA and 2.5 % NaCl for 15 min.

17. Stain cells with secondary antibodies according to the manufacturer's instructions for 2 h at room temperature. Cover cells to protect from light.

18. Stain cell nuclei with TO-PRO®-3 Stain for 5 min according to the manufacturer's instructions (*see* **Note 22**).

19. Wash twice with PBS containing 2.5 % NaCl and 0.1 % BSA, 5 min each.

20. Wash with PBS alone for 5 min.

21. Mount slides with mounting medium. Coverslip and seal with nail polish.

22. Image with fluorescence microscope.

---

4 Notes

1. Sodium deoxycholate is a lung irritant and should be handled under a chemical fume hood. Peracetic acid is corrosive to the skin and eyes and hazardous in case of inhalation. Peracetic acid waste should be disposed according to Environmental, Health, and Safety Guidelines.

2. Warming the solution to 37 °C aids solubility. Return solution to room temperature before use.

3. Peracetic acid solution should be prepared fresh on the day of use.

4. Drain caps can be made by drilling a concentric pattern of 1 mm diameter holes through 50 mL tube caps. Alternatively, tubes may be uncapped and solutions poured out carefully to avoid loss of tissue.

5. StemPro 34 contains a basal liquid medium and frozen supplement. If combined, it should be consumed within 2 weeks. Otherwise, store the medium in 50 mL aliquots and the supplements in 1.3 mL aliquots. Combine right before using.

6. Warm 200Å~ L-glutamine in 37 °C water bath until it is dissolved completely. Store in 5 mL aliquots at –20 °C. Once thawed, use within 3 weeks.

7. Prepare a stock solution of 5 mg/mL ascorbic acid in cold distilled water. Reconstitute on ice, and vortex periodically until completely dissolved. Filter, aliquot, and store at –20 °C. Once opened, discard the excess.

8. Store monothioglycerol in 1 mL aliquots at 4 °C. It is strongly recommended to aliquot monothioglycerol to minimize the amount of oxidation due to repeated opening of the stock bottle.

9. Weigh appropriate amount of trypsin-EDTA and add to warm PBS (without $Ca2^+$ and $Mg2^+$). Allow to dissolve for 15 min. Once dissolved, filter and aliquot. Store at –20 °C.

10. Freeze–thaw cycles will damage the enzyme solution. Recommended maximum is one freeze–thaw cycle.

11. The orbital shaker should be maintained at approximately 200 rpm throughout decellularization.

12. When snap-freezing, flatten tissue sections out as much as possible to aid in the pulverization process. Sections frozen in a globular form are significantly more difficult to pulverize.

13. A Kimwipe is used to cover the uncapped tube during lyoph-ilization so that during retrieval electrostatic interactions do not result in the loss of any ECM powder.

14. A bicinchoninic acid assay (BCA assay) may be performed to determine the total solubilized protein concentration.

15. Different concentrations of ECM hydrogels (ranging from 2 to 6 mg/mL) can be prepared by varying the volume of ECM digest. For concentrations below 6 mg/mL, the volume of 1Å~PBS should be increased to obtain the final desired volume. Alternatively, cell culture media or a cell suspension may be used in place of the 1Å~PBS reagent for encapsulation of cells within the cardiac ECM hydrogel.

16. Gentle closing and opening of the incubator door are critical during this period to prevent disturbing the aggregation process.

17. This step is critical because the cytokines BMP4 and activin A, if not completely washed out, will affect the next stage induction even at very low concentrations. The washing must be gentle enough so as not to disturb the aggregates, as they are relatively easy to break apart. Aggregation is key for successful differentiation. Harvest for FACS analysis at days 4–7 to ensure mesoderm induction and cardiovascular progenitor specification in the hEBs (Subheading 3.5).

18. Cut off the tip of a 200 µL pipette tip for easier transportation of hEBs.

19. Gently add medium along the sidewall of the culture well, and try not to disturb the newly formed gel. Here we provide two types of culture medium that can be used: stage induction medium (with growth factors) and base medium alone (without growth factors). Note that stage induction medium is optimized for maximizing the percentage of cardiac progenitor cells without the addition of native hydrogel. We find that hydrogel/collagen without the addition of growth factors achieves a significantly better outcome than collagen hydrogels with the supplement of growth factors [3].

20. Avoid using vacuum for medium aspiration because of the fragility of the hEB/hydrogel mixture. Gently remove medium with a 1,000 µL pipette.

21. Ensure that slides are stored in a moist environment to prevent desiccation of staining solution overnight.

22. If fluorescence microscope has ultraviolet excitation, use mounting medium containing 4′,6-diamidino-2-phenylindole (DAPI) to label cell nuclei.

Acknowledgements

This work was supported by the NIH (grants HL076485, EB 17103 and EB002520), NYSTEM (grant C028119), and starter funding from Columbia University Technology Ventures Office.

References

1. Freytes DO, Godier-Fournemont A, Duan Y, O'Neill J, Vunjak-Novakovic G (2013) Native heart matrix as a source of scaffolds for cardiac regeneration. In: Ren-Ke L, Richard W (eds) Cardiac regeneration and repair: biomaterials and tissue engineering. Woodhead Publishing, Cambridge, pp 201–224, Chapter 8

2. Godier-Furnemont A, Martens T, Koeckert M, Wan LQ, Parks J, Zhang G, Hudson J, Vunjak-Novakovic G (2011) Composite scaffold provides a cell delivery platform for cardiovascular repair. PNAS 108(19):7974–7979

3. Duan Y, Liu Z, O'Neill J, Wan L, Freytes DO, Vunjak-Novakovic G (2011) Hydrogel derived from native heart matrix induces cardiac differentiation of human embryonic stem cells without supplemental growth factors. J Cardiovasc Transl Res 4(5):605–615

4. Yang L, Soonpaa MH, Adler ED, Roepke TK, Kattman SJ, Kennedy M, Henckaerts E, Bonham K, Abbott GW, Linden RM, Field LJ, Keller GM (2008) Human cardiovascular progenitor cells develop from a KDR+embryonic-stem-cell-derived population. Nature 453(7194):524–528

Chapter 8

Magnetically Actuated Alginate Scaffold: A Novel Platform for Promoting Tissue Organization and Vascularization

Yulia Sapir, Emil Ruvinov, Boris Polyak, and Smadar Cohen

Abstract

Among the greatest hurdles hindering the successful implementation of tissue-engineered cardiac patch as a therapeutic strategy for myocardial repair is the know-how to promote its rapid integration into the host. We previously demonstrated that prevascularization of the engineered cardiac patch improves cardiac repair after myocardial infarction (MI); the mature vessel networks were generated by including affinity-bound angiogenic factors in the patch and its transplantation on the blood vessel-enriched omentum. Here, we describe a novel in vitro strategy to promote the formation of capillary-like networks in cell constructs without supplementing with angiogenic factors. Endothelial cells (ECs) were seeded into macroporous alginate scaffolds impregnated with magnetically responsive nanoparticles (MNPs), and after pre-culture for 24 h under standard conditions the constructs were subjected to an alternating magnetic field of 40 Hz for 7 days. The magnetic stimulation per se promoted EC organization into capillary-like structures with no supplementation of angiogenic factors; in the non-stimulated constructs, the cells formed sheets or aggregates.

This chapter describes in detail the preparation method of the MNP-impregnated alginate scaffold, the cultivation setup for the cell construct under magnetic field conditions, and the set of analyses performed to characterize the resultant cell constructs.

Key words Alginate scaffolds, Endothelial cells, Magnetic nanoparticles, Vessel-like networks, Tissue engineering, Magnetic alginate composite

1 Introduction

Prevascularization of a tissue-engineered cardiac patch prior to its implantation on an infarcted myocardium is a scientific and technological challenge in the clinical realization of cardiac tissue engineering for myocardial repair. We previously showed that prevascularization of a tissue-engineered cardiac patch containing angiogenic factors on the omentum for 7 days prior to its implantation on the infarcted myocardium improves its therapeutic outcome on cardiac repair [1]. The omentum, a blood vessel-enriched tissue, provided a continuous source of endothelial cells (ECs), pericytes, and angiogenic factors leading to formation of mature

Milica Radisic and Lauren D. Black III (eds.), *Cardiac Tissue Engineering: Methods and Protocols*, Methods in Molecular Biology, vol. 1181, DOI 10.1007/978-1-4939-1047-2_8, © Springer Science+Business Media New York 2014

vasculature within the patch [2–6]. Although successful, prevascularization of the cardiac patch by this strategy is currently not a clinical option for patients with heart failure.

Other strategies to promote vessel-like network formation in cardiac patches currently focus on the incorporation of angiogenic factors in the scaffolds for prolonged presentation [7–12] and/or co-culture of ECs with the target tissue cells [13–17]. These strategies are still under development, and additional studies are required to optimize the type and dosing of angiogenic factors, the cell ratio of ECs to cardiomyocytes (CMs) and other target cells, in order to achieve functional vasculature in the tissue-engineered patch.

We recently developed a novel strategy to promote vessel-like formation in EC constructs by employing a magnetically responsive alginate scaffold and applying an external alternating magnetic field [18]. The scaffold was impregnated with iron oxide (magnetite) magnetic nanoparticles (MNPs), seeded with bovine aortic ECs, and pre-cultured for 24 h for cell adhesion, and then the cell construct was exposed to an alternating magnetic field of 40 Hz for 7 days. No angiogenic factors were added to the cell constructs. We hypothesized that magnetic stimulation of the MNP-impregnated scaffold would provide an appropriate physical signal to induce the organization of ECs into capillary-like structures. Such a strategy enables actuation at a distance on the nanoscale and cellular levels. The desired process can be actuated within a target cell by magnetically responsive materials due to its coupling to a magnetic field, regardless of the presence of any intervening structures such as tissue between the cell and the magnetic field source. Furthermore, the magnetic field can penetrate deep into tissues, reaching a single cell and acting directly on its organelles, unlike an electric field, which is shielded by the membrane potential. Stress parameters can also be varied dynamically simply by varying the properties (e.g., strength, frequency) of the applied field.

In our hands, application of the magnetic stimulation strategy led to EC organization into capillary-like structures without supplementation of any angiogenic growth factors, while in the non-stimulated scaffolds the cells were mainly organized as sheets or aggregates. We previously showed that magnetic stimulation combined with cell cultivation in MNP-impregnated scaffolds was effective in promoting the organization of cardiomyocytes into anisotropically organized striated cardiac fibers [19]. While the exact mechanism of action of the magnetic composite biomaterial and magnetic field is currently under investigation, this strategy clearly has been proven to be capable of "priming" the seeded cells to organize into their native structures, as endothelial cells were organized into a primitive vascular network.

2 Materials

Prepare all solutions using ultrapure water (prepared by purifying deionized water to attain a resistivity of 18.2 MΩ cm at 25 °C) and analytical grade reagents. Prepare and store all reagents at room temperature (RT) (unless indicated otherwise).

2.1 Magnetically Responsive Alginate Scaffolds

1. Materials for magnetite preparation: Iron (III) chloride hexahydrate and iron (II) chloride tetrahydrate.

2. 2 N NaOH solution in water for magnetite preparation: Into a beaker placed on ice and containing 70 mL of water, slowly add while stirring 4 g of NaOH and mix until dissolved. Make up to 100 mL solution. Store at RT.

3. Alginate solution for coating the magnetite crystals for stabilization: Dissolve sodium alginate (Pronova™ UP LVG, NovaMatrix FMC Biopolymers, Drammen, Norway) in water to a final concentration of 2.4 % (w/v). The suspension is dissolved by stirring for 2 h until achieving a clear solution. Filter (0.22 μm) the solution into a sterile 50 mL Falcon tube and store at 4 °C (*see* **Note 1**).

4. 1.2 % (w/v) solution of alginate modified with cell adhesion peptides (glycine$_4$-arginine-glycine-aspartate-tyrosine (RGD) and glycine$_4$-serine-proline-proline-arginine-arginine-alanine-arginine-valine-threonine-tyrosine (HBP)) (Alg-PEP): The peptides were covalently attached to alginate *via* carbodiimide chemistry [20] with modification [21], thus creating an amide bond between the terminal amine of the peptide and the alginate carboxylic group. Briefly 10 mL of 1.2 % (w/v) alginate solution was prepared in 0.2 M MES buffer and 0.3 M NaCl, at pH 6.5. The carboxylic group on the alginate polymer backbone was activated using 0.2 mmol 1-ethyl (dimethylaminopropyl) carbodiimide (EDAC), and 0.1 mmol of the co-reactant, *N*-hydroxysulfosuccinimide (sulfo-NHS), was added to stabilize the reactive intermediate. The mixture was stirred at room temperature for 3 h, and 3.84 μmol of GGGGSPPRRARVTY (HBP) or GGGGRGDY (RGD) was added and stirred for additional 12 h to ensure peptide coupling. Then the two modified alginates were mixed together. The peptide-attached alginate products were purified by dialysis (3500 MWCO) against DDW for 3 days, lyophilized until dry, and stored desiccated. Dissolve the Alg-PEP in water, at room temperature, while stirring for 2 h (*see* **Note 2**).

5. Cross-linker solution for alginate gelation: D-Gluconic acid/hemi-calcium salt in water to a final concentration of 1.32 % (w/v). The suspension is dissolved by stirring for 1–2 h until achieving a clear solution. Filter (0.22 μm) the solution into a sterile Falcon tube and store at 4 °C.

6. Magnetic separation plate (LifeSep™, Dexter Magnetic Technologies, Elk Grove Village, Illinois, USA).

2.2 Magnetic Stimulation Setup

1. Magnetic coils (HHS 5201-98, Schwarzbeck Mess-Elektronik, Schonau, Germany).

2. Power supply (Associated Power Technologies 5000 series power supply) (DiamondBar, CA).

3. Magnetometer (Model 410 Hand-held Gaussmeter, Lake Shore Cryotronics, Inc., Westerville, OH).

2.3 Materials for Cell Culture and Analyses of Cell Constructs

1. Complete growth medium for bovine aortic endothelial cells (BAECs): High-glucose DMEM, supplemented with 1 % (v/v) penicillin–streptomycin, 2 % (v/v) L-glutamine, and 10 % (v/v) fetal bovine serum (FBS) (*see* **Note 3**).

2. Solution for scaffold dissolution: 4 % (w/v) Sodium citrate (trisodium dihydrate) in PBS.

3. AlamarBlue™, PrestoBlue™, or similar for evaluation of cell metabolic activity.

4. Hoeschst 33258, Picogreen™, or similar to measure DNA content.

5. DMEM buffer: Into 900 mL water, add 0.265 g $CaCl_2 \times 2H_2O$, 0.4 g KCl, 0.2 g $MgSO_4 \times 7H_2O$, 6.4 g NaCl, 3.7 g $NaHCO_3$, and 0.109 g NaH_2PO_4, and mix until dissolution. Titrate to pH 7.2–7.4, complete to 1 L, and filter (0.22 μm) into a sterile bottle. Store at 2–8 °C for 1 month (*see* **Note 4**).

6. Fixation buffer: 10 % Formalin (v/v) in DMEM buffer (*see* **Note 5**).

7. Blocking buffer: 1 % (w/v) Bovine serum albumin (BSA) in DMEM buffer.

8. Permeabilization buffer: Prepare 0.2 % (v/v) Triton-X100 in DMEM buffer.

9. Primary antibodies to label and follow cell organization in the construct. The antibody solution is prepared according to the manufacturer's instructions.

10. Secondary antibodies, depending on the primary antibody choice and laser configuration of confocal system. The antibody solution is prepared according to the manufacturer's instructions.

11. Nuclear dye, e.g., DAPI.

3 Methods

3.1 Preparation of Alginate-Stabilized Magnetite Nanoparticles

The procedure below is described for a magnetite sample prepared in one 20 mL vial. Batch size is a total of four vials prepared using this procedure. The general scheme of preparation is depicted in Fig. 1.

1. Iron oxide preparation

$$2FeCl_3 \times 6H_2O + FeCl_2 \times 4H_2O + 8NaOH \rightarrow 8NaCl + 20H_2O + Fe_3O_4$$

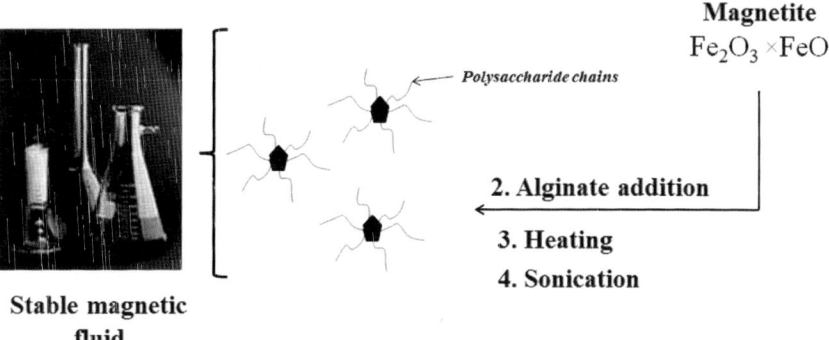

Magnetite

$Fe_2O_3 \times FeO$

Polysaccharide chains

2. Alginate addition

3. Heating

4. Sonication

Stable magnetic fluid

Fig. 1 A general scheme for preparation of alginate-stabilized magnetic nanoparticles (MNPs)

1. Weigh an empty 20 mL glass vial before use.

2. Add into the vial 520 mg of iron (III) chloride hexahydrate and 192 mg of iron (II) chloride tetrahydrate.

3. Immediately add 6.2 mL of water, and mix gently by hand until fully dissolved (*see* **Note 6**).

4. Quickly add 3.8 mL NaOH 2 N, and mix the vial manually (*see* **Note 6**).

5. Place the vial on magnetic separation plate, and wait for the complete sedimentation of the magnetites (the supernatant becomes transparent).

6. Remove supernatant by vacuum suction while keeping the vial on the magnet plate.

7. Wash the pelleted magnetite with 5 mL of water to remove any unreacted salts.

8. Remove supernatant by vacuum suction.

9. Repeat **steps** 7 and **8** (*see* **Note 7**).

10. Repeat suction on magnet every 10 min for a total of 60 min (*see* **Notes 7** and **8**).

11. Combine the magnetite batches from four vials (theoretical total of 890 mg) in one 50 mL Falcon tube.

12. Disperse the magnetites in 5.5 mL of 2.4 % (w/v) alginate by pipetting. Final alginate concentration should be 1.2 % (w/v), and final volume of the magnetite-stabilized suspension is 11 mL (*see* **Note 9**).

13. Place the tube with the magnetite dispersed in alginate in 90 °C water bath for 20 min, and mix periodically.

14. Sonicate for 5 min (*see* **Note 10**).

15. Repeat **steps 13** and **14**.

16. Flush with Argon and store at 4 °C.

17. The size of the alginate-stabilized magnetite particles can be determined by photon correlation spectroscopy (PCS) or other standard methods (*see* **Note 11**). The alginate coating of the MNPs can be tested by contact angle measurements of the MNPs before and after coating.

3.2 Fabrication of Magnetically Responsive Alginate Scaffolds

1. Mix desired amount of Alg-PEP 1.2 % (w/v) solution with alginate-stabilized magnetite solution prepared according to Subheading 3.1 (at volume ratio 5:1), and stir for 15 min in glass vial.

2. Add D-gluconic acid/hemi-calcium salt (cross-linker) into alginate mixture from the previous step to reach 0.22 % (w/v) and 1 % (v/v) final cross-linker and alginate concentrations, respectively. Stir for 3 h.

3. Pour 100 μL of a cross-linked solution per well into 96-well plates.

4. Place the plates at –20 °C overnight.

5. Lyophilize (–56 °C, 0.05–0.08 mbar) the scaffolds for at least 48 h.

6. The internal morphology and pore surface topography of the dry MNP-impregnated scaffolds could be analyzed by scanning electron microscopy (SEM) (Fig. 2). The magnetic properties of the scaffold could be analyzed from hysteresis curves measured by an alternating gradient magnetometer (Fig. 3a). The dynamic viscoelastic behavior could be determined using a stress-control rheometer (Fig. 3b).

3.3 Cell Seeding and Culture

1. Seed BAECs (or other endothelial cell types) at an initial cell density of 2.23×10^7 cells/cm^3 scaffold by dropping 15 μL of the cell suspension in culture medium onto the MNP-impregnated scaffold placed in a 96-well plate.

2. Centrifuge the cell constructs in plate for 2 min at $100 \times g$ to achieve homogenous cell distribution in the scaffold.

3. Add 200 μL/well of culture medium, and incubate for 15 min (37 °C, 5 % CO_2).

4. Gently transfer the cell constructs with small forceps to 48-well plates supplemented with 1 mL of culture medium. Cultivate for 24 h, and proceed to stimulation.

3.4 Magnetic Stimulation of the Cell Constructs

1. Place two 48-well plates with the cell constructs in a gap between two Helmholtz coils (Fig. 4) (*see* **Notes 12** and **13**).

2. Turn on the system to generate an AC magnetic field of 10–15 Gauss by passing an electrical current through serially connected coils at a frequency of 40 Hz.

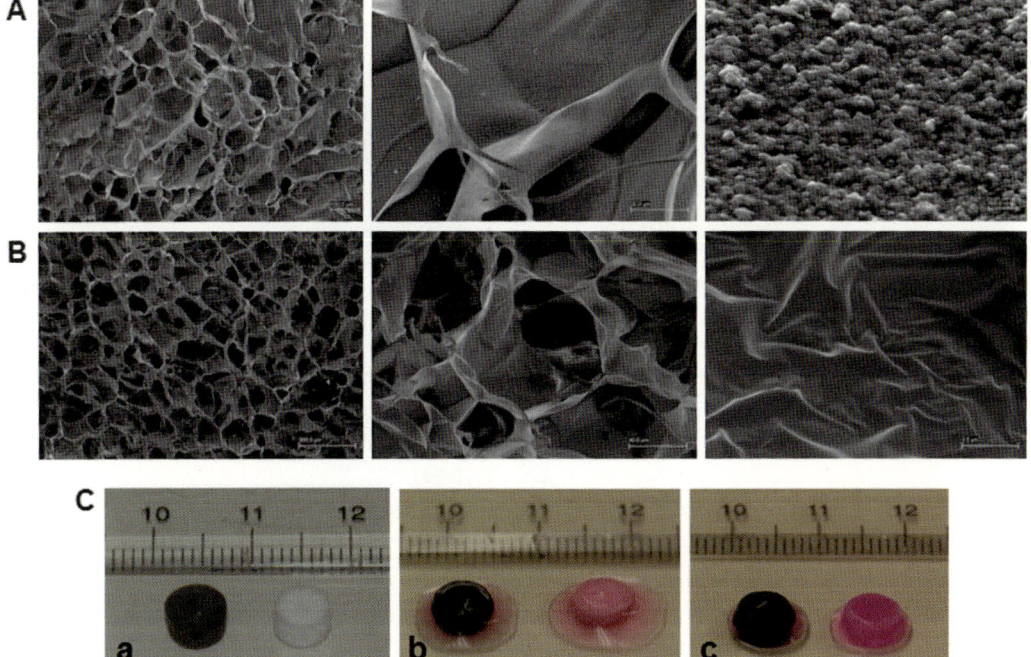

Fig. 2 The internal morphology and pore surface topography of the alginate scaffolds: Scanning electron microscope (SEM) images of MNP-impregnated (**a**) and nonmagnetic (**b**) scaffolds. (**c**) Macroscopic views of the scaffolds (a) dry and (b, c) wetted with culture medium for 30 min (b) and 24 h (c). The *black* scaffold is MNP–alginate composite, and the *white/pink* is without MNP inclusion. Reprinted with permission from [18]. Typical pore size is 100–150 μm

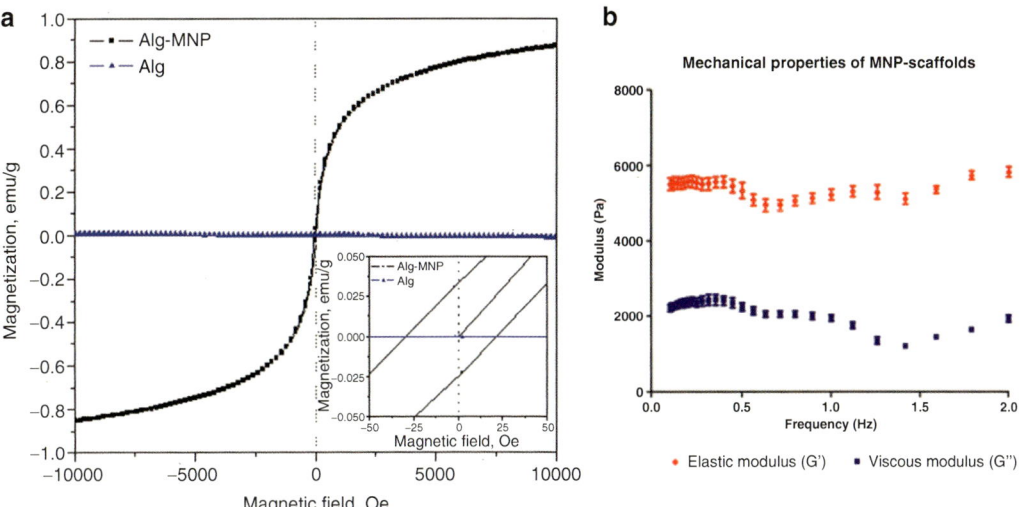

Fig. 3 Magnetic and mechanical properties of the MNP-impregnated alginate scaffold. (**a**) Magnetization curves of MNP scaffold; M_r (0 Oe) = 0.4 emu/g (remanent magnetization); M_f (15 Oe) = 1.25 emu/g (magnetization at the experimental field condition); M_s (5 kOe) = 24 emu/g (saturation magnetization). Inset graph depicts magnified region near the center of coordinates. (**b**) Mechanical spectra of the wetted scaffolds (10 mm diameter and 2 mm thickness). The scaffolds were wetted with culture medium for 2 days in a humidified atmosphere of 5 % CO_2 and 95 % air at 37 °C. G′—elastic modulus, G″—viscous modulus; their measurements were conducted at a frequency sweep of 0.1–2 Hz at 37 °C. We used stress-control Bohlin CS-10 Rheometer (Malvern Instruments, Westborough, MA), operated in the plate–plate mode with 8 mm diameter

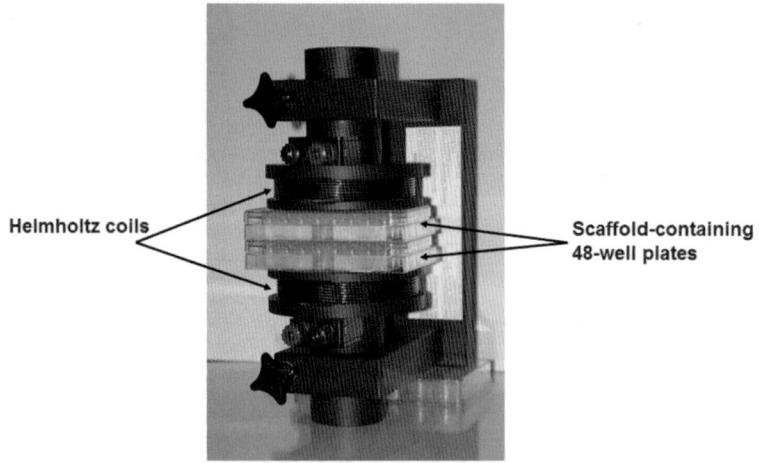

Fig. 4 AC magnetic field-generating setup

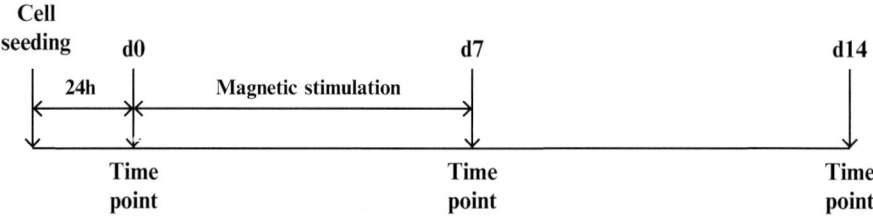

Fig. 5 Experiment timeline. Seed endothelial cells (ECs) and allow to properly adhere to the scaffold for 24 h. At time zero (d0), subject the cell constructs to an alternating magnetic field for 7 days, while control cell constructs are cultivated in a similar incubator with no stimulation. After stimulation, turn the field off and cultivate "stimulated" and "non-stimulated" groups for an additional 7 days. The setup design could be changed based on the experimental needs

3. Validate the strength of the magnetic field by a handheld magnetometer.

4. Stimulate according to the scheme in Fig. 5.

3.5   Cell Construct Imaging by Laser Scanning Confocal Microscopy and Assessment of Cell Organization (See Note 14)

1. Fixation of the cell constructs: Transfer the cell constructs to a new 96-well plate, gently add 300 μL of warm (~37 °C) DMEM buffer, repeat twice to wash, and discard gently. Add 200 μL of warm (~37 °C) 10 % (v/v) formalin in DMEM buffer at room temperature for 7 min, and wash with DMEM buffer three times, 3 min each wash.

2. Permeabilization of cell constructs: Add 200 μL of 0.2 % (v/v) Triton-X100 in DMEM buffer for 5 min, and wash with DMEM buffer three times, 3 min each wash.

3. Blocking: Add 500 μL of 1 % BSA (w/v) in DMEM buffer, incubate for 1 h at room temperature, and wash with DMEM buffer once for 1 min.

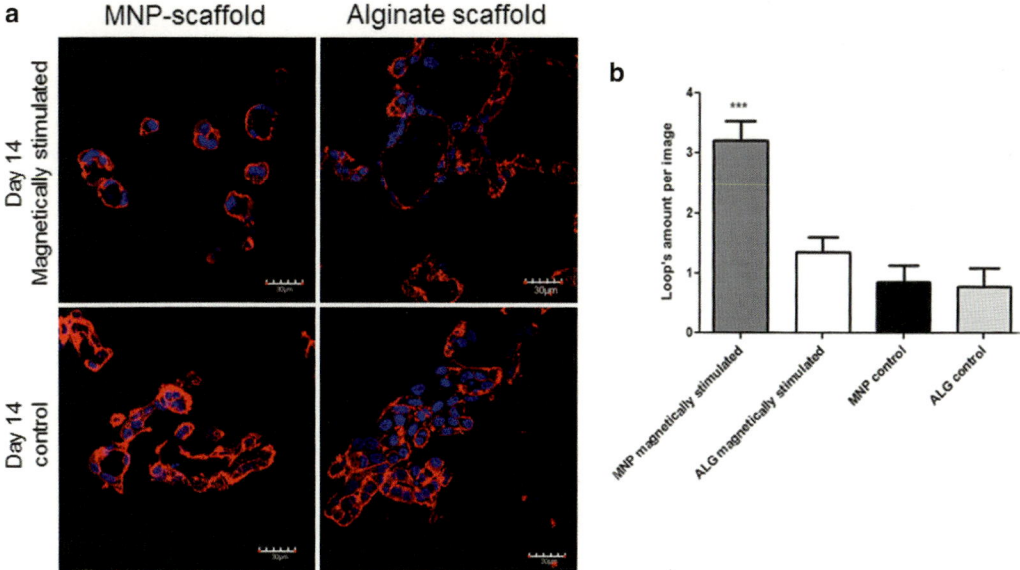

Fig. 6 In vitro vessel-like formation in scaffolds under magnetic stimulation. (**a**) Endothelial cell organization in magnetically responsive or pristine alginate scaffolds, on day 14 post-cell seeding, with or w/o magnetic stimulation. The cells are stained with Alexa-Fluor 546-conjugated phalloidin (Life Technologies), which stains F-actin for cell visualization; for nuclei detection cells were stained with DAPI (Life Technologies) (bar: 30 μm). (**b**) Average loop number per image field counted on day 14 post-cell seeding. A total of 25 randomly selected fields were analyzed per each group. Loop-like structures were quantified per image. *Asterisks* denote significant difference (by two-way ANOVA), ***$p < 0.005$ (Bonferroni's post hoc test was used for comparison between the groups). Reprinted with permission from [18]

4. Add primary antibody in blocking solution, and incubate overnight at 4 °C (*see* **Note 15**).

5. Wash with DMEM buffer three times, 15 min each wash.

6. Add secondary antibody in blocking solution, and incubate for 1 h at RT. Protect from light from now on.

7. Wash with DMEM buffer three times, 15 min each wash.

8. Add nuclear stain of choice; incubate for 5 min at RT.

9. Wash with DMEM buffer three times, 15 min each wash.

10. Discard the liquid, but maintain the cell construct wet.

11. Visualize the cell construct using laser scanning confocal imaging system mounted on an inverted microscope.

12. Assessment of EC organization: F-actin-labeled cells are identified by their red color (Fig. 6a). The number of loop-like structures was determined in various fields (Fig. 6b).

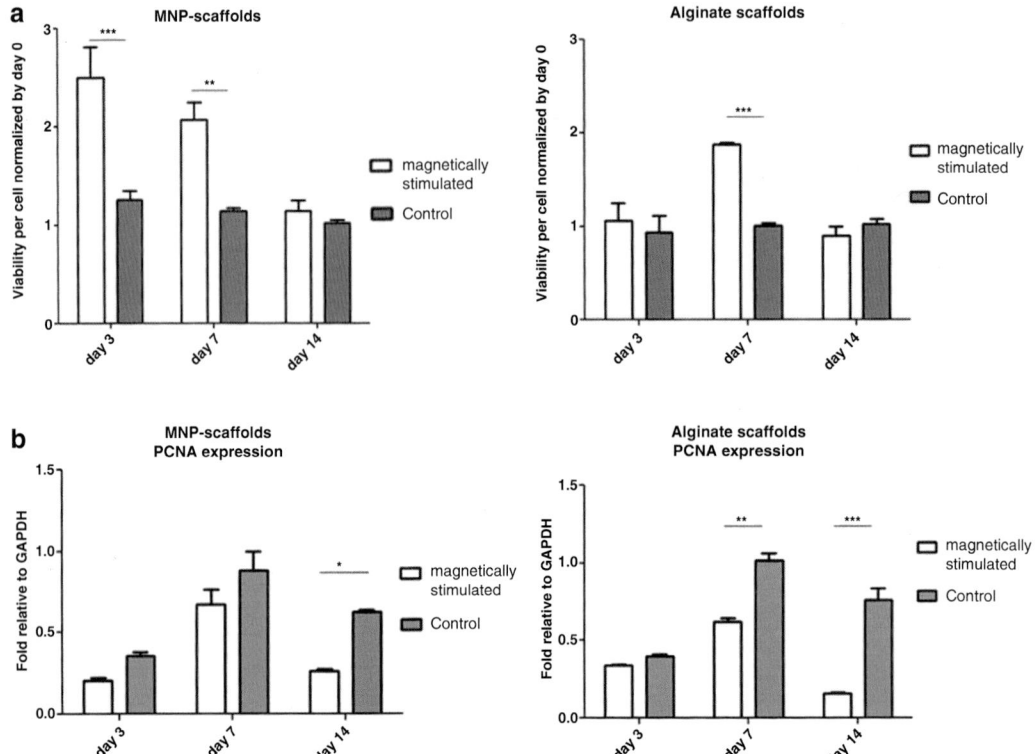

Fig. 7 Cell metabolic activity and proliferation in the different constructs. (**a**) Metabolic activity per cell in the different constructs. Data ($n = 3$–5 constructs per tested group) presented are normalized to the DNA content in the construct, relative to day 0. *Asterisks* denote significant difference (by two-way ANOVA), *$p < 0.05$, **$p < 0.01$, and ***$p < 0.005$ (Bonferroni's post hoc test was used for comparison between the groups). (**b**) Densitometric analysis of Western blots for expression of proliferating cell nuclear antigen, PCNA. Protein expression is normalized by GAPDH protein, for MNP scaffolds and nonmagnetic alginate scaffolds. *Asterisks* denote significant difference (by two-way ANOVA), *$p < 0.05$, **$p < 0.01$, and ***$p < 0.005$ (Bonferroni's post hoc test was used for comparison between the groups). Reprinted with permission from [18]

3.6 Cell Viability and Proliferation

1. Cell viability and/or metabolic activity could be measured using various metabolic assays (e.g., XTT, AlamarBlue™, PrestoBlue™, and many others) by direct incubation of the scaffolds in working reagent solution and following the manufacturer's instructions for measurement (*see* **Note 16**).

2. The cell constructs in the scaffolds could be subjected to DNA content, Western immunoblotting, qPCR, and other downstream analyses (Fig. 7).

3. Transfer the scaffolds to a clean Eppendorf tube, and dissolve by adding 200 μL of 4 % (w/v) sodium citrate in PBS (*see* **Note 17**). Centrifuge at $2,400 \times g$ for 10 min at RT. Discard the liquid, and use the cell pellet for Western immunoblotting or other subsequent analyses of choice.

4 Notes

1. Store alginate solutions only in plastic containers. Glass containers contain calcium ions that may leach out and gradually cross-link the alginate.

2. Dissolve Alg-PEP just prior to scaffold preparation.

3. We used BAECs. The readers are encouraged to test the system with additional endothelial cell types, both primary and cell lines.

4. Alginate scaffolds are sensitive to most commonly used buffers since these may disturb the calcium cross-linking, leading to scaffold dissociation. DMEM buffer is appropriate for all staining procedures.

5. Check primary/secondary antibody instructions for the requirements for formalin preparations, specifically the need for methanol-free formulations.

6. These steps should be accomplished rapidly in order to avoid oxidation of the components.

7. Remove as much liquid as possible.

8. The magnetite particles are prepared by the "Massart method" [22] *via* alkaline precipitation of ferrous and ferric chloride. Diameter distribution of the MNPs (obtained by alkaline precipitation) is in the range of 5–20 nm, optionally analyzed by TEM.

9. The theoretic yield of this procedure is 890 mg of magnetite in total. Remaining water content is determined by weighing the vial prior to and after magnetite synthesis after subtraction of the theoretic magnetite weight. The total batch of all four magnetite vials usually has 5.5 mL of the remaining water. Therefore the magnetite is resuspended by adding 5.5 mL of 2.4 % (w/v) alginate in order to reach a final concentration of 1.2 % (w/v).

10. The sonication step is performed to improve alginate coating and decrease magnetite crystal aggregation. We used 550 Sonic Dismembrator, Fisher Scientific, equipped with 0.5 in. horn (probe) with 0.5 in. diameter and 5 in. length probe extender, operated at 35–40 % power (corresponding to 190–220 W).

11. According to our experience, the alginate-stabilized MNPs have a diameter range of between 416 and 776 nm with a polydispersity index of 0.225.

12. Make sure that the distance between the coils is optimal for producing a uniform magnetic field.

13. Ensure that the constructs are placed in the magnetic field between the coils.

14. This general protocol is suited for most primary antibodies. Refer to specific manufacturer's instructions for each antibody, and modify the protocol as necessary.

15. Overnight incubation at 4 °C results in more specific signal and background reduction.

16. We use PrestoBlue™ reagent from Life Technologies. This reagent has a good linear response range with short (30 min or less, depending on the cell number) incubation periods.

17. Optimization of scaffold number per sample (e.g., pooling) could be required to get optimal results. In general, use at least $1-2 \times 10^6$ cells per sample for Western immunoblotting.

Acknowledgements

Yulia Sapir gratefully acknowledges the generous fellowship from the late Mr. Daniel Falkner and his daughter Ms. Ann Berger and thanks the Azrieli Foundation for the award of an Azrieli Fellowship supporting her PhD program. This study was partially supported by the Louis and Bessie Stein Family foundation through the Drexel University College of Medicine, American Associates of the Ben-Gurion University of the Negev, Israel, the European Union FWP7 (INELPY) and USA Award Number 5R01HL107771 from the National Heart, Lung and Blood Institute, and Israel Science Foundation Grant 793/04. Smadar Cohen holds the Claire and Harold Oshry Professor Chair in Biotechnology.

References

1. Dvir T, Kedem A, Ruvinov E et al (2009) Prevascularization of cardiac patch on the omentum improves its therapeutic outcome. Proc Natl Acad Sci U S A 106(35):14990–14995. doi:10.1073/pnas.0812242106, 0812242106 [pii]

2. Lee H, Cusick RA, Utsunomiya H et al (2003) Effect of implantation site on hepatocytes heterotopically transplanted on biodegradable polymer scaffolds. Tissue Eng 9(6):1227–1232. doi:10.1089/10763270360728134

3. Zhang QX, Magovern CJ, Mack CA et al (1997) Vascular endothelial growth factor is the major angiogenic factor in omentum: mechanism of the omentum-mediated angiogenesis. J Surg Res 67(2):147–154. doi:10.1006/jsre.1996.4983, S0022-4804(96)94983-5 [pii]

4. Zhou Q, Zhou JY, Zheng Z et al (2010) A novel vascularized patch enhances cell survival and modifies ventricular remodeling in a rat myocardial infarction model. J Thorac Cardiovasc Surg 140(6):1388–1396, e1381–1383. doi:10.1016/j.jtcvs.2010.02.036

5. Shimizu T, Sekine H, Yang J et al (2006) Polysurgery of cell sheet grafts overcomes diffusion limits to produce thick, vascularized myocardial tissues. FASEB J 20(6):708–710. doi:10.1096/fj.05-4715fje

6. Suzuki R, Hattori F, Itabashi Y et al (2009) Omentopexy enhances graft function in myocardial cell sheet transplantation. Biochem Biophys Res Commun 387(2):353–359. doi:10.1016/j.bbrc.2009.07.024, S0006-291X(09)01358-8 [pii]

7. Richardson TP, Peters MC, Ennett AB et al (2001) Polymeric system for dual growth factor delivery. Nat Biotechnol 19(11):1029–1034. doi:10.1038/nbt1101-1029

8. Perets A, Baruch Y, Weisbuch F et al (2003) Enhancing the vascularization of three-

dimensional porous alginate scaffolds by incorporating controlled release basic fibroblast growth factor microspheres. J Biomed Mater Res A 65(4):489–497. doi:10.1002/jbm.a.10542

9. Freeman I, Kedem A, Cohen S (2008) The effect of sulfation of alginate hydrogels on the specific binding and controlled release of heparin-binding proteins. Biomaterials 29(22):3260–3268. doi:10.1016/j.biomaterials.2008.04.025, S0142-9612(08)00278-0 [pii]

10. Freeman I, Cohen S (2009) The influence of the sequential delivery of angiogenic factors from affinity-binding alginate scaffolds on vascularization. Biomaterials 30(11):2122–2131. doi:10.1016/j.biomaterials.2008.12.057, S0142-9612(08)01060-0 [pii]

11. Ruvinov E, Leor J, Cohen S (2010) The effects of controlled HGF delivery from an affinity-binding alginate biomaterial on angiogenesis and blood perfusion in a hindlimb ischemia model. Biomaterials 31(16):4573–4582. doi:10.1016/j.biomaterials.2010.02.026, S0142-9612(10)00255-3 [pii]

12. Ruvinov E, Leor J, Cohen S (2011) The promotion of myocardial repair by the sequential delivery of IGF-1 and HGF from an injectable alginate biomaterial in a model of acute myocardial infarction. Biomaterials 32(2):565–578. doi:10.1016/j.biomaterials.2010.08.097, S0142-9612(10)01135-X [pii]

13. Caspi O, Lesman A, Basevitch Y et al (2007) Tissue engineering of vascularized cardiac muscle from human embryonic stem cells. Circ Res 100(2):263–272. doi:10.1161/01. RES.0000257776.05673.ff, 01. RES.0000257776.05673.ff [pii]

14. Levenberg S, Rouwkema J, Macdonald M et al (2005) Engineering vascularized skeletal muscle tissue. Nat Biotechnol 23(7):879–884. doi:10.1038/nbt1109, nbt1109 [pii]

15. Iyer RK, Chui J, Radisic M (2009) Spatiotemporal tracking of cells in tissue-engineered cardiac organoids. J Tissue Eng Regen Med 3(3):196–207. doi:10.1002/term. 153

16. Iyer RK, Chiu LL, Radisic M (2009) Microfabricated poly(ethylene glycol) templates enable rapid screening of triculture conditions for cardiac tissue engineering. J Biomed Mater Res A 89(3):616–631. doi:10.1002/jbm.a.32014

17. Iyer RK, Chiu LL, Vunjak-Novakovic G et al (2012) Biofabrication enables efficient interrogation and optimization of sequential culture of endothelial cells, fibroblasts and cardiomyocytes for formation of vascular cords in cardiac tissue engineering. Biofabrication 4(3):035002. doi:10.1088/1758-5082/4/3/035002

18. Sapir Y, Cohen S, Friedman G et al (2012) The promotion of in vitro vessel-like organization of endothelial cells in magnetically responsive alginate scaffolds. Biomaterials 33(16):4100–4109. doi:10.1016/j.biomaterials.2012.02.037

19. Sapir Y, Polyak B, Cohen S (2014) Cardiac tissue engineering in magnetically actuated scaffolds. Nanotechnology 25(1):014009

20. Rowley JA, Madlambayan G, Mooney DJ (1999) Alginate hydrogels as synthetic extracellular matrix materials. Biomaterials 20(1):45–53

21. Tsur-Gang O, Ruvinov E, Landa N et al (2009) The effects of peptide-based modification of alginate on left ventricular remodeling and function after myocardial infarction. Biomaterials 30(2):189–195. doi:10.1016/j.biomaterials.2008.09.018, S0142-9612(08)00687-X [pii]

22. Massart R (1981) Preparation of aqueous magnetic liquids in alkaline and acidic media. IEEE Trans Magnetics 17(2):1247–1248. doi:10.1109/Tmag.1981.1061188

Chapter 9

Shrink-Induced Biomimetic Wrinkled Substrates for Functional Cardiac Cell Alignment and Culture

Nicole Mendoza, Roger Tu, Aaron Chen, Eugene Lee, and Michelle Khine

Abstract

The anisotropic alignment of cardiomyocytes in native myocardium tissue is a functional feature that is absent in traditional in vitro cardiac cell culture. Microenvironmental factors cue structural organization of the myocardium, which promotes the mechanical contractile properties and electrophysiological patterns seen in mature cardiomyocytes. Current nano- and microfabrication techniques, such as photolithography, generate simplified cell culture topographies that are not truly representative of the multifaceted and multiscale fibrils of the cardiac extracellular matrix. In addition, such technologies are costly and require a clean room for fabrication. This chapter offers an easy, fast, robust, and inexpensive fabrication of biomimetic multi-scale wrinkled surfaces through the process of plasma treating and shrinking prestressed thermoplastic. Additionally, this chapter includes techniques for culturing stem cells and their cardiac derivatives on these substrates. Importantly, this wrinkled cell culture platform is compatible with both fluorescence and bright-field imaging; real-time physiological monitoring of CM action potential propagation and contraction properties can elucidate cardiotoxicity drug effects.

Key words Stem cells, Cardiomyocytes, Cell mechanics, Alignment, Shrink film, Biomaterials, Biomimetic topography

1 Introduction

Heart disease, the number one cause of death in developed countries [1], is particularly difficult to treat due to the post-mitotic nature of adult cardiomyocytes (CMs). CMs have limited regenerative potential; therefore, damaged cells are not replaced nor repaired. In recent years, human pluripotent stem cells (hPSCs) have become an attractive cell source for basic and translational cardiac studies. These cells are capable of self-renewal [2–4] and differentiation into CMs [5–7], which provides an unlimited supply of cardiac cells.

Such cardiac cells from hPSCs have promising applications in high-throughput drug screening [8–10]. Current preclinical drug testing methods fail to effectively recapitulate human cardiac

Milica Radisic and Lauren D. Black III (eds.), *Cardiac Tissue Engineering: Methods and Protocols*, Methods in Molecular Biology, vol. 1181, DOI 10.1007/978-1-4939-1047-2_9, © Springer Science+Business Media New York 2014

responses, thus contributing to cardiac toxicity. In fact, cardiac toxicity is the leading cause for drug removal from the market [11]. Since the discovery of hPSC-derived CMs (hPSC-CMs) over a decade ago [12], optimization of cardiac cell culture has been the objective of thousands of research reports in pursuit of a functional in vitro cardiac model.

Despite their benefits, hPSC-CMs cultured in vitro commonly display embryonic-like phenotypes, which currently prevents their adoption both for transplantation and even as an acceptable in vitro model [13]. Their dysfunctional features, including immature structural development, electrical propagation, and contractile properties [14–17], differ significantly compared to mature cardiac cells. Exposure to both biological and mechanical factors drives cardiac maturation in vivo. In the native human heart, layers of anisotropic extracellular matrix (ECM) fibrils, spanning from nano- to microscale, naturally align cardiac cells [18]. Such organization facilitates the proper development of physiological, mechanical, and electrical functions of CMs [19].

Common in vitro cell culture devices (e.g., petri dishes) fail to recreate the native microenvironment of the heart. Yet studies have shown that tuning various components of the in vitro cellular environment can induce alignment of sarcomere structures, improved action potential propagations, and greater contractile forces in CMs [20, 21]. Several micropatterning and microfabrication techniques have been implemented to provide CMs with necessary mechanical cues [22]. However, there are significant trade-offs between the cost and effectiveness of each technique. Microcontact printing is a micropatterning technique, in which proteins are directly printed onto soft substrates, aligning CMs that are cultured directly on these proteins [23–25]. While this process is relatively cheap and suitable for large culture surface areas, the proteins lack the mechanical structure of the native microenvironment. Microfabrication techniques, such as photolithography, are traditionally used to generate surface topographies composed of uniform nano- to micro-sized grooves or ridges (Fig. 1a) [18, 26–28]. These features are typically arranged in repetitive patterns that are limited by inability to capture the complex, varying structures of the ECM. Replicating the finer features (at the nm scale) of the ECM usually requires more advanced fabrication techniques, such as electro-beam lithography and nanoprint technology [29]. Using advanced technologies to produce biomimetic features over substantial areas is a timely process and requires expensive equipment not accessible to most cell culture laboratories.

Shrink film technologies provide both an inexpensive and effective method for achieving multi-scale, biomimetic wrinkles that have similar size and structure to those of native collagen bands (Fig. 1b) [19]. Shrink film is composed of polyethylene (PE), a prestressed thermoplastic that shrinks by 90 % by area when heated in a conventional oven [30]. Prior to shrinking, the surface

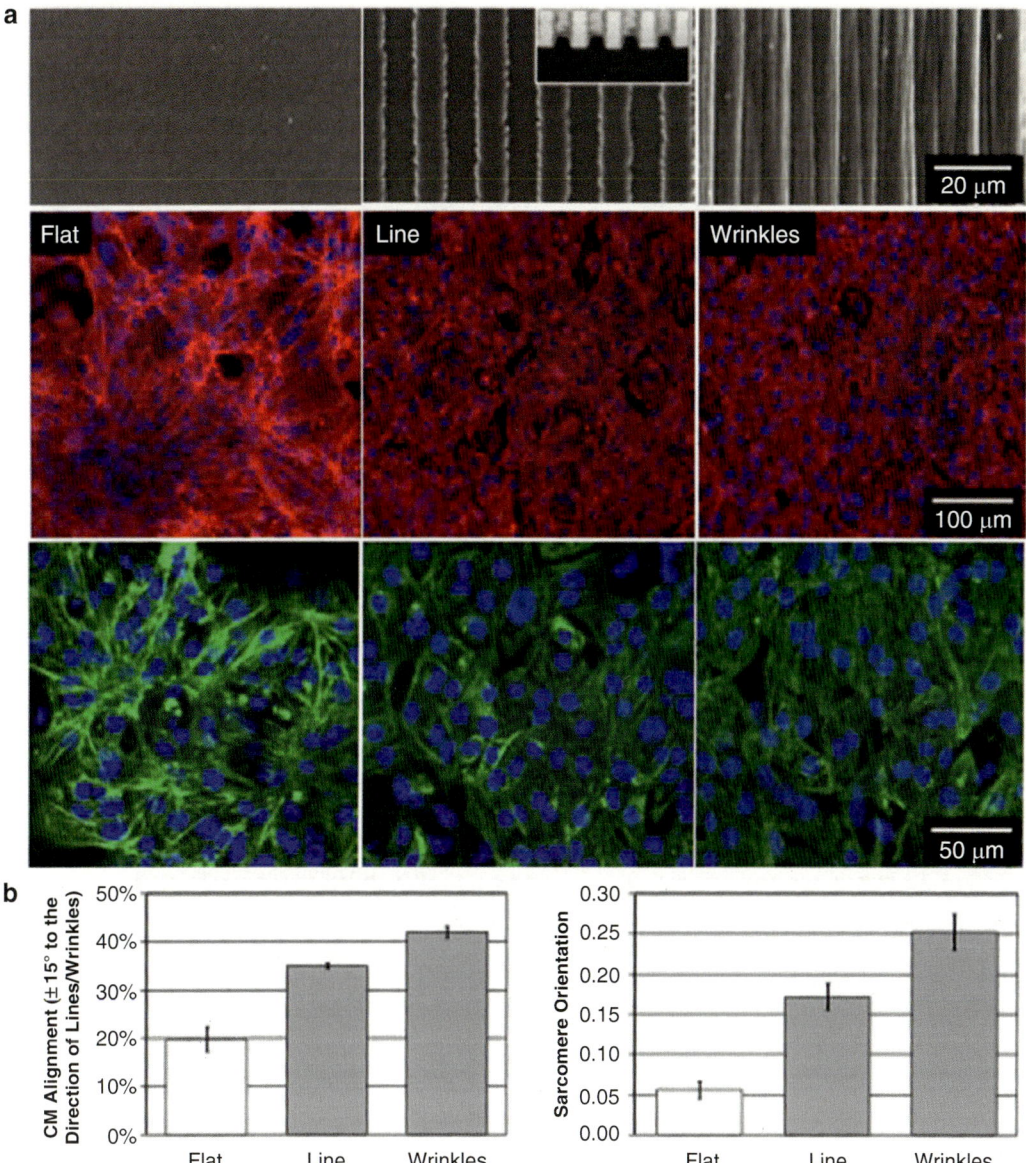

Fig. 1 Alignment of hESC-CMs. (**a**) Comparison of (*left*) flat, (*middle*) lined, and (*right*) wrinkled substrate topographies. (*Top panel*) Scanning electron micrographs (SEM) of topographies at ×1,000 magnification. Inset is the tilted view of the cross section of lines. Inset scale bar is 20 μm. (*Middle panel*) Fluorescent images taken at ×20 magnification of f-actin. (*Lower panel*) Fluorescent images taken at ×40 magnification of α-actinin, showing hESC-CM alignment. (**b**) Bar graphs quantifying (*left*) f-actin alignment by percent aligned and (*right*) sarcomere alignment using orientation organization parameter. This figure was reproduced with permission from Biomaterials [33]

of the PE film is oxidized by a plasma machine. The formation of wrinkles depends on the stiffness mismatch between the oxidized surface layer and the shrink film. During the shrinking process the oxidized layer buckles on itself, causing multi-scale wrinkles to self-assemble on the surface [31]. These self-similar wrinkles are aligned

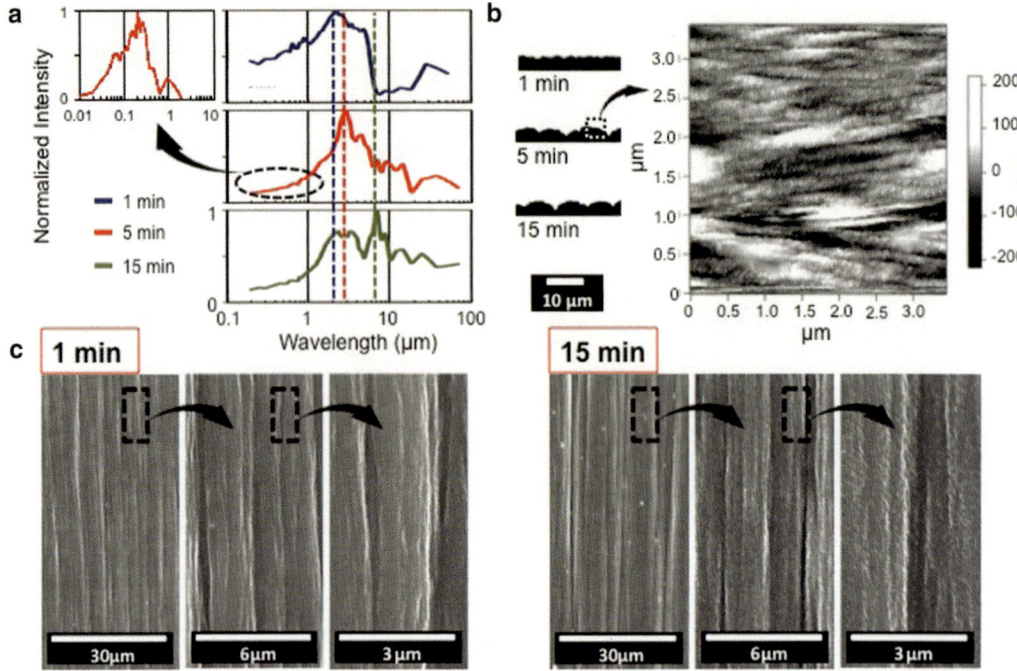

Fig. 2 Characterization of wrinkles. (**a**) Fast Fourier transform (FFT) was performed on SEM images to determine the characteristic wavelengths of wrinkles of three different plasma conditions: 1, 5, and 15 min. Inset shows smaller wrinkles at ×20,000 magnification. (**b**) (*Left*) Bright-field image at ×150 magnification shows cross section of wrinkles at the three plasma time conditions. (*Right*) AFM of matrigel-coated wrinkles of 5-min plasma time condition. (**c**) SEM images with progressive zoom to illustrate self-similar wrinkles and wrinkle bundling of (*left*) 1-min plasma time condition and (*right*) 15-min plasma time condition

and layered in such a way that they resemble the bundled structure of the ECM. The average predominant wavelengths of the smaller wrinkles (minor wavelengths) range from 60 to 380 nm, while average wavelengths of larger wrinkles (major wavelengths) range from 1 to 7 μm (Fig. 2). This scale is consistent with natural ECM, in which individual collagen fibers measure about 300 nm and bundles of fibrils reach micrometers in size [32]. These feature sizes are easily tunable with longer oxidation periods yielding greater wrinkle bundling [30].

These biomimetic substrates have served as a platform for functional monitoring of hPSC-CMs [33]. Capturing time-varying mechanical and electrical properties of contracting CMs is crucial for in vitro characterization and detection of changes in CM behavior. Wrinkled substrates are compatible with bright-field and fluorescence-based microscopy, which provides a noninvasive option for real-time physiological monitoring and cardiotoxicity assays of hPSC-CMs. Functional anisotropic alignment of human embryonic stem cell-derived ventricular CMs (hESC-VCMs) using the multi-scale wrinkles represents a more accurate model for efficacious drug discovery and development as well as arrhythmogenicity

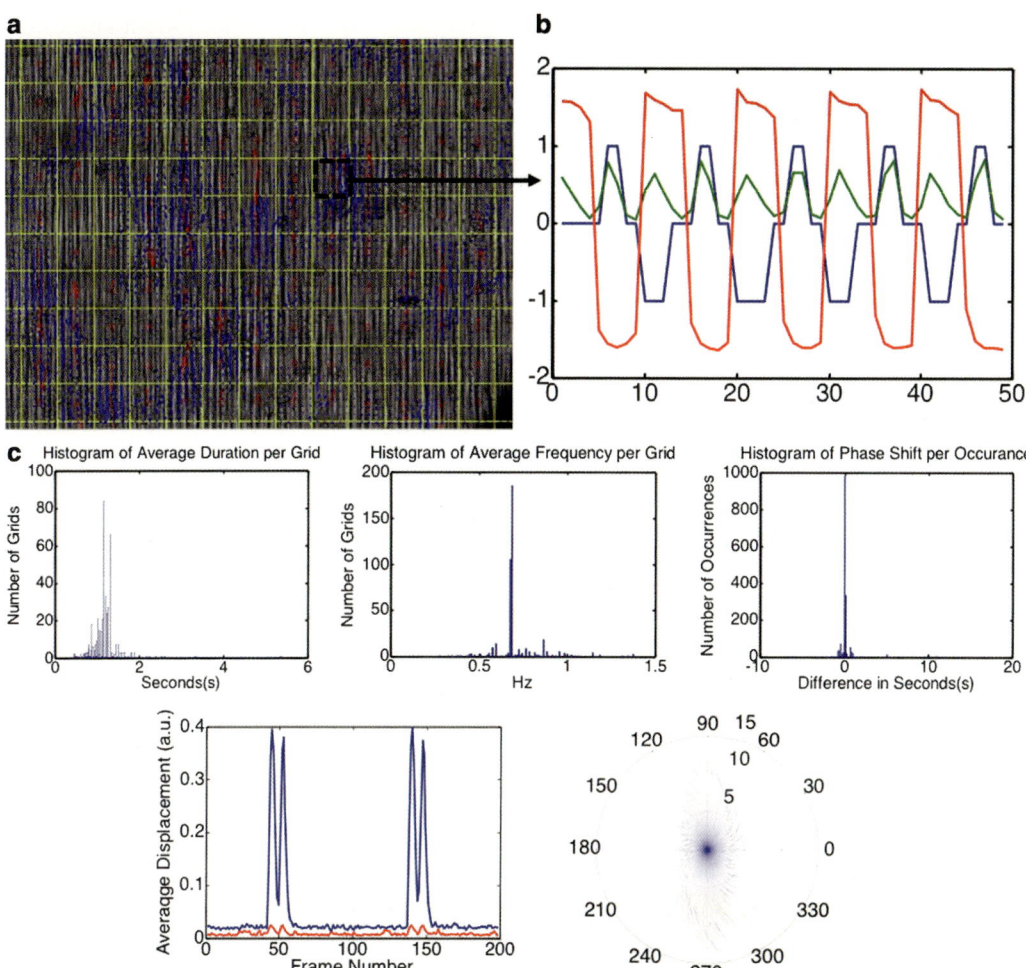

Fig. 3 Optical flow hPSC-CM contraction analysis. (**a**) Time-lapse images are gridded (*yellow boxes*), and motion vectors (*blue lines*) are generated using the optical flow code. Average motion vectors are calculated for each grid (*red arrows*). (**b**) Contraction analysis plot. The *green line* represents displacement, and the *red line* is direction of motion vector in radians. The *blue line* is the categorical plot, with values of 1 indicating contractions and −1 indicating relaxations. (**c**) Types of analysis that can be generated by the optical flow code include histograms of contractile duration, frequency, and synchronicity (labeled as phase shift). Contractile orientation can be represented in two graphs: *x*- and *y*-motion vector trances (*bottom left*, *blue* is *y*, and *red* is *x*) and a compass plot (*bottom right*). This figure was reproduced with permission from Biomaterials [33]

screening when combined with action potential measurements [34]. When combined with time-lapse bright-field imaging and optical flow vector analysis, these biomimetic substrates provide an integrated platform capable of determining cardiotoxicity drug effects by simply characterizing CM contraction properties [33]. Optical flow analysis generates a series of motion vectors, which can be analyzed to give information regarding CM contraction duration, frequency, synchronicity, and orientation (Fig. 3).

This chapter provides the easy, inexpensive method for fabricating biomimetic plastic substrates, which takes less than 10 min [30]. The PE-wrinkled surface can be either directly cultured on or used as a master mold for transferring wrinkled features on to any other cell culture plastic via hot embossing (also detailed in this chapter). In addition, techniques for culturing stem cell-derived CMs on these surfaces are provided.

2   Materials

2.1   Shrink Film

1. Prestressed PE shrink film (Cryovac® D-film, LD935, Sealed Air Corporation) is used to create biomimetic-wrinkled surfaces.

2. Polydimethylsiloxane (PDMS) (Sylgard 184) can be used to make a stamp to transfer wrinkled features onto other cell culture plastics.

3. These methods describe hot embossing features onto a tissue culture-compatible polystyrene sheet (Grafix Clear Shrink Film), but any tissue culture plastic substrate can be substituted.

4. Equipment needed includes a plasma machine (Plasma Prep II, SPI Supplies) for oxygen plasma treatment of surfaces and a conventional oven for shrinking prestressed plastics.

2.2   Cell Culture

This protocol is specific to the culture and alignment of hESC-CMs. Wrinkled shrink substrates have also been proven to successfully align mouse embryonic fibroblasts (MEFs), aortic smooth muscle cells (AoSMCs), and human embryonic stem cells (hESCs) [30]. Many different protocols for the culture and differentiation of hESC-CMs exist [2, 5, 6, 12, 35–38] and are likely compatible with wrinkled shrink substrates.

2.2.1   Cell Culture Media and Factors

1. Maintain hESC cells with mTeSR1 medium.

2. For differentiation procedure, use RPMI-B27 with and without insulin.

3. Supplements include 100 ng/mL activin A (R&D Systems) and 10 ng/mL BMP4 [39].

2.2.2   hESC-CM Seeding

1. Use 0.25 % trypsin–EDTA to lift cells from plate.

2. Prepare plating media consisting of Dulbecco's modified Eagle medium (D-MEM) high glucose supplemented with 10 % fetal bovine serum (FBS), 1× nonessential amino acids, and 2 mM Glutamax.

3. Maintain hESC-CMs in RPMI-B27 culture media with insulin.

3 Methods

3.1 Fabrication of Biomimetic Wrinkles

3.1.1 Wrinkled PE Master Mold

1. Treat a PE film with oxygen plasma for 5 min (Fig. 4I) (*see* **Note 1**).

2. On top of a glass plate, place a piece of paper that is slightly bigger than the size of the treated PE film (*see* **Note 2**).

3. Place the PE film on top of the paper-covered glass slide, and secure the shorter ends with binder clips.

4. Place the construct (PE film, paper, and glass plate) in an oven that is preheated to 150 °C for 3 min. This process will shrink the film and generate the wrinkled surface features on the PE (Figs. 1 and 4.II) (*see* **Note 3**).

5. The resulting PE-wrinkled substrate can be used directly for cell culture (skip to Subheading 3.1.4) or used as a master mold for transferring features via hot embossing (continue to Subheading 3.1.2).

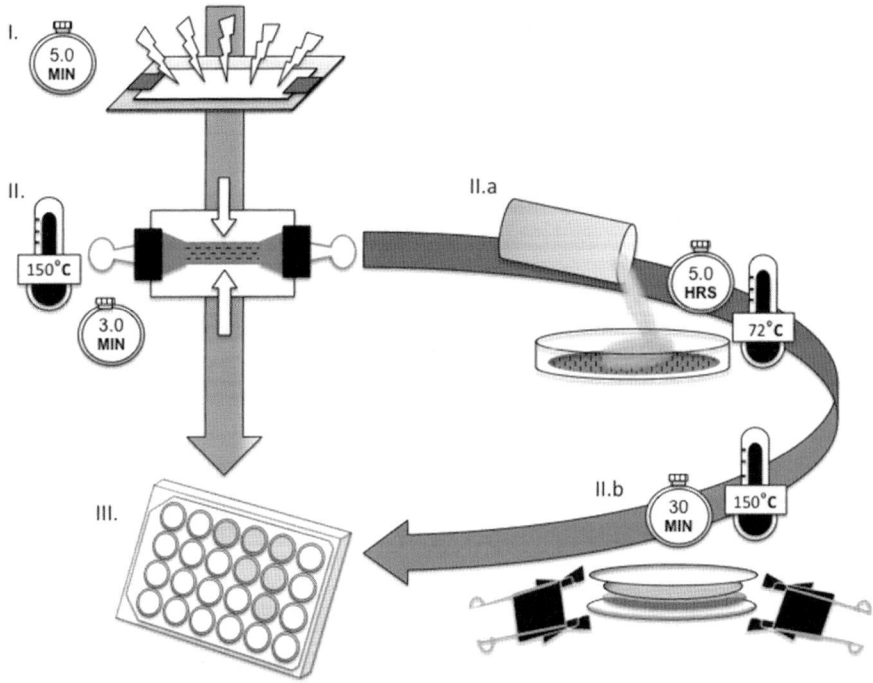

Fig. 4 Fabricating wrinkled cell culture substrates. (I) Treat PE with oxygen plasma for 5 min. (II) Bind both ends of PE and shrink uniaxially by placing in an oven at 150 °C for 3 min [30]. (II.a) Pour PDMS on top of featured side of PE in a petri dish. (II.b) Hot emboss PS by placing on featured side of PDMS, sandwich together with glass slides and binder clips, and heat at 150 °C for 30 min. (III) Use either the original wrinkled PE substrate or the hot embossed PS substrate for cell culture

3.1.2 PDMS-
Wrinkled Stamp

1. Inspect wrinkles under a microscope for micro-tears (*see* **Note 3**).

2. Cut out the parallel portion of wrinkled PE, and use double-sided tape to adhere it to a petri dish, featured side up.

3. Mix PDMS monomer and curing agent at a 1:10 ratio.

4. Spin PDMS mixture in a centrifuge at 2,060 × g for 5 min to degas.

5. Pour the degassed PDMS mixture over the wrinkled PE in the petri dish (Fig. 4.II.a). The volume of the mixture is dictated by the desired thickness of the stamp.

6. Place the petri dish under a vacuum for 3 h. Then allow it to sit at room temperature for an additional 5 h.

7. Transfer the petri dish to an oven at 72 °C for at least 5 h.

8. Remove from oven, and cut out the PDMS stamp.

3.1.3 Hot Embossing

1. Place a sheet of PS (or any other tissue culture plastic) atop the featured side of the PDMS stamp (*see* **Note 4**).

2. Sandwich the PS and PDMS stamp with two pieces of glass and secure together with binder clips (Fig. 4.II.b).

3. Preheat a conventional oven to 150 °C (*see* **Note 5**).

4. Heat the construct in the oven for 30 min. Features on the PDMS stamp will be hot embossed onto the PS sheet (*see* **Note 6**).

5. Remove construct from the oven, and allow it to cool to room temperature.

3.1.4 Substrate
Mounting and Sterilization

1. Inspect wrinkles under a microscope for micro-tears.

2. Cut wrinkled substrate to desired size/shape (15 mm diameter circles are described here).

3. Pipette 500 μL of PDMS onto a 15 mm glass cover slip. Place the wrinkled substrate directly on top of PDMS to adhere it to the 15 mm glass cover slip (*see* **Note 7**).

4. Sterilize the wrinkled substrate by submerging in 70 % ethanol and treating with UV light for at least 30 min.

5. Submerge wrinkled substrate again in double-distilled water and treat with UV light for a minimum of 30 min.

3.2 hESC Expansion

1. Feeder-free hPSC culture has been described elsewhere [39].

2. Begin hESC cell culture in standard tissue culture plates coated with 1:200 Matrigel®.

3. Feed cells daily with mTeSR1 medium.

4. Passage when cells reach 80–90 % confluency; during passage, plate cells in a 1:10 ratio (*see* **Note 8**).

3.3 Cardiac Differentiation

Day 2: Passage a confluent well of hPSC into three wells of a tissue culture well plate (*see* **Note 9**) [5].

Day 1: Exchange media with mTeSR1. Cells should be very confluent.

Day 0: Exchange media using RPMI-B27 (without insulin), and add a supplement of 100 ng/mL activin A for the first 24 h.

Day 1: Exchange media using fresh RPMI-B27 (without insulin), and add a supplement of BMP4 (10 ng/mL). Wait for 4 days.

Day 5: Exchange media with RPMI-B27 (with insulin). Do not add any cytokines. Following this, exchange media every 3 days using RPMI-B27 with insulin.

Days 8, 11, 14, etc.: Exchange media. Spontaneous contractions usually start occurring.

Day 20: Start the experiment (*see* **Note 10**).

3.4 Cell Seeding onto Wrinkled Substrates

1. Oxidize wrinkled cell culture substrates by UV/ozone for 8 min.

2. Coat surface-activated substrates with 50 µg/mL fibronectin and incubate overnight at 2–8 °C.

3. Lift cardiac cells by incubating them in 0.25 % trypsin–EDTA for 5 min.

4. Resuspend cells in "plating medium" (DMEM with 10 % FBS and 2 mM Glutamax).

5. Load cells onto fibronectin-coated wrinkled substrate at a seeding density of 5×10^5 cells/mL.

6. The following day, replace culture media using RPMI-B27 (with insulin).

7. Change media every 3–4 days.

4   Notes

1. This plasma procedure is specifically developed for Plasma Prep II. Optimization is required when using different plasma machines. Settings are as follows: Turn vacuum pump on, and wait for 10 min for vacuum to warm up. Turn oxygen gas valve open and bring to 7 PSI. With the sample in the chamber, turn on machine vacuum and pump down to 200 mTorr. Begin plasma treatment at a power level of 60–70 mA. Wait for 5 min (plasma chamber should glow a light blue color). Oxygen will be introduced into the chamber continuously over the 5 min. Stop plasma treatment, de-vacuum the plasma machine, and remove the sample from the chamber.

2. The paper acts as a non-sticking surface so that the film can retract freely during shrinkage. It is fine to substitute any other non-stick surface for paper.

3. When film is completely shrunk, it should resemble the shape of a bowtie, with the center being relatively narrow and parallel (Fig. 4.II). Depending on the quality and handling of the PE film, as well as plasma treatment condition, micro-tears sometimes do occur after shrinking. Micro-tears can be identified as discoloration within the parallel portion of the wrinkles. If tears do occur, the piece is deemed unusable.

4. Depending on the desired thickness of final wrinkled substrates, thicker or multiple PS sheets may be used.

5. If using a different type of thermal plastics, make sure that the oven temperature is higher than the glass transition temperature.

6. Depending on the quality and placement of PS sheets, bubbles can form and get trapped during hot embossing. If this occurs, multiple layers of PS sheets should be exchanged for a single sheet of thick, high-quality PS.

7. In general, be gentle with the cells. Aspirate and pipette media slowly. Do not place media in warm water bath; instead, slowly bring to room temperature by placing in a dark drawer.

8. The PDMS acts as glue that fastens the wrinkled substrate to the glass. Be careful not to get PDMS on top of the plastic surface or wrinkles will be compromised. It is okay to use double-sided tape instead of PDMS as long as the tape does not affect cell culture.

9. Make sure that cells are confluent in the three new wells. Many cells will die during differentiation.

10. On days 0, 1, 2, 5, and 8, do not use an aspirator when exchanging media. Use a pipette tip and exchange medium SLOWLY. Cells are very loosely attached during differentiation. It is usually best to wait before starting cardiac experiments until after day 20; however, cardiac experiments may technically begin any time after day 8.

References

1. Roger VL, Go AS, Lloyd-Jones DM et al (2011) Heart disease and stroke statistics–2011 update: a report from the American Heart Association. Circulation 123:e18–e209

2. Thomson JA, Itskovitz-Eldor J, Shapiro SS et al (1998) Embryonic stem cell lines derived from human blastocysts. Science 282:1145–1147

3. Reubinoff BE, Pera MF, Fong CY et al (2000) Embryonic stem cell lines from human blastocysts: somatic differentiation in vitro. Nat Biotechnol 18:399–404

4. Amit M, Carpenter MK, Inokuma MS et al (2000) Clonally derived human embryonic stem cell lines maintain pluripotency and proliferative potential for prolonged periods of culture. Dev Biol 227:271–278

5. Laflamme MA, Chen KY, Naumova AV et al (2007) Cardiomyocytes derived from human

embryonic stem cells in pro-survival factors enhance function of infarcted rat hearts. Nat Biotechnol 25:1015–1024

6. Yang L, Soonpaa MH, Adler ED et al (2008) Human cardiovascular progenitor cells develop from a KDR+ embryonic-stem-cell-derived population. Nature 453:524–528

7. Zhang J, Klos M, Wilson GF et al (2012) Extracellular matrix promotes highly efficient cardiac differentiation of human pluripotent stem cells: the matrix sandwich method. Circ Res 111:1125–1136

8. Chow MZ, Boheler KR, Li RA (2013) Human pluripotent stem cell-derived cardiomyocytes for heart regeneration, drug discovery and disease modeling: from the genetic, epigenetic, and tissue modeling perspectives. Stem Cell Res Therapy 4(4):97

9. Himmel HM (2013) Drug-induced functional cardiotoxicity screening in stem cell-derived human and mouse cardiomyocytes: Effects of reference compounds. J Pharmacol Toxicol Methods 68:97–111

10. Pet L et al (2013) Drug screening using a library of human induced pluripotent stem cell-derived cardiomyocytes reveals disease-specific patterns of cardiotoxicity. Circulation 127(16):1677–1691

11. Lexchin J (2005) Drug withdrawals from the Canadian market for safety reasons, 1963–2004. Can Med Assoc J 172:765–767

12. Xu CH, Inokuma MS, Denham J et al (2001) Feeder-free growth of undifferentiated human embryonic stem cells. Nat Biotechnol 19:971–974

13. Robertson C, Tran DD, George SC (2013) Concise review: maturation phases of human pluripotent stem cell-derived cardiomyocytes. Stem Cells 31:829–837

14. Lieu DK, Liu J, Siu CW et al (2009) Absence of transverse tubules contributes to non-uniform Ca(2+) wavefronts in mouse and human embryonic stem cell-derived cardiomyocytes. Stem Cells Dev 18:1493–1500

15. Satin J, Itzhaki I, Rapoport S et al (2008) Calcium handling in human embryonic stem cell-derived cardiomyocytes. Stem Cells 26:1961–1972

16. Hoffman BF (1962) Electrophysiology of the conducting system of the heart. Trans NY Acad Sci 24:886–890

17. Peng S, Lacerda AE, Kirsch GE et al (2010) The action potential and comparative pharmacology of stem cell-derived human cardiomyocytes. J Pharmacol Toxicol Methods 61:277–286

18. Kim D-H, Lipke EA, Kim P et al (2010) Nanoscale cues regulate the structure and function of macroscopic cardiac tissue constructs. Proc Natl Acad Sci U S A 107:565–570

19. Pins GD, Christiansen DL, Patel R et al (1997) Self-assembly of collagen fibers. Influence of fibrillar alignment and decorin on mechanical properties. Biophys J 73:2164–2172

20. Grosberg A, Alford PW, McCain ML et al (2011) Ensembles of engineered cardiac tissues for physiological and pharmacological study: heart on a chip. Lab Chip 11:4165–4173

21. Au HTH, Cheng I, Chowdhury MF et al (2007) Interactive effects of surface topography and pulsatile electrical field stimulation on orientation and elongation of fibroblasts and cardiomyocytes. Biomaterials 28:4277–4293

22. Luna JI, Ciriza J, Garcia-Ojeda ME et al (2011) Multiscale biomimetic topography for the alignment of neonatal and embryonic stem cell-derived heart cells. Tissue Eng Part C Methods 17:579–588

23. Grosberg A, Nesmith AP, Goss JA et al (2012) Muscle on a chip: In vitro contractility assays for smooth and striated muscle. J Pharmacol Toxicol Methods 65:126–135

24. Badie N, Satterwhite L, Bursac N (2009) A method to replicate the microstructure of heart tissue in vitro using DTMRI-based cell micropatterning. Ann Biomed Eng 37:2510–2521

25. Bauwens CL, Peerani R, Niebruegge S et al (2008) Control of human embryonic stem cell colony and aggregate size heterogeneity influences differentiation trajectories. Stem Cells 26:2300–2310

26. Dalby MJ, Riehle MO, Yarwood SJ et al (2003) Nucleus alignment and cell signaling in fibroblasts: response to a micro-grooved topography. Exp Cell Res 284:274–282

27. Chen CS, Mrksich M, Huang S et al (1997) Geometric control of cell life and death. Science 276:1425–1428

28. Au HTH, Cui B, Chu ZE et al (2009) Cell culture chips for simultaneous application of topographical and electrical cues enhance phenotype of cardiomyocytes. Lab Chip 9:564–575

29. Engelmayr GC Jr, Cheng M, Bettinger CJ et al (2008) Accordion-like honeycombs for tissue engineering of cardiac anisotropy. Nat Mater 7:1003–1010

30. Chen A, Lieu DK, Freschauf L et al (2011) Shrink-film configurable multiscale wrinkles for functional alignment of human embryonic stem cells and their cardiac derivatives. Adv Mater 23(48):5785–5791

31. Fu C-C, Grimes A, Long M et al (2009) Tunable nanowrinkles on shape memory polymer sheets. Adv Mater 21:4472–4476

32. Pamula E, De Cupere V, Dufrene YF et al (2004) Nanoscale organization of adsorbed collagen: Influence of substrate hydrophobicity and adsorption time. J Colloid Interface Sci 271:80–91

33. Chen A, Lee E, Tu R et al (2013) Integrated Platform for Functional Monitoring of Biomimetic Heart Sheets Derived From Human Pluripotent Stem Cells. Biomaterials 35(2):675–83

34. Wang J, Chen A, Lieu DK et al (2013) Effect of engineered anisotropy on the susceptibility of human pluripotent stem cell-derived ventricular cardiomyocytes to arrhythmias. Biomaterials 34:8878–8886

35. Li Y, Powell S, Brunette E et al (2005) Expansion of human embryonic stem cells in defined serum-free medium devoid of animal-derived products. Biotechnol Bioeng 91:688–698

36. Ludwig TE, Bergendahl V, Levenstein ME et al (2006) Feeder-independent culture of human embryonic stem cells. Nat Methods 3:637–646

37. Lee JB, Lee JE, Park JH et al (2005) Establishment and maintenance of human embryonic stem cell lines on human feeder cells derived from uterine endometrium under serum-free condition. Biol Reprod 72:42–49

38. Ma J, Guo L, Fiene SJ et al (2011) High purity human-induced pluripotent stem cell-derived cardiomyocytes: electrophysiological properties of action potentials and ionic currents. Am J Physiol 301:H2006–H2017

39. Eremeev AV, Svetlakov AV, Polstianoy AM et al (2009) Derivation of a novel human embryonic stem cell line under serum-free and feeder-free conditions. Dokl Biol Sci 426: 293–295

Chapter 10

Injectable ECM Scaffolds for Cardiac Repair

Todd D. Johnson, Rebecca L. Braden, and Karen L. Christman

Abstract

Injectable biomaterials have been developed as potential minimally invasive therapies for treating myocardial infarction (MI) and heart failure. Christman et al. first showed that the injection of a biomaterial alone into rat myocardium can improve cardiac function after MI (Christman et al. Tissue Eng 10:403–409, 2004). More recently, hydrogel forms of decellularized extracellular matrix (ECM) materials have shown substantial promise. Here we present the methods for fabricating an injectable cardiac specific ECM biomaterial shown to already have positive outcomes in small and large animal models for cardiac repair (Singelyn et al. Biomaterials 30:5409–5416, 2009; Singelyn et al. J Am Coll Cardiol 59:751–763, 2012; Seif-Naraghi et al. Sci Transl Med 5:173ra25, 2013). Also covered are the methods for the injection of a biomaterial into rat myocardium using a surgical approach through the diaphragm. Although the methods shown here are for injection of an acellular biomaterial, cells or other therapeutics could also be added to the injection for testing other regenerative medicine strategies.

Key words Injectable, Extracellular matrix, Hydrogel, Cardiac repair, Rat, Decellularization

1 Introduction

A minimally invasive approach for cardiac repair has numerous positive benefits including decreasing local tissue trauma, surgery times, risk due to surgery, hospital stays, and recovery times. These positive attributes have led to the investigation of injectable therapies for treating myocardial infarction (MI). Christman et al. showed that injection of a biomaterial alone directly into the myocardium could lead to beneficial outcomes for cardiac repair post-MI [1, 5]. Since this initial study numerous naturally derived biomaterials including alginate, collagen, chitosan, decellularized tissues, fibrin, hyaluronic acid (HA), keratin, and Matrigel, along with several synthetic biomaterials composed of polyethylene glycol (PEG) or poly(N-isopropylacrylamide) (PNIPAAm), have been investigated [6, 7]. Ideally these injectable biomaterials would be injected utilizing current catheter technology for quicker translation to the clinic. But this mode of delivery provides unique challenges and design parameters for the biomaterial such as

Milica Radisic and Lauren D. Black III (eds.), *Cardiac Tissue Engineering: Methods and Protocols*, Methods in Molecular Biology, vol. 1181, DOI 10.1007/978-1-4939-1047-2_10, © Springer Science+Business Media New York 2014

incorporating the ability to pass through a 27G needle and the appropriate kinetics to not gel at body temperature for up to an hour due to the duration of these procedures [6]. One material recently developed by Singelyn et al. is a decellularized biomaterial derived from porcine myocardial tissue, which provides a tissue-specific material for cardiac repair [2]. In brief, fresh porcine myocardium is decellularized by spinning the chopped tissue in detergents, and then the decellularized tissue is lyophilized and milled into a fine powder. This powder is then partially digested in acidic conditions by pepsin into a liquid form that once brought to physiological conditions (salt, pH, and temperature) gels with the appropriate kinetics for catheter delivery [3, 4]. This myocardial matrix hydrogel was initially tested by injection into rat myocardium post-MI and was shown to maintain cardiac function, increase the size of cardiomyocyte islands within the infarcted region, and even recruit cardiac stem cells into the region of repair [3]. The matrix was also shown to be deliverable through numerous endocardial injections via catheter delivery in a porcine model [3]. Later studies in a porcine MI model demonstrated that the myocardial matrix hydrogel led to increasing cardiac function, decreased infarct fibrosis, and increased cardiac muscle at the endocardium [4]. In this chapter, the methods for decellularization, material digestion, and processing of the matrix into an injectable liquid form are presented. Also, detailed instructions for injecting a biomaterial into rat myocardium with a surgical approach through the diaphragm are included. Here the injection is occurring into a healthy rat heart, but several methods for modeling myocardial infarction could be applied before the injection with total coronary occlusion, coronary occlusion followed by reperfusion, or cryo-injury. Although the specific approach is for a biomaterial alone both growth factors and/or cells could be included in this procedure for further study options.

2   Materials

Use ultrapure water for all solutions and rinsing steps. All materials and supplies for the material processing should be sterile or as clean as possible to prevent contamination. Any surgical supplies or tools that come in contact with the animal during surgery should be autoclaved and/or sterilized.

2.1   Decellularization Materials

1. Sharp knife and cutting board.
2. Decellularization solution: 1 % SDS, 1× PBS. Dissolve 80 g of SDS powder (*see* **Note 1**) in 800 mL of water to make a 10 % stock solution of SDS. In an autoclaved 4 L beaker combine 3,400 mL of water, 400 mL of the 10 % SDS stock, and 200 mL of a 20× PBS stock solution. Stir until dissolved.

3. Plastic cryomolds and OCT compound.

4. Autoclaved 1 L beakers with $3/8'' \times 2\ 1/2''$ stir bars.

5. Stir plate that can be set to 125 rpm and can hold a 1 L beaker.

6. Penicillin/streptomycin or PenStrep (PS): 10,000 U/mL Penicillin and 10,000 μg/mL streptomycin.

7. Autoclaved fine mesh metal strainer.

8. Autoclaved 1 L bottles.

9. Sterile 50 mL plastic conicals.

2.2 Digestion and Injection Preparation Materials

1. Lyophilizer and Wiley® Mini-Mill (Thomas Scientific, Swedesboro, NJ).

2. Digestion solution: 0.1 M HCl, 1 mg/mL pepsin from porcine gastric mucosa (2,500 U/mg protein). Fully dissolve the pepsin in acid by vortexing or shaking the solution. Then sterile filter through a 0.22 μm pore size filter.

3. Autoclaved 20 mL scintillation vial and $5/16'' \times 1/2''$ stir bar.

4. Sterile filtered solutions for pH and salt adjustments: 1 M NaOH, 0.1 M HCl, and 10× PBS.

2.3 Cardiac Surgical Injection Materials

1. Isoflurane anesthesia system.

2. Ventilator with temperature probe.

3. Heating pump connected to warmed surgical table.

4. Electric razor and vacuum.

5. Sterile syringes: 1, 3, and 10 mL.

6. Sterile needles: 25, 27, and 30G.

7. IV Catheter $14G \times 2''$.

8. Aspiration tube constructed from a 20G Intramedic tip with 10 cm of PE 100 tubing attached.

9. Autoclaved surgical tools: Scalpel, scissors, fine forceps, standard forceps, needle holder, and towel clamp.

10. Topical: Betadine, isopropyl alcohol, artificial tears, Surgilube, and triple-antibiotic ointment.

11. Injectable: Lactated ringers, 1 % lidocaine, and Buprenex.

12. Suture: Vicryl 4-0 FS-1, Vicryl 5-0 FS-2, and Prolene 5-0 RB-1.

13. Lab tape or masking tape.

3   Methods

3.1 Tissue Processing and Decellularization

1. Starting with a fresh unfrozen (*see* **Note 2**) porcine heart (Fig. 1a), isolate only the left ventricle and septal wall by removing the thinner right ventricular free wall tissue, both atria, and the valves.

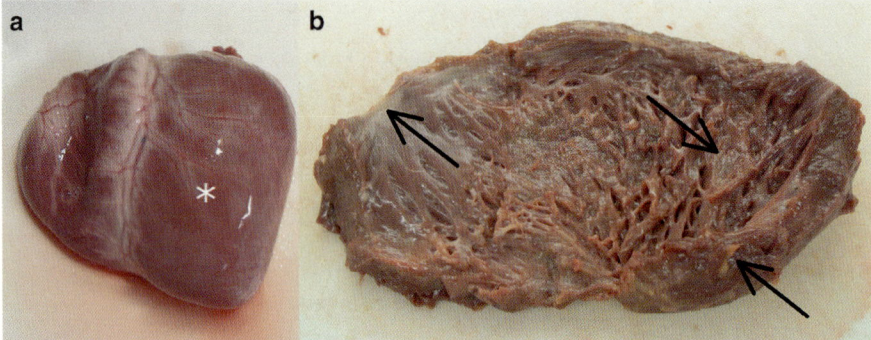

Fig. 1 A fresh harvested porcine heart before processing. Labeled with a (*) is the left ventricular myocardium used in the protocol (**a**). The fresh porcine heart is trimmed down such that only the myocardial tissue of the left ventricle remains. As indicated with *arrows*, the valves, superficial fat, fascia, chordae tendineae, and papillary muscles were all removed leaving behind the red myocardial tissue (**b**)

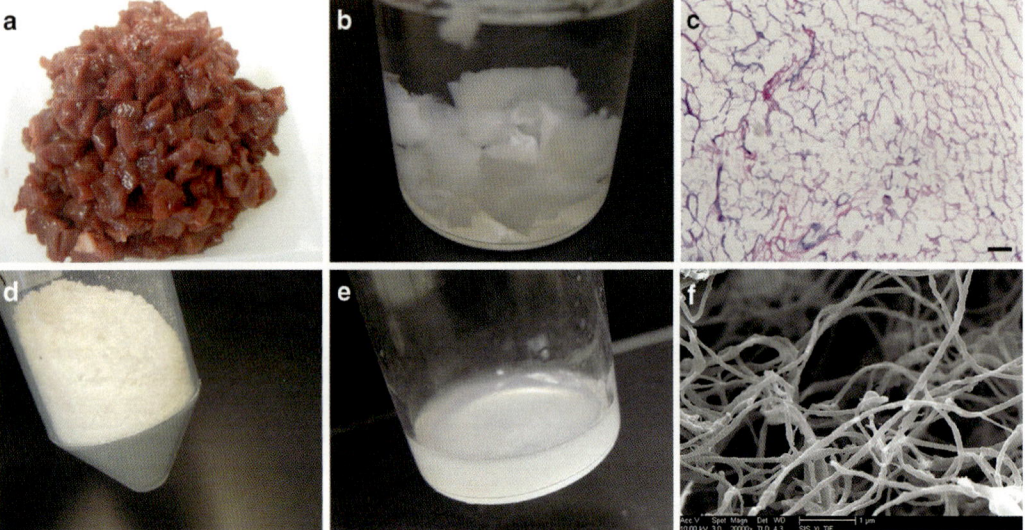

Fig. 2 Fresh myocardial tissue chopped into fine pieces (**a**). Fully decellularized, white myocardial tissue (**b**). An H&E-stained section of fully decellularized myocardial tissue showing no remaining nuclei (**c**). Lyophilized and milled porcine myocardial matrix (**d**). Once partially digested in acid by pepsin, the material is now in a liquid form (**e**). When brought to physiological conditions including salt, temperature, and pH the material forms a nanofibrous gel as viewed by scanning electron microscopy (SEM) (**f**). Modified and reprinted from [3] with permission from Elsevier

2. Remove any superficial fat, fascia, chordae tendineae, and papillary muscles just leaving behind only the red myocardial tissue (Fig. 1b).

3. Cut the remaining tissue into small regular cubed pieces about $2 \times 2 \times 2$ mm (*see* **Note 3**) (Fig. 2a).

4. Divide cut tissue into 1 L autoclaved beakers with 20–35 g of tissue per beaker (*see* **Note 4**).

5. Add 800 mL of water and a 3/8″ × 2 1/2″ stir bar to each beaker.

6. Stir beakers at 125 rpm for 30–45 min. Cover beaker with parafilm (*see* **Note 5**).

7. Strain tissue through the autoclaved fine mesh metal strainer and rinse with water.

8. Place tissue back into beaker with stir bar. Add 800 mL of decellularization solution and 4 mL of PS.

9. Stir beakers at 125 rpm for 2 h. Cover beaker with parafilm.

10. Repeat **steps 7** and **8**. Then stir beakers at 125 rpm for 24 h.

11. Repeat **step 10** two to four times (for a total of 3–5 days in the decellularization solution) until the tissue has become completely decellularized or turned fully white in color (*see* **Note 6**) (Fig. 2b).

12. Repeat **step 7**, and then place tissue back into beaker with stir bar and 800 mL of water. Stir at 125 rpm for 24 h.

13. Repeat **step 12**, but stir in water for only 30–45 min.

14. Repeat **step 7**, and then transfer decellularized tissue into 1 L autoclaved bottles with 800 mL of water. Vigorously shake for 1 min.

15. Repeat **step 14**, and then check for bubble formation after shaking. Lack of bubbles indicates removal of SDS solution. If bubbles remain then keep rinsing and shaking in the 1 L bottle with water to remove SDS from tissue.

16. Strain tissue with fine mesh metal strainer, transfer each into a 50 mL conical, and freeze at –80 °C (*see* **Note 7**).

3.2 Digestion and Injection Preparation

1. Using a lyophilizer, freeze dry the decellularized tissue.

2. Once fully dry, with a Wiley® Mini Mill process the material through a #40 mesh filter (pore size of 0.422 mm) into an autoclaved 20 mL vial (Fig. 2d). Before using the mill, make sure to fully clean all surfaces of the mill and mesh with 70 % ethanol to minimize contamination of the decellularized tissue.

3. In a sterile environment (such as a tissue culture hood), mass out 15–25 mg of milled ECM into an autoclaved 20 mL vial with a 5/16″ × 1/2″ stir bar.

4. Still in a sterile environment, add an appropriate amount of sterile-filtered digestion solution to the 20 mL vial until the ECM is at 10 mg/mL. Stir at 60–85 rpm for 48–56 h (*see* **Note 8**).

5. Once digested (Fig. 2e), the ECM in liquid form is brought to physiological conditions on ice with ice-chilled solutions. Add 1 M NaOH, and thoroughly mix until at pH 7.4 as confirmed with a pH strip.

6. Calculate the new total volume of liquid in the vial, and add 1/9 this volume of 10× PBS to bring the solutions up to 1× PBS salt concentration.

7. Dilute the solution to the desired ECM concentration of 6 mg/mL for the injection with 1× PBS.

8. The liquid form of the ECM can either be kept on ice for immediate injection preparation or frozen and re-lyophilized for long-term storage at −80 °C. If stored in a lyophilized state, the material can simply be resuspended with the appropriate amount of sterile water back to 6 mg/mL ECM.

9. Liquid form of the ECM is then loaded into a sterile syringe and kept on ice until injected into the myocardium. Upon injection into the animal the material will gel or self-assemble into a nanofibrous structure (Fig. 2f).

3.3 Cardiac Surgical Injection

All animal work should be done under an approved animal protocol as governed by your institution. For our studies, all experiments were performed in accordance with the guidelines established by the Committee on Animal Research at the University of California, San Diego, and the American Association for Accreditation of Laboratory Animal Care.

1. Anesthetize a female Sprague Dawley rat (225–250 g) for 3–5 min with 5 % isoflurane.

2. Intubate with a 14G×2″ IV catheter. Secure the rat in a supine position on the surgical table, and connect to the ventilator where anesthesia is maintained with 2.5 % isoflurane.

3. Apply ophthalmic ointment (artificial tears) to the eyes.

4. Insert the temperature probe coated in Surgilube.

5. Administer two 1.5 mL subcutaneous injections of lactated ringers through a 25G needle away from the incision region.

6. With the electric razor, shave and vacuum the abdomen and chest region free of hair. Disinfect the incision region with Betadine and wipe clean with 70 % isopropyl alcohol swabs.

7. Use a 25G needle to inject 3–4 beads of 1 % lidocaine subcutaneously along the length of the initial incision region.

8. Use the scalpel to make a 3–4 cm cutaneous incision beginning posterior to the xiphoid process and continuing lateral left approximately 1 cm caudal to the ribcage (Fig. 3).

9. Carefully cut through the muscle to expose the xiphoid process without damaging the liver. Once the xiphoid process is exposed, cut through the muscle along the length of the cutaneous incision.

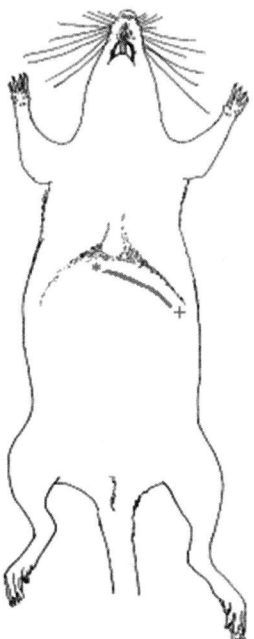

Fig. 3 The initial cutaneous incision beginning posterior to the xiphoid process (*) and continuing lateral left approximately 1 cm caudal to the ribcage (+). Modified and reprinted from [8] with permission from Elsevier

10. Expose the diaphragm by lifting and anchoring the xiphoid process. This can be accomplished by running half the length of an appropriate suture (4-0) through the xiphoid process and taping the free ends of the suture to a nearby high point (*see* **Note 9**).

11. Use fine forceps to hold the diaphragm dorsal to the heart apex, and slightly pull the diaphragm caudal in order to prevent damaging lung or heart tissue. Cut a small opening in the diaphragm at a point dorsal to the heart apex. From this opening make a ventral incision that extends 1–1.5 cm being careful not to cut the lungs (Fig. 4).

12. Tear the pericardium at the apex without damaging the lungs to expose the left ventricle free wall (*see* **Note 10**).

13. Hold the heart lightly enough to provide stability but do not interfere with its regular beating. Inject 75 μL of the biomaterial through a 27 or a 30G needle at a steady rate with the bevel or the bore hole oriented toward the left ventricle lumen. The needle tip should enter the left ventricle free wall almost parallel to the epicardial surface to ensure that the injection does not enter the lumen (*see* **Note 11**) (Fig. 5).

14. Close the diaphragm with a taper needle suture (5-0). Anchor the suture using a double knot at the dorsal edge of the incision, and close 3/4 of the incision with a running stitch (*see* **Note 12**).

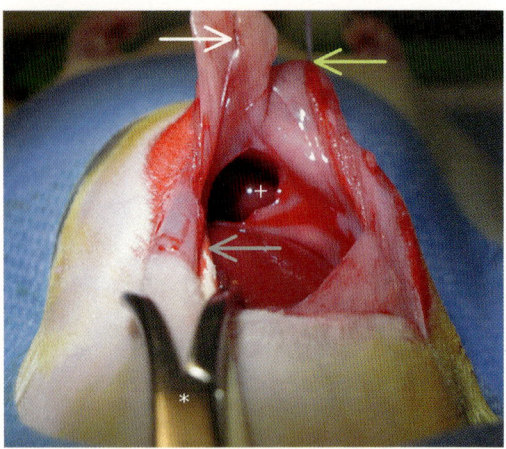

Fig. 4 The diaphragm is exposed by lifting xiphoid process by running half the length of an appropriate suture (Vicryl 4-0 FS-1) through the xiphoid process (*white arrow*) and taping the free ends of the suture to a nearby high point. Diaphragm visibility can be increased by repeating this process with another suture placed approximately 1 cm lateral left from the xiphoid process (*yellow arrow*). The heart apex is visible through the diaphragm (+). The diaphragm incision is made beginning dorsal to the heart apex and extends ventral 1–1.5 cm. A sterile cotton swab (*grey arrow*, cotton head of swab not visible) anchored with a towel clamp (*) is used to keep the right lung from obscuring the heart apex

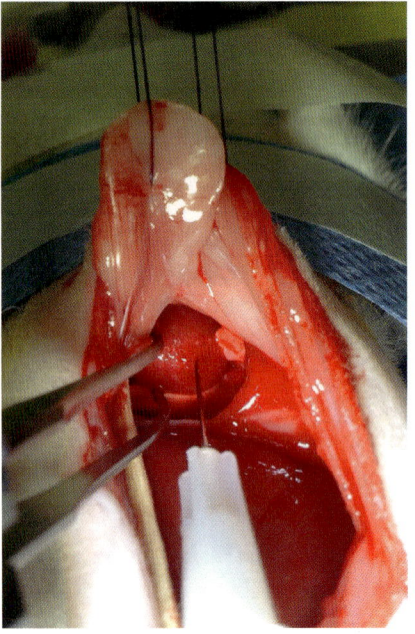

Fig. 5 The material is then injected into the left ventricle of the heart as shown. A 27G needle is used to deliver the material into the wall of the heart. If too much force is used the needle can fully puncture into the lumen of the ventricle, thus leading to a failed delivery

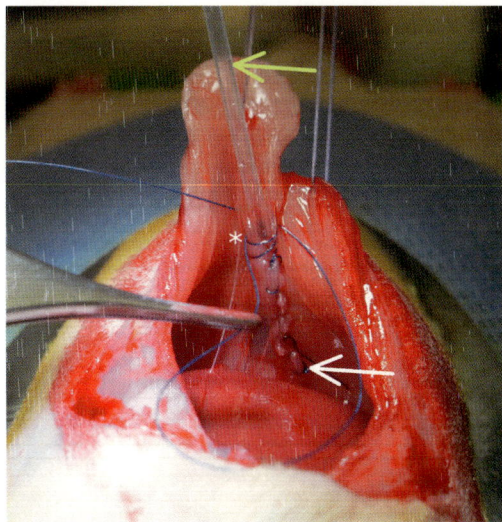

Fig. 6 The diaphragm is sutured closed beginning at the base of the incision using a running stitch (*white arrow*). The running stitch extends around the aspiration tubing (*yellow arrow*) and is held tight while leaving the last stitch loose (*) for the closing knot once the aspiration tube is removed

15. When 3/4 of the opening is closed, insert an aspiration tube into the cavity along the incision and continue the running stitch around the tube until the incision is completely closed (*see* **Note 13**).

16. Place one last stitch around the tube, but leave the loop large enough to use in a final double knot (Fig. 6).

17. Attach a 10 mL syringe to the aspiration tube.

18. With one hand, hold the running stitch tight while leaving the last stitch loose (Fig. 6). Use the other hand to suction the air from the thoracic cavity while simultaneously pulling the tubing free (*see* **Note 14**).

19. Continue to hold the suture, and watch the diaphragm once the aspiration tubing is free. If the diaphragm remains tight and concave the sutures can be closed with a double knot (*see* **Note 15**) (Fig. 7).

20. Reduce the isoflurane level to 1 % after closing the diaphragm.

21. Release and remove the anchored suture used to make the diaphragm visible.

22. Suture the muscle layer closed with an appropriate suture (4-0) using close-spaced intermittent stitches.

23. Close the skin with an appropriate suture (5-0) using intermittent stitches and seal with tissue adhesive.

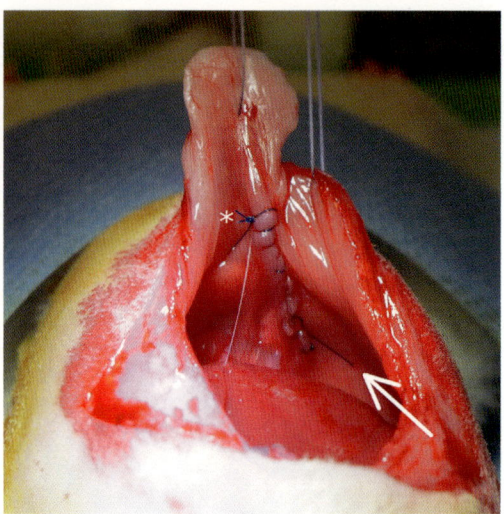

Fig. 7 The airtight running stitch closing the diaphragm (*) produces a concave diaphragm (*white arrow*) after aspirating air from the thoracic cavity

24. Apply triple-antibiotic ointment over the wound.

25. Reduce the isoflurane level to 0 %, and monitor breathing until the rat can breathe independent of the ventilator.

26. Administer an appropriate dose of postoperative analgesic such as Buprenex. Check with your animal care facility to determine which analgesic, administrative route, and dosage are recommended for your facility (*see* **Note 16**).

4   Notes

1. The quality of SDS powder can vary greatly between brands and can even have variation from the same brand. The SDS powder should be white and very fine. When dissolved, the SDS solution should turn completely clear. Our lab has previously received batches of SDS that had gathered into large aggregates due to exposure to water before storage and a batch that dissolved into a yellow solution. Double check the quality of SDS since this could have significant impact on the decellularization of the tissue. We have had consistent results with ordering Fisher Scientific, #S529–500.

2. After the heart is harvested it should be kept on ice to minimize degradation of the ECM, which is an issue due to the immediate release of matrix metalloproteinases (MMPs) from cell death.

3. The size of tissue pieces for this process has been optimized for the type of tissue being decellularized. If the pieces are too large, the core of the tissue will not decellularize properly and

can even begin to decompose. If the tissue pieces are too small, the tissue pieces will shred and fall apart in the decellularization solution. Denser tissues decellularize slower and require smaller pieces. Weaker tissues, or those with little ECM, can decellularize more quickly and should be cut into larger pieces.

4. It is helpful to remove a couple of pieces of fresh tissue to freeze for analysis against the decellularized tissue. Some of the fresh tissue should also be frozen in OCT compound in a cryomold for cryo-sectioning and histological staining.

5. The tissue is first spun in water to not only rinse off the blood, but it is also a hypotonic solution, which can actually begin the decellularization process by rupturing the exposed cells.

6. The length of time needed to decellularize the tissue varies and is dependent on the tissue type and size of the cut tissue pieces. Larger pieces take longer than smaller ones, and thus, after 24 h in the decellularization solutions the larger pieces can be cut in half if needed. While spinning in the decellularization solution the tissue will release the cellular content, as seen by a cloudy haze that accumulates over time in the beaker. Eventually the tissue will turn completely white, leaving behind only the ECM scaffold of the tissue.

7. Similar to the fresh tissue, some of the decellularized tissue should be kept for analysis and frozen in OCT compound in a cryomold for cryo-sectioning and histological staining to confirm complete decellularization of the tissue (Fig. 2c).

8. The digest should be checked regularly throughout 48–56 h. The ECM material can creep up the walls of the glass vial and out of the digestion solution due to the motion of the stir bar. If this happens, the material should be scraped back down into the solution using a small autoclaved metal spatula. Once digested, the material should have an increased viscosity and be homogeneous but it will not be transparent.

9. To increase visibility, a second suture can be used in the same manner but placed along the ribcage edge approximately 1 cm lateral left from the xiphoid process. With the diaphragm exposed, the location of the heart apex will be visible through the diaphragm between the lungs.

10. Once the heart apex is free from the pericardium, visibility can be aided by anchoring a sterile cotton swab with a towel clamp to keep the incision open and the right lung from blocking visibility.

11. Blanching of the left ventricular free wall during the injection is a visible indication of injection success but may not always be visible.

12. The stitch placement should be close set to make an airtight closure to allow the lungs to properly inflate. If air is present in

the thoracic cavity the animal will not be able to breathe. Mouth breathing or gasping post-surgery may be an indication of air in the thoracic cavity.

13. The aspiration tubing can be taped to a light in order to keep the aspiration tube positioned next to the sutures. The placement of the tube along the diaphragm closure is necessary to limit the chances of aspirating lung tissue or pericardium.

14. Between 1 and 3 mL of total volume may be removed from the cavity. If resistance is felt, the suction may be blocked by pericardium or lung tissue. If this occurs, stop suctioning and adjust the aspiration tubing. Keep the aspiration tubing next to the suture line to reduce the chances of aspirating any lung tissue.

15. If there is a leak in the sutures the diaphragm will not remain tight and will balloon. The sutures can be loosened to reinsert the aspiration tube and repeat the process.

16. Buprenorphine is a controlled substance commonly sold under the brand name Buprenex and may require special licensing to purchase. When a 0.05 mg/kg dose is subcutaneously administered using a 27G needle the animal will be mobile, sternal, and alert during its immediate recovery. Within an hour the animal's behavior and coat appearance should be normal. There should be no signs of distress such as hunched back, ruffled fur, or mouth breathing.

References

1. Christman KL, Fok HH, Sievers RE et al (2004) Fibrin glue alone and skeletal myoblasts in a fibrin scaffold preserve cardiac function after myocardial infarction. Tissue Eng 10(3–4):403–409

2. Singelyn JM, DeQuach JA, Seif-Naraghi SB et al (2009) Naturally derived myocardial matrix as an injectable scaffold for cardiac tissue engineering. Biomaterials 30(29):5409–5416

3. Singelyn JM, Sundaramurthy P, Johnson TD et al (2012) Catheter-deliverable hydrogel derived from decellularized ventricular extracellular matrix increases endogenous cardiomyocytes and preserves cardiac function post-myocardial infarction. J Am Coll Cardiol 59:751–763

4. Seif-Naraghi SB, Singelyn JM, Salvatore MA et al (2013) Safety and efficacy of an injectable extracellular matrix hydrogel for treating myocardial infarction. Sci Transl Med 5:173ra25

5. Christman K, Vardanian A, Fang Q et al (2004) Injectable fibrin scaffold improves cell transplant survival, reduces infarct expansion, and induces neovasculature formation in ischemic myocardium. J Am Coll Cardiol 44:654–660

6. Johnson TD, Christman KL (2012) Injectable hydrogel therapies and their delivery strategies for treating myocardial infarction. Expert Opin Drug Deliv 10(1):59–72

7. Rane AA, Christman KL (2011) Biomaterials for the treatment of myocardial infarction: a 5-year update. J Am Coll Cardiol 58:2615–2629

8. Yang CH, Lee BB, Jung HS et al (2002) Effect of electroacupuncture on response to immobilization stress. Pharmacol Biochem Behav 72(4): 847–855

Chapter 11

Generation of Strip-Format Fibrin-Based Engineered Heart Tissue (EHT)

Sebastian Schaaf, Alexandra Eder, Ingra Vollert, Andrea Stöhr, Arne Hansen, and Thomas Eschenhagen

Abstract

This protocol describes a method for casting fibrin-based engineered heart tissue (EHT) in standard 24-well culture dishes. In principle, a hydrogel tissue engineering method requires cardiomyocytes, a liquid matrix that forms a gel, a casting mold, and a device that keeps the developing tissue in place. This protocol refers to neonatal rat heart cells as the cell source; the matrix of choice is fibrin, and the tissues are generated in rectangular agarose-casting molds (12 × 3 × 3 mm) prepared in standard 24-well cell culture dishes, in which a pair of flexible silicone posts is suspended from above. A master mix of freshly isolated cells, medium, fibrinogen, and thrombin is pipetted into the casting mold and, over a period of 2 h, polymerizes and forms a fibrin cell block around two silicone posts. Silicone racks holding four pairs of silicone posts each are used to transfer the fresh fibrin cell blocks into new 24-well dishes with culture medium. Without further handling, the cells start to remodel the fibrin gel, form contacts with each other, elongate, and condense the gel to approximately ¼ of the initial volume. Spontaneous and rhythmic contractions start after 1 week. EHTs are viable and relatively stable for several weeks in this format and can be subjected to repeated measurements of contractile function and final morphological and molecular analyses.

Key words Engineered heart tissue, Cardiomyocytes, Neonatal rat heart cells, Fibrin, Tissue engineering

1 Introduction

When working with cardiomyocytes in cell culture, the sensitivity of these cells is an issue. Quick degeneration and cessation of contractility limit the time window for experiments, e.g., to a few hours to 1–2 days in adult cardiomyocytes from mice or rat. Furthermore, contractile function is strongly affected by loading conditions. Freely floating cells experience essentially no load and contract isotonically; cells attached to plastic surfaces face a very peculiar situation in that the cell basis is essentially fixed and cannot

Milica Radisic and Lauren D. Black III (eds.), *Cardiac Tissue Engineering: Methods and Protocols*, Methods in Molecular Biology, vol. 1181, DOI 10.1007/978-1-4939-1047-2_11, © Springer Science+Business Media New York 2014

deform the substrate (isometric), whereas the rest of the cell can freely contract (isotonic). Both conditions are highly unphysiological and limit the validity of these force measurements.

Some of these limitations can be overcome by working with 3D engineered heart tissue (EHT) instead of single cells [1–3]. Advantages include the following: (a) the tissue structure as such; that is, cells are not isolated but embedded in a 3D tissue context similar to the native heart; (b) stability over weeks; that is, EHTs can be used for experiments repetitively (e.g., measurement of reversible effects of substances); (c) defined longitudinal orientation allowing directed force measurement; and (d) higher degree of cardiac myocyte maturation than newborn myocytes in 2D culture [4]. EHTs also possess promising features useful in applications of regenerative medicine [5].

Compared to earlier models developed by our group such as the lattice [2] or the ring model [1] the new 24-well fibrin EHT method [6] offers a number of advantages. (a) A key feature of this protocol is the flexible silicone posts. They are aligned in parallel at the beginning of EHT development when the fibrin gel is formed, and they slightly bent over time when cells remodel the matrix, which leads to shortening of the muscle strip. This configuration of bent flexible silicone posts exposes the growing tissue to a continuous strain (diastolic or resting tension). With each contraction twitch, the EHT has to overcome this resting elastic force, then moves the posts towards each other, and finally relaxes. This creates an auxotonic force cycle, similar to the native heart. Static stretch (as in the lattice or the ring EHT), in contrast, is not physiological and, as shown recently, promotes a pathological form of cardiac hypertrophy [7]. (b) The diastolic tension and the continuous contractile work against the elastic resistance of the posts induce a high degree of organization and maturation of the cells [6, 8], which resembles a heart more than a cell culture in a dish does. (c) The 24-well format allows medium-throughput studies without manual handling steps except the initial transfer from the casting molds. (d) It is easily accessible to video-optical recordings of contractile function [6]. Knowledge of the elastic properties of the silicone and the length and diameter of the posts allows simple transformation of µm post-deflection to force in mN.

As a source for heart cells, this protocol refers to cells of neonatal rat origin. However, it can be easily adapted to cells of other origin. For example, it is possible to generate EHTs from neonatal mouse heart cells. This is attractive for the vast number of genetically modified mouse models [9]. A disadvantage of mice is the small size of the heart, requiring quite a number of pups for one cell preparation. In addition, rat-based EHTs have proven themselves to be more robust. Even more interesting is the option to create EHTs from cardiomyocytes derived from pluripotent mouse or human cells. With an adequate protocol for the differentiation

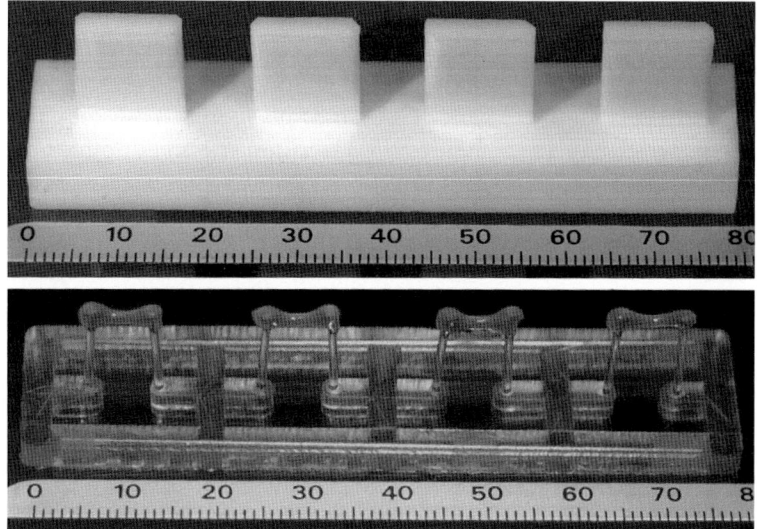

Fig. 1 *Top*: Teflon spacer for the generation of four casting molds. *Bottom*: Silicone rack with four EHTs suspended between two posts each

of these cells (minimum 30–40 % cardiomyocytes), it is possible to generate human EHTs using essentially the identical EHT protocol [8]. In combination with the induced pluripotent stem cell technology this opens exciting possibilities for modeling human cardiac diseases in vitro.

The method described in this protocol starts with freshly isolated cells obtained from neonatal rat hearts, meaning a mixture of cardiac cells is used and not pure cardiomyocytes. A protocol for the preparation of these cells needs to be obtained elsewhere. In our experience all methods that give rise to good-quality neonatal rat (or murine or human) cardiomyocytes (including commercially available such as the Worthington or Miltenyi kits) are feasible for generating EHTs according to the protocol described herein. Moreover, anybody with a basic training in cell culture can easily learn this method in a few days. It does not require special equipment with the exception of the teflon spacer and silicone racks (Fig. 1). The latter have been manufactured at large numbers and can be obtained from the authors.

The next step of the protocol is the preparation of the casting molds. Liquid agarose solution is poured into the wells of a 24-well cell culture plate, and a custom-made teflon spacer is then placed in the agarose, creating casting molds as the agarose solidifies. The spacers are removed, and the custom-made silicone racks are placed on the 24-well plate, with two silicone posts being lowered into each mold. The heart cells are then mixed with fibrinogen and diluted to the desired cell number of 4.1×10^6 per ml, i.e., 0.4×10^6 cells per EHT. Just before transfer into the casting mold, each 100 µl pipet load is mixed with an aliquot of thrombin, thus starting the polymerization process. After 90 min, a fibrin gel has formed, with heart cells evenly dispersed in it and suspended between the silicone posts. The racks can now be transferred to a new cell culture plate with fresh medium. In order to replace the medium for feeding, the silicone racks with the EHTs are placed into a new cell culture plate. After roughly 2 weeks, the cardiomyocytes have formed a network throughout the construct, beating synchronously and generating force. Each contraction visibly bends the silicone posts towards each other.

2 Materials

All reagents and tools needed for this protocol are standard cell culture equipment. Exceptions are the spacers needed to produce the agarose-casting molds and the silicone racks. Also, an advanced setup is needed to measure the contractility of EHTs. Contact the authors for further information.

1. Aprotinin: Dissolve 1 g aprotinin in 30.3 ml aqua ad injectabilia (33 mg/ml), aliquot 250 μl, and store at –20 °C.

2. Fibrinogen: Pre-warm 0.9 % NaCl solution to 37 °C, dissolve fibrinogen (bovine) in appropriate volume to a final concentration of 200 mg/ml, add 72.1 μl of aprotinin (33 mg/ml stock) per 25 ml of fibrinogen solution, aliquot 200 μl, and store at –20 °C for short term and –80 °C for long term (more than 4 weeks; *see* **Note 4**).

3. Thrombin: Dissolve 250 U of thrombin (bovine) in 1.5 ml of phosphate buffer saline (PBS) and 1.0 ml aqua ad injectabilia (100 U/ml), mix thoroughly, make aliquots of 450 μl for storage, prepare 3 μl aliquots for EHT generation in small reaction tubes (e.g., 0.2 ml PCR tubes), and store at –20 °C.

4. Agarose 2 %: Add agarose to the appropriate volume of PBS to a final concentration of 2 %, and autoclave. Agarose solution need to be stored at 60 °C.

5. 10× Dulbecco's modified Eagle medium (DMEM): Dissolve 670 mg 10× DMEM powder in 5 ml aqua ad injectabilia, filter through a sterile filter with 0.22 μm pores, and store at 4 °C.

6. 2× DMEM: Mix 10× DMEM solution (2 ml), inactivated horse serum (2 ml), chick embryo extract (0.4 ml, centrifuge at $650 \times g$ for 10 min before use), penicillin/streptomycin (0.2 ml, Gibco), and aqua ad injectabilia ad 10 ml (=5.4 ml), and store at 4 °C.

7. EHT medium: Add penicillin/streptomycin (1 %) to DMEM, heat-inactivated horse serum (10 %, always thaw fresh from –20 °C), chick embryo extract (2 %, centrifuge at $650 \times g$ for 10 min before use), insulin (10 μg/ml, Sigma-Aldrich), and

aprotinin (0.1 %, equals 33 μg/ml), and filter through a sterile 0.22 μm filter; *see* **Note 1** for further details.

8. Non-cardiomyocyte (NCM) medium: Add penicillin/streptomycin (1 %) to DMEM, heat-inactivated fetal calf serum (10 %, Gibco), and l-Glutamine (1 %, Gibco).

9. Cell culture microplates, 24-well (Nunc/Thermo Scientific).

10. Ca^{2+}- and bicarbonate-free Hank's buffer (CBFHH): NaCl 136.8 mM, KCl 5.4 mM, $MgSO_4$ 0.81 mM, KH_2PO_4 0.44 mM, Na_2HPO_4 0.35 mM, glucose 5.6 mM, 4-(2-hydroxyethyl)-1-piperazineethanesulfonic acid (HEPES) 20 mM, pH 7.4.

11. Chick embryo extract: Use fertilized chicken eggs at days 7–9, open eggs, transfer embryos into a glass bottle filled with CBFHH (150 ml for 60 eggs), homogenize, fill volume to 300 ml with CBFHH, centrifuge ($60 \times g$, 15 min), collect supernatant, homogenize pellet again, centrifuge pellet ($60 \times g$, 15 min), pool supernatants, aliquot, and store at –20 °C. Everything is done under sterile conditions and on ice.

12. Spacer: The spacers are used in combination with the agarose in order to create the casting molds (Fig. 1). They were cut from a block of teflon using a milling cutter and fit into 4 wells of a 24-well plate; that is, 6 spacers are needed for one complete plate. The dimensions of the casting molds are $12 \times 3 \times 4$ mm. For further information contact the authors.

13. Silicone racks: Silicone racks are custom-made (Siltec GmbH & Co, KG) pieces that consist of a reinforced backbone and four pairs of silicone posts (Fig. 1). EHTs are suspended between these posts, meaning that four EHTs are attached to one rack. For a 24-well cell culture plate, six racks are needed. For further information contact the authors.

3 Methods

3.1 Preparations

1. Prepare all solutions described in Subheading 2.

2. Prepare enough 3 μl aliquots of thrombin; one for each EHT is needed.

3. Make sure that teflon spacers and silicone racks (Fig. 1) are sterile and ready to use. Boil them twice in deionized water and autoclave afterwards. Spacers and silicone racks are supposed to be reused. Do not use soap/detergents for cleaning.

4. Prepare neonatal rat heart cells, count them, and place them on ice. Instructions for this step are not provided in this protocol. *See* **Note 2** for further details.

5. Good timing is essential, as the heart cells should not be kept on ice for extended periods of time.

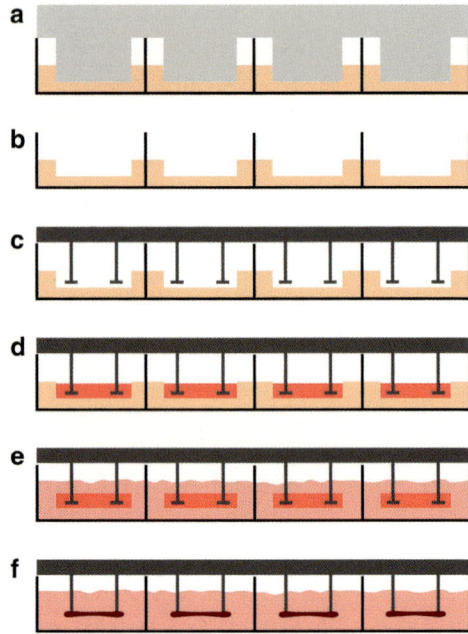

Fig. 2 (**a**) Teflon spacer in liquid agarose, (**b**) casting molds after solidification, (**c**) silicone racks with two posts in each casting mold, (**d**) master mix of cells, fibrinogen, and thrombin inside casting molds, (**e**) cells in a block of fibrin transferred to fresh medium after polymerization, (**f**) EHTs after 2 weeks of development

3.2 Step 1: Agarose-Casting Molds and Preparation of the Master Mix

1. In a 24-well cell culture plate, pipet 1.6 ml of agarose solution in each well.

2. Place a teflon spacer in each column of wells, i.e., 6 spacers in total per 24-well plate (Fig. 2a).

3. Allow 15 min for solidification (*see* **Note 3**).

4. During solidification prepare the master mix (*see* **Note 4**). Calculate the desired amount of cells (0.4×10^6 cells per EHT), and prepare the master mix (on ice) as follows (*see* **Note 5**):
   Cell suspension (89 µl per EHT), 2× DMEM (5.5 µl), and fibrinogen (2.5 µl).

5. Remove the spacers from the plate by moving them strictly in an upward direction (Fig. 2b). Do not move the spacers back and forth or tilt them, as this might damage the casting molds.

6. Place the silicone racks on the plate, one rack for each column of wells (Fig. 2c). Make sure that the posts are correctly placed in the casting molds (two in each, check for bent posts).

3.3 Step 2: Generation of Engineered Heart Tissue

1. Make sure that there are enough pipet tips and thrombin aliquots; one of each is needed per EHT.

2. Pipet 97 µl of master mix into a tube with a 3 µl thrombin aliquot, take it up again, and immediately pipet the mixture

into an agarose-casting mold (Fig. 2d). Use a new tip for each EHT, and resuspend the master mix after four to eight EHTs (*see* **Note 6**).

3. Place the freshly cast EHTs in a cell culture incubator (37 °C, CO_2 7 %, O_2 40 %) for 80 min.

4. Pipet a small amount of NCM medium (approx. 0.3 ml) in each well and incubate for another 10 min (*see* **Note 7**).

5. Prepare a fresh 24-well plate with 1.5 ml of EHT medium per well.

6. Transfer the silicone racks with the EHTs from the casting plate to the plate with fresh medium (Fig. 2e).

3.4 Maintenance

1. Place EHTs in a cell culture incubator at 37 °C, 7 % CO_2, and 40 % O_2.

2. Feed every other day (e.g., Monday, Wednesday, Friday) with freshly prepared EHT medium (*see* **Notes 1** and **8**).

3. Check the development of EHT regularly. Contractions of single cells should be visible from day 1 or 2 on. Larger areas should show contractions after 2–5 days, and after 5–10 days synchronized contraction and bending of the post should be clearly visible (*see* **Note 9** and Fig. 2f).

4 Notes

1. EHT medium: Always use fresh EHT medium for feeding. Especially the horse serum should be initially aliquoted, frozen, and then thawed for every preparation of EHT medium. Do not use horse serum which has been stored at 4 °C.

2. Preparation of cells: After the preparation of cells, centrifuge and resuspend. A cell number of 0.4×10^6 per 89 µl of medium (equals 4.1×10^6/ml) is ideal. It is very important to only use media containing heat-inactivated serum for resuspension. Otherwise the polymerization of fibrinogen into fibrin might start already in the master mix. Keep the cells on ice until they are needed.

3. Preparation of casting molds: Do not prepare more than 8 or 12 wells at a time; otherwise the agarose might start to solidify before the spacers are in place. Solidification times should be at least 15 min; otherwise pieces of agarose might stick to the spacers. The casting molds should not be prepared too much in advance, as too long periods might lead to small cracks on the surface due to drying, making the agarose leaky and unsuitable for EHT generation.

4. Fibrinogen/handling the master mix: Due to its viscosity, fibrinogen can be tricky to handle. At the concentrations described, it needs to be at least at room temperature, better

at 37 °C, to be pipettable. This is especially important when preparing the master mix for the generation of EHTs, since the fibrinogen solution will clot the pipet tip immediately if it gets cold. So do not place the tip of the pipet into the master mix, but instead hold it roughly 0.5 cm above the liquid and let the warm fibrinogen drop into the mix. Then mix without disturbing the cardiomyocytes too much, so use either a serological pipet for trituration or a 1 ml Eppendorf-style pipet. Note that it might take a few minutes until the fibrinogen is completely dissolved. The master mix should be kept on ice at all times.

As an alternative, mix fibrinogen, NCM, and 2× DMEM without the cells. By doing so, extended trituration can be used to dissolve the fibrinogen without any risk of harming the cells. With this pre-master mix, take up the pelleted cells after centrifugation (usually the last step of the preparation).

Problems: Clotting of the master mix usually has two possible reasons. First, serum has been used that has not been heat inactivated. This might start the polymerization process. If the problem remains, the cells might have been treated with active serum (e.g., for stopping of the dissociation process). Try washing the cells with medium containing heat-inactivated serum as the last step of the preparation. Second, contamination can lead to clotting. Make sure to use a new pipet tip/new serological pipet for every step.

5. Calculating the master mix: As for every EHT a new pipet tip is used, there is a rather large discrepancy between calculated volume and actually needed volume. Calculate at least 10–20 % more master mix than theoretically needed.

6. Casting EHTs: When pipetting the master mix (with thrombin) into the casting molds, do not empty the pipet tip completely, but leave 1–2 μl. By doing so, extensive amounts of air bubbles in the freshly cast EHT can be avoided. There still might be some air bubbles, but they usually disappear after a few days and are not harmful to the EHT. Use a new pipet tip for each EHT. Even small remnants of thrombin will lead to polymerization of the master mix.

7. Transfer of EHTs: Adding medium to the freshly cast EHTs helps to separate EHTs from the agarose-casting mold and increases the number of successfully transferred EHTs. Do this 80 min after casting and 10 min before transfer.

8. Maintenance: For feeding, prepare a second 24-well plate for each plate with EHT. On feeding days, remove the medium from the second plate and replace with fresh medium. Then transfer EHTs from the first plate into the second plate. Keep the first plate, and use it for feeding next time.

9. Mature EHTs: EHTs from neonatal rat cardiomyocytes usually have a stable beating pattern from day 14 to day 28 and should be used for experiments within this time frame. Contractions appear either in isolated, single beats or in bursts, i.e., 10–30-s periods of rapid contractions followed by pauses of 20–40 s.

References

1. Zimmermann W-H, Schneiderbanger K, Schubert P, Didié M, Münzel F, Heubach JF, Kostin S, Neuhuber WL, Eschenhagen T (2002) Tissue engineering of a differentiated cardiac muscle construct. Circ Res 90(2):223–30

2. Eschenhagen T, Fink C, Remmers U, Scholz H, Wattchow J, Weil J, Zimmermann W-H, Dohmen HH, Schäfer H, Bishopric N, Wakatsuki T, Elson EL (1997) Three-dimensional reconstitution of embryonic cardiomyocytes in a collagen matrix: a new heart muscle model system. FASEB J 11(8):683–94

3. Eschenhagen T, Eder A, Vollert I, Hansen A (2012) Physiological aspects of cardiac tissue engineering. Am J Physiol 303(2):H133–43

4. Tiburcy M, Didié M, Boy O, Christalla P, Döker S, Naito H, Karikkineth BC, El-Armouche A, Grimm M, Nose M, Eschenhagen T, Zieseniss A, Katschinksi DM, Hamdani N, Linke WA, Yin X, Mayr M, Zimmermann W-H (2011) Terminal differentiation, advanced organotypic maturation, and modeling of hypertrophic growth in engineered heart tissue. Circ Res 109(10):1105–14

5. Zimmermann W-H, Melnychenko I, Wasmeier G, Didié M, Naito H, Nixdorff U, Hess A, Budinsky L, Brune K, Michaelis B, Dhein S, Schwoerer A, Ehmke H, Eschenhagen T (2006) Engineered heart tissue grafts improve systolic and diastolic function in infarcted rat hearts. Nat Med 12(4):452–8

6. Hansen A, Eder A, Bönstrup M, Flato M, Mewe M, Schaaf S, Aksehirlioglu B, Schwoerer AP, Schwörer A, Uebeler J, Eschenhagen T (2010) Development of a drug screening platform based on engineered heart tissue. Circ Res 107(1):35–44

7. Hirt MN, Sörensen NA, Bartholdt LM, Boeddinghaus J, Schaaf S, Eder A, Vollert I, Stöhr A, Schulze T, Witten A, Stoll M, Hansen A, Eschenhagen T (2012) Increased afterload induces pathological cardiac hypertrophy: a new in vitro model. Basic Res Cardiol 107(6):307

8. Schaaf S, Shibamiya A, Mewe M, Eder A, Stöhr A, Hirt MN, Rau T, Zimmermann W-H, Conradi L, Eschenhagen T, Hansen A (2011) Human engineered heart tissue as a versatile tool in basic research and preclinical toxicology. PloS One 6(10):e26397

9. Stöhr A, Friedrich F, Flenner F, Geertz B, Eder A, Schaaf S, Hirt MN, Uebeler J, Schlossarek S, Carrier L, Hansen A, Eschenhagen T (2013) Contractile abnormalities and altered drug response in engineered heart tissue from / Mybpc3/-targeted knock-in mice. J Mol Cell Cardiol 63:189–98

Chapter 12

Cell Tri-Culture for Cardiac Vascularization

Ayelet Lesman, Lior Gepstein, and Shulamit Levenberg

Abstract

Poor graft survival is a critical obstacle toward production of clinically relevant engineered tissues. Here we utilize a multicellular culturing approach for induction of vascular networks embedded within cardiac tissue constructs. The construct is composed of human cardiomyocytes, endothelial cells (ECs), and embryonic fibroblast cells co-seeded onto highly porous three-dimensional (3D) scaffolds. The resulting vascularized cardiac constructs showed microstructural details characteristic of cardiomyocytes and nascent vessels and exhibited synchronous beating activity in vitro. Upon implantation, stable grafts were formed presenting intense vascularization, with evidence of anastomosis between the pre-formed endothelial capillaries and host neovessels.

Key words Tissue engineering, Vascularization, Cardiomyocytes, Endothelial cells, Scaffold, Co-culture

1 Introduction

Bioengineering strategies utilizing functional cells seeded onto biomaterials to generate cardiac tissue constructs represent a promising strategy for regenerating infarcted hearts. Our group has focused on generating vascularized constructs in order to enhance the survival of the graft following implantation, a critical obstacle in the design of clinically relevant engineered tissues.

The native myocardium is a highly metabolic tissue. Cardiomyocytes are concertedly aligned and in close proximity to a rich vasculature indispensable for conforming to the high metabolic demands of this essential tissue. It has been reported that a capillary resides next to almost every cardiomyocyte, and ECs outnumber cardiomyocytes by 3:1 [1]. Beyond the essential role of endothelial capillaries in oxygen and nutrient delivery, cross-talk signaling between ECs and muscle cells has been shown to act as an important regulator of early myocardium development and growth of cardiomyocytes [2]. Incorporation of functional blood vessel networks into engineered tissues can potentially improve the

Milica Radisic and Lauren D. Black III (eds.), *Cardiac Tissue Engineering: Methods and Protocols*, Methods in Molecular Biology, vol. 1181, DOI 10.1007/978-1-4939-1047-2_12, © Springer Science+Business Media New York 2014

maturation of the tissue in vitro, its blood supply, and integration following implantation.

For induction of vascularized constructs, we have utilized an in vitro multicellular prevascularization strategy. The multicellular approach entails co-seeding of ECs and vascular mural cells together with tissue-specific cells within 3D porous scaffolds. The cells are then allowed to self-assemble to form vascular networks embedded within the engineered tissue of choice. The vascular mural cells are perivascular cells which provide physical support to ECs and release angiogenic growth factors which induce vascularization. Embryonic fibroblasts and mesenchymal precursor cells are extensively used in tissue prevascularization techniques due to their ability to differentiate into mural cells [3].

We have shown that the multicellular approach is successful in vascularization of a variety of engineered tissues including the heart [4], skeletal muscle [5], and pancreas [6]. Importantly, we found that extended in vitro incubation allows for the formation of a more structured vascular network, which further promotes anastomosis and vascular integration in vivo [7]. We also dynamically traced the formation of the networks in vitro and found that it followed a series of defined steps. Initially the ECs were sparse and isolated; then they gradually developed into group of cells which united and organized into an array of interconnected 3D vascular bed [8].

Other promising procedures to induce vascularization of engineered constructs are under extensive investigation. For example, neovascularization originating from the host animal can be used for providing the adequate vascularization of the construct. The engineered tissue is implanted onto a rich vascular area in the host animal, such as the omentum [9]. After allowing sufficient time for penetration of blood vessels into the graft, the construct is detached and implanted in the desired organ. Such technique has been shown to be successful in improving cardiac function after myocardial infraction [9]. Also, casting of patterned well-organized vascular networks has recently become possible by printing 3D filament networks of carbohydrate glass and using them as a template to generate tube networks that could be lined with ECs and perfused with blood under pulsatile flow [10].

Here, we describe the formation of synchronously contracting engineered cardiac tissue containing endothelial vessel networks [4]. The 3D muscle tissue consists of cardiomyocytes, ECs, and embryonic fibroblasts seeded into poly-L-lactic acid/polylactic-glycolic acid (PLLA/PLGA) porous scaffolds (Fig. 1). We have shown that the engineered human vascularized cardiac muscle is functional in vitro and can be transplanted to rat hearts to form stable grafts. The presence of vessel networks increased biograft vascularization following implantation, and the preexisting human vessels became functional and contribute to tissue perfusion [11].

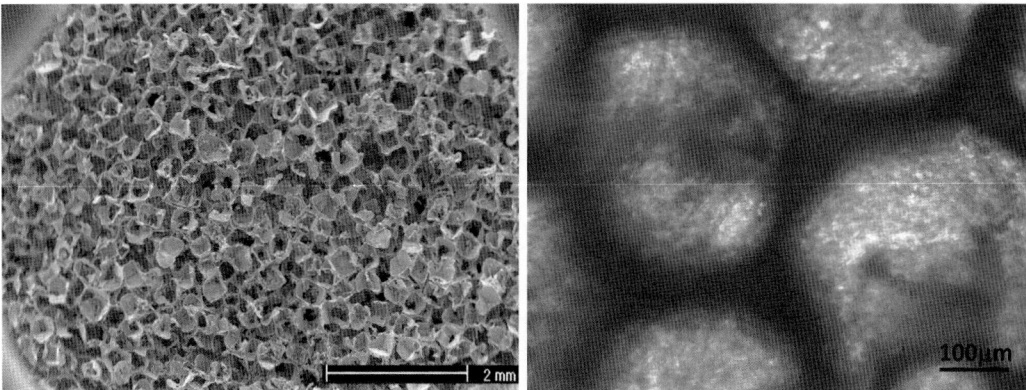

Fig. 1 The PLLA/PLGA porous scaffolds. *Left*: Low-zoom image showing the porous structure of the scaffolds. *Right*: Higher zoom image of the scaffold (*block areas*) seeded with the tri-culture cells

2 Materials

2.1 Cell Culture Components

1. Dulbecco's modified Eagle medium (DMEM).
2. Knockout DMEM.
3. Fetal bovine serum.
4. EGM-2 medium and bullet kit.
5. Nonessential amino acids (NEAA).
6. Mercaptoethanol collagenase B.
7. Collagenase IV.
8. Growth factor-reduced Matrigel.

2.2 Cell Types and Culture Conditions

1. Human umbilical vein ECs (HUVECs):
   HUVECs (passages 3–7) are grown on tissue culture plates in EGM-2 medium supplemented with 2 % FBS and EGM-2 bullet kit. HUVEC-GFP and HUVEC-RFP (passages 4–6), grown as above, are also used in some experiments to monitor the dynamic development of vascular networks in vitro and in vivo.

2. Mouse embryonic fibroblast (MEF) cells:
   MEF cells are derived from mouse embryos at day 13 and cultured in DMEM with 10 % FBS (used until passage 3).

3. Human foreskin fibroblast (HFF) cells:
   Primary culture of HFF cells were prepared in our laboratory from newborn's foreskin and used until passage 20. Protocol by Diecke et al. in this book can also be used for HFF isolation. HFF cells were cultured in medium consisting of DMEM supplemented with 10 % FBS, 1 % NEAA, and 0.2 % mercaptoethanol.

4. Human embryonic-derived cardiomyocytes (hES-CMs): Pluripotent hESC of the H9.2 clone [12] (passages 30–60) are grown in the undifferentiated state on top of MEF feeder layer as previously described [13]. The culture medium consists of 20 % FBS, 79 % knockout DMEM supplemented with 1 mM L-glutamine, 0.1 mM mercaptoethanol, and 1 % NEAA. To induce differentiation, the hESC are dispersed to small clumps using 1 mg/mL collagenase IV. Then, they are transferred to plastic petri dishes and cultured in suspension for 7–10 days where they form embryoid bodies (EBs). Hereupon, the EBs are plated on gelatin-coated culture dishes. Spontaneously beating areas can be noted in some of the EBs after 5–20 days of plating. The contracting areas within the EBs can be isolated with a curved 23G needle after 25–30 days of in vitro differentiation. The contracting areas are then dissociated into small cell clusters (20–100 cells) by incubation with 1 mg/ml of collagenase B for 45 min.

2.3 PLLA/PLGA Scaffold Components

1. PLLA.
2. PLGA.
3. Teflon cylinders (diameter = 21.5 mm, height = 25 mm).
4. Sodium chloride.
5. Sieve with mesh size of 212–600 μm.
6. Chloroform.

3 Methods

3.1 Creating PLLA/PLGA Scaffolds for Cell Seeding

1. Weight 0.5 g PLGA, and dissolve it in 10 ml chloroform in a small glass tube to yield a solution of 5 % polymer (w/v).

2. Weight 0.5 g of PLLA, and dissolve it in 10 ml chloroform in a small glass tube to yield a solution of 5 % polymer (w/v). Place it on a heated magnetic stirring plate and stir until the chloroform begins to bubble. Then, the heating should be stopped and repeated until the polymer is completely dissolved.

3. Create a new stock solution by mixing 50 % PLGA solution (**step 1**) and 50 % PLLA solution (**step 2**).

4. Load 0.24 ml solution of the 50/50 PLLA/PLGA stock solution (**step 3**) into a Teflon cylinder packed with 0.4 g sodium chloride particles sieved to size between 212 and 600 μm.

5. Cover the container with a lid, and let it sit for 1 h at room temperature so that the chloroform will fill the entire space between the salt particles. Then, take off the lid in the chemical hood, and let it sit air exposed overnight to allow the chloroform to evaporate.

6. After overnight incubation, remove the scaffold from its mold by tapping the containers upside down.

7. Place each scaffold in a histology cassette, and put it in a 3–4 L beaker filled with distilled water to leach out the salt.

8. Change the water each hour for 6–8 h.

9. Remove the scaffolds from the histology cassettes, dry them on a kimwipe, and freeze at –80 °C for at least 12 h.

10. Place the scaffolds in the lyophilizer (freeze-dryer) overnight and keep in a vacuum container until use (*see* **Note 1**). The appearance and structure of the scaffolds can be seen in Fig. 1.

3.2 Generating 3D Vascularized Cardiac Constructs

To create the 3D constructs, we seed into the PLLA/PLGA sponges the following cell types:

1. hES-CMs: Their differentiation and purification methods are detailed elsewhere [4, 12, 13].

2. HUVECs.

3. Mouse embryonic or human foreskin fibroblast cells.

Over the course of 3–5 days in culture, the vascular cells self-assemble into vessel-like networks that are arranged next to groups of contracting cardiomyocytes. Occasionally, a uniform beating of the whole construct appears after 3 days of culture [4].

3.2.1 Early Preparations Before Cell Seeding

1. Slice the PLLA/PLGA 50/50 sponges into squares using scissors (to dimensions of about 5 mm × 5 mm × 1 mm, volume of ~25 mm^3). Place the desired number of square sponges in 70 % ethanol and incubate overnight for sterilization. Wash them three times with phosphate-buffered saline before use, each time for 15 min.

2. Place a –20 °C frozen aliquot of Matrigel in an ice basket, and put it in the 4 °C freezer so that it will thaw slowly overnight.

3. Prepare media for the multicellular constructs: Combine 50 % volume HUVEC media with 50 % volume hES-CM media.

3.2.2 Seeding Cells in PLLA/PLGA Sponges to Generate Vascularized Cardiac Constructs

1. Prepare cells: Trypsinize and count to achieve 4×10^5 hESC-CMs, 4×10^5 HUVECs, and 2×10^5 fibroblasts, and place them together inside a 1.7 ml centrifuge tube (*see* **Notes 2** and **3**).

2. Centrifuge the cell mixture: 1,000 rpm for 4–5 min.

3. While centrifuging, place each sterilized sponge on the bottom of a 6-well plate. Cover each scaffold with a drop of media (~50 μl). After 5–10 min, aspirate out the media, keeping the scaffold as dry as possible. This step helps the sponges to absorb more easily the cell suspension in **step 5**.

4. Take out the supernatant (with a pipet tip), and suspend the cell pellet in 8 μl of a 1:1 mixture of culture medium and Matrigel.

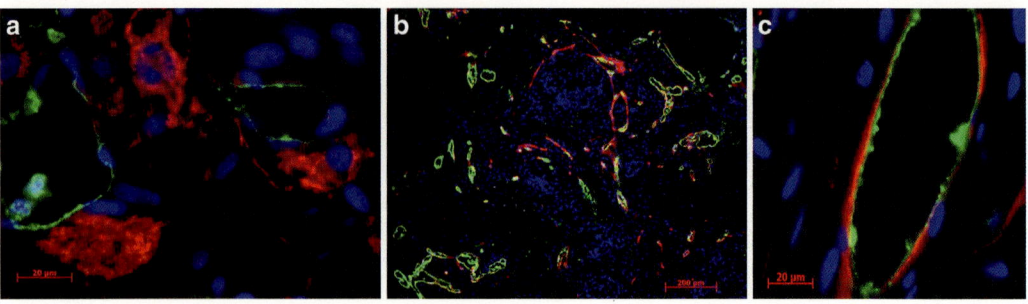

Fig. 2 Immunofluorescent staining of the vascularized cardiac constructs (paraffin-embedded slices). (**a**) hES-CMs labeled with anti-troponin I (*red*) and ECs labeled with anti-vWF (*green*). (**b**) ECs labeled with both anti-vWF (*green*) and anti-CD31 (*red*). (**c**) High-zoom image describing a vessel with an EC layer (anti-vWF, *green*) surrounded by smooth muscle cell layer (anti-αSMA, *red*). Nuclei are stained with DAPI

5. Transfer the cell suspension (total volume of ~10 μl, use a pipette) gently into the sterilized sponge sitting on the bottom of a 6-well plate (*see* **Note 1**). Pipet up and down a few times to disperse cells equally throughout the volume of the sponge. Try to avoid air bubbles.

6. Incubate at 37 °C for 20–30 min to allow solidification of the Matrigel and attachment of the cells to the scaffold.

7. Fill (slowly) the well with 4 ml of cell media (HUVEC/hES-CMs 1:1 ratio).

8. Detach the sponges from the bottom of the plate by gently tapping it from the side with a sterilized pipette tip.

9. Incubate at 37 °C on an XYZ shaker (to achieve better media penetration into the scaffolds). Every other day the medium should be changed. Culture the constructs for about 2 weeks in vitro.

10. Following 2 weeks of culture, the constructs can proceed to in vitro assessment studies [4] or for in vivo implantation in rodent hearts [11]. For in vitro assessment, constructs can be fixed in 10 % formalin and processed by either whole-mount or paraffin sectioning and then proceed for immunofluorescent staining (Fig. 2). Alternatively, constructs can follow DNA extraction for evaluation of gene expression using RT-PCR [8].

4   Notes

1. Although in this protocol we have focused on using sponge scaffolds made of PLLA/PLGA, we note that other types of biomaterials can be used in a similar way to support the tri-culture cell system as we showed previously in the case of tri-culture of myoblasts, ECS, and fibroblast seeded onto fibrin gels [8].

2. In some of the experiments, in order to track the hESC-CMs following in vivo transplantation, we pre-labeled them with the lipophilic fluorescent cell tracker CM-DiO (2.5 µg/ml for 45 min, Molecular Probes) just before use in the cell seeding procedure.

3. HUVEC-RFP or HUVEC-GFP cells were also used frequently to allow their trace in vitro or following implantation in vivo [8].

Acknowledgement

This research was supported by the European Union 7th Framework Programme as part of the project NanoCARD, grant agreement no. 229294.

References

1. Brutsaert DL (2003) Cardiac endothelial-myocardial signaling: its role in cardiac growth, contractile performance, and rhythmicity. Physiol Rev 83:59–115. doi:10.1152/physrev.00017.2002

2. Hsieh PC, Davis ME, Lisowski LK, Lee RT (2006) Endothelial-cardiomyocyte interactions in cardiac development and repair. Annu Rev Physiol 68:51–66. doi:10.1146/annurev.physiol.68.040104.124629

3. Ding R, Darland DC, Parmacek MS, D'Amore PA (2004) Endothelial-mesenchymal interactions in vitro reveal molecular mechanisms of smooth muscle/pericyte differentiation. Stem Cells Dev 13:509–520. doi:10.1089/1547328042417336

4. Caspi O et al (2007) Tissue engineering of vascularized cardiac muscle from human embryonic stem cells. Circ Res 100:263–272

5. Levenberg S et al (2005) Engineering vascularized skeletal muscle tissue. Nat Biotechnol 23:879–884

6. Kaufman-Francis K, Koffler J, Weinberg N, Dor Y, Levenberg S (2012) Engineered vascular beds provide key signals to pancreatic hormone-producing cells. PloS One 7:e40741. doi:10.1371/journal.pone.0040741

7. Koffler J et al (2011) Improved vascular organization enhances functional integration of engineered skeletal muscle grafts. Proc Natl Acad Sci U S A 108:14789–14794

8. Lesman A, Koffler J, Atlas R, Blinder YJ, Kam Z, Levenberg S (2011) Engineering vessel-like networks within multicellular fibrin-based constructs. Biomaterials 32:7856–7869

9. Dvir T et al (2009) Prevascularization of cardiac patch on the omentum improves its therapeutic outcome. Proc Natl Acad Sci U S A 106:14990–14995

10. Miller JS et al (2012) Rapid casting of patterned vascular networks for perfusable engineered three-dimensional tissues. Nat Mat 11:768–774. doi:10.1038/nmat3357

11. Lesman A et al (2010) Transplantation of a tissue-engineered human vascularized cardiac muscle. Tissue Eng Part A 16:115–125

12. Amit M et al (2000) Clonally derived human embryonic stem cell lines maintain pluripotency and proliferative potential for prolonged periods of culture. Dev Biol 227:271–278. doi:10.1006/dbio.2000.9912

13. Kehat I, Kenyagin-Karsenti D, Snir M, Segev H, Amit M, Gepstein A, Livne E, Binah O, Itskovitz-Eldor J, Gepstein L (2001) Human embryonic stem cells can differentiate into myocytes with structural and functional properties of cardiomyocytes. J Clin Invest 108:407–414

Chapter 13

Cell Sheet Technology for Cardiac Tissue Engineering

Yuji Haraguchi, Tatsuya Shimizu, Katsuhisa Matsuura,
Hidekazu Sekine, Nobuyuki Tanaka, Kenjiro Tadakuma,
Masayuki Yamato, Makoto Kaneko, and Teruo Okano

Abstract

In this chapter, we describe the methods for the fabrication and transfer/transplantation of 3D tissues by using cell sheet technology for cardiac tissue regeneration. A temperature-responsive culture surface can be fabricated by grafting a temperature-responsive polymer, poly(N-isopropylacrylamide), onto a polystyrene cell culture surface. Cells cultured confluently on such a culture surface can be recovered as an intact cell sheet, and functional three-dimensional (3D) tissues can then be easily fabricated by layering the recovered cell sheets without any scaffolds or complicated manipulation. Cardiac cell sheets, myoblast sheets, mesenchymal stem cell sheets, cardiac progenitor cell sheets, etc., which are prepared from temperature-responsive culture surfaces, can be easily transplanted onto heart tissues of animal models, and those cell sheet constructs enhance the cell transplant efficiency, resulting in the induction of effective therapy.

Key words Cardiomyocytes, Cell sheet engineering, Extracellular matrix, Temperature-responsive culture surface, Three-dimensional tissue, Tissue engineering, Transplantation

1 Introduction

Recently, cell-based regenerative therapy has been utilized to develop some of the most hopeful treatment strategies for curing cardiovascular disease. To date most of these have involved the direct injection of cells, and autologous myoblasts, bone marrow cells, and cardiac progenitor cells for regenerating damaged heart tissues have already been used clinically [1–5]. However, many injected cells die and vanish after transplantation, and few cells remain in the infarcted heart tissues, while a large percentage of the cells are found in the liver and spleen [6–8]. Recently, methodologies

Electronic supplementary material The online version of this article (doi:10.1007/978-1-4939-1047-2_13) contains supplementary material. This video is also available to watch on http://www.springerimages.com/videos/978-1-4939-1046-5. Please search for the video by the article title.

Milica Radisic and Lauren D. Black III (eds.), *Cardiac Tissue Engineering: Methods and Protocols*, Methods in Molecular Biology, vol. 1181, DOI 10.1007/978-1-4939-1047-2_13, © Springer Science+Business Media New York 2014

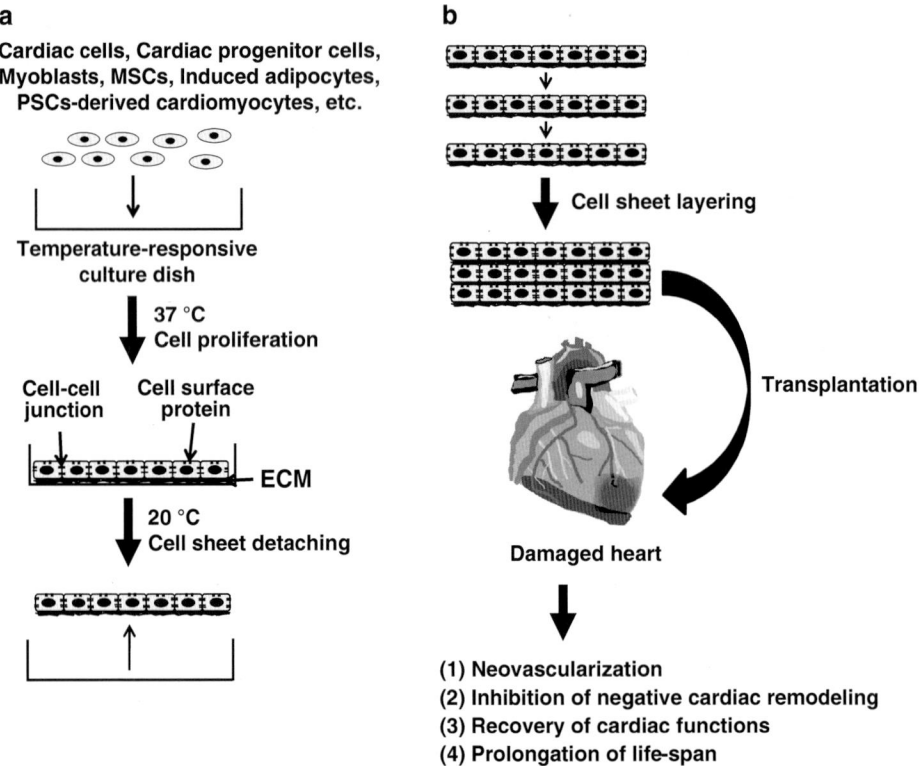

Fig. 1 Schematic illustrations of the fabrication of a cell sheet and a cell-dense 3D tissue. By lowering the temperature, confluently cultured cells on a temperature-responsive culture surface are harvested as a contiguous cell sheet; preserving cell–cell junctions, cell surface proteins, and the extracellular matrix (ECM) (**a**). Cell-dense 3D tissue can be fabricated by layering single-cell sheets, and the transplantation of these fabricated cell sheets contributes to significant therapeutic effects (**b**). MSCs: mesenchymal stem cells; PSCs: pluripotent stem cells

for creating engineered tissue have been developed and are expected to be the basis of a second generation of cell-based regenerative therapy for cardiovascular disease. Using this new methodology, synchronously beating three-dimensional (3D) cardiac tissues have already been successfully fabricated by seeding cardiac cells into 3D scaffolds, hydrogels, or decellularized native tissues [9–12]. Bioengineered tissues fabricated by using biodegradable 3D collagen type I matrixes and autologous bone marrow cells for regenerating damaged heart tissues have also been used clinically [13, 14].

Our laboratory developed a scaffold-free tissue-engineered methodology, "cell sheet engineering," using a temperature-responsive culture surface, which can be fabricated by covalent immobilization of a temperature-responsive polymer, poly(N-isopropylacrylamide) (PIPAAm) onto a conventional cell culture surface [15, 16]. The culture surface alternates between hydrophobicity and hydrophilicity simply by changing the temperature from the PIPAAm's lower critical solution temperature (LCST) of 32 °C. Cells cultured confluently on the surface can then be

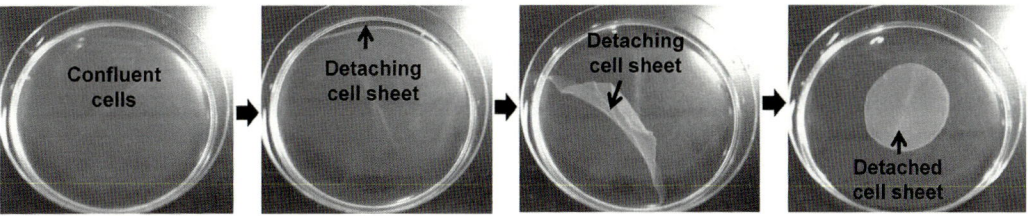

Fig. 2 A cell sheet detaching spontaneously from a temperature-responsive culture surface. Confluent C2C12 mouse myoblasts on a 100-mm temperature-responsive culture surface are spontaneously detached as a cell sheet by decreasing culture temperature to 20 °C

detached spontaneously as an intact cell sheet from the surface by decreasing the temperature from 37 to 20 °C (Figs. 1a and 2, 17–20). Because cell sheets can be recovered from culture surfaces without using any protease treatments, the cell sheets retain their own unique deposited extracellular matrix (ECM) on the surfaces of the sheets which were produced during cultivation [21–23]. The presence of ECM allows cell sheets easily to adhere to each other without any additional scaffolds, and the resulting cell sheet constructs can be transplanted directly onto host tissues without suturing or cell loss. Layered cardiac cell sheets will couple electrically and rapidly via gap junction formation and show spontaneous, synchronous, and macroscale beating [24–27]. Several kinds of cell sheets have been fabricated including (1) cardiac cell sheets, (2) myoblast sheets, (3) adipose tissue-derived mesenchymal stem cell (MSC) sheets, (4) endometrial gland-derived MSC sheets, (5) cardiac progenitor cell sheets, and (6) induced adipocyte cell sheets using cell sheet technology; and their transplantation showed (1) the induction of neovascularization, (2) the inhibition of negative cardiac remodeling, (3) the recovery of cardiac function, and (4) the prolongation of life-span in various heart-damaged animal models (Fig. 1b, 28–40). It has been shown that the transplantation of cell sheets yields significantly greater cell survival and induces better therapeutic effects than the injection of dissociated cells [28, 37]. Clinical trials using autologous myoblast sheets are also now under way [41]. More recently, the fabrication of cardiac cell sheets and engineered cardiac tissues using mouse as well as human pluripotent stem cell (PSC)-derived cardiomyocytes has been carried out [42–46]. The transplantation of PSC-derived cardiac cell sheets into a rat myocardial infarction model and a porcine ischemic cardiomyopathy model induced significant therapeutic effects, including (1) high capillary density in the infarction area, (2) an improvement in cardiac function, and (3) the inhibition of negative cardiac remodeling [44, 45].

In this chapter, we describe the methods to fabricate functional 3D tissues by layering cell sheets and the transfer/transplantation of these cell sheets onto animal tissues for use in cardiac tissue engineering and regenerative medicine.

2   Materials

2.1   Temperature-Responsive Culture Surface

Temperature-responsive culture surfaces are available in several sizes [Upcell (6-, 12-, 24-, 48-, and 96-well plates, and 35-, 60-, and 100-mm dishes), CellSeed, Tokyo, Japan] that can be selected based on intended use. For example, in mouse, rat, and hamster damaged heart models, small cell sheets prepared from 12-well plates or 35-mm dishes are used [30, 32, 34, 35, 37, 44]. On the other hand, in canine and porcine models, and clinical trials, larger cell sheets prepared onto 60- or 100-mm culture dishes are used [29, 36, 40, 45]. Many cells can be effectively transplanted by using large cell sheets prepared on a 100-mm dish (Fig. 3a). Figure 3b shows double-layered PSC-derived cardiac cell sheets recovered from 24-well temperature-responsive culture surfaces on a MED probe (AlphaMED Scientific, Osaka, Japan) containing multiple microelectrodes, which detected the field potentials of the cardiac cell sheets, and the action potentials of each cardiac cell sheet showed that the layered cell sheets also coupled electrically (Fig. 3c, 42). For (1) in vitro drug screenings, (2) toxicologic tests, (3) molecular biological analyses, and (4) electrophysiological analyses using cardiac tissues, small 3D cardiac tissue fabricated by using 96-, 48-, or 24-well plates may be advantageous because of the reduced number of cardiac cells required in the fabrication of small-size cell sheets and the assay system using small cell sheets enables random screening.

2.2   Cells and Culture Media

See Table 1. All cells are cultured at 37 °C in a humidified atmosphere with 5 % CO_2.

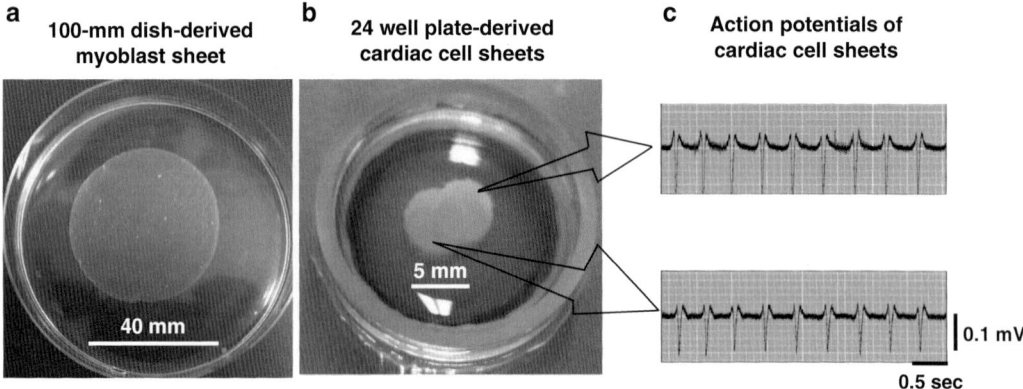

a 100-mm dish-derived myoblast sheet

b 24 well plate-derived cardiac cell sheets

c Action potentials of cardiac cell sheets

40 mm

5 mm

0.1 mV

0.5 sec

Fig. 3 A C2C12 mouse myoblast sheet on a 100-mm temperature-responsive culture surface (**a**) and double-layered cardiac cell sheets including mouse embryonic stem (ES) cell-derived cardiomyocytes and mouse cardiac fibroblasts, which detached from a 24-well temperature-responsive culture surface, using a multiple-electrode probe (**b**). The electrograms show the spontaneous action potentials of the double-layered cardiac cell sheets that are coupled electrically (**c**)

Table 1
Cell sources and medium for generating cell sheets

Cell type	Media	Cell sources
Rat neonatal cardiac cells	6 % Fetal bovine serum (FBS, Japan Bio Serum, Nagoya, Japan), containing 40 % Medium 199 (Invitrogen, Carlsbad, CA, USA), 0.2 % penicillin–streptomycin (Sigma-Aldrich, St. Louis, MO, USA), 2.7 mM glucose, and 54 % balanced salt solution containing (in mM) 116 NaCl, 1.0 NaH$_2$PO$_4$, 0.8 MgSO$_4$, 1.18 KCl, 0.87 CaCl$_2$, and 26.2 NaHCO$_3$ (*see* **Note 1**) [24, 26, 27, 32, 37, 47]	Cardiac cells are isolated from Wistar or Sprague–Dawley neonatal rats (Nisseizai, Tokyo, Japan). Recently, a living-cell isolation device was developed, which can semiautomatically isolate beating cardiomyocytes from neonatal rat heart tissues [48]
Mouse ES cell-derived cardiomyocytes and neonatal mouse cardiac fibroblasts	Dulbecco's modified Eagle medium (DMEM, Sigma-Aldrich) supplemented with 15 % FBS [34]	Mouse embryonic stem (ES) cells differentiate efficiently into cardiomyocytes [40]. The existence of fibroblasts expressing ECM seems to be important for fabricating cardiac cell sheets. Mouse cardiac fibroblasts are isolated from neonatal mouse heart tissues as described in previous reports [42, 49] (*see* **Note 1**)
Human iPS cell-derived cardiomyocytes	Differentiation is carried out via activin A, bone morphogenetic protein 4, and fibroblast growth factor-2 treatment under serum-free cultivation [43, 46]. Differentiated cells are cultured in DMEM, supplemented with 10 % FBS and 1 % penicillin–streptomycin	Human iPS cells are provided by the RIKEN BRC through the National Bio-Resources Project of the MEXT, Japan. The cells differentiate efficiently into beating cardiomyocytes after treatment with three cytokines (Activin A, bone morphogenetic protein 4, and fibroblast growth factor-2) under serum-free cultivation [43, 46]
Mouse cardiac progenitor cells	Cells are cultured on 1 % gelatin-coated dishes with Iscove's Modified Dulbecco's Medium (IMDM, Sigma-Aldrich) supplemented with 10 % FBS and 1 % penicillin–streptomycin [34] (*see* **Note 1**)	Mouse cardiac progenitor cells (Sca1-positive cells) are isolated from the heart tissues of C57BL/6 mice
Mouse C2C12 myoblasts	DMEM supplemented with 10 % FBS and 1 % penicillin–streptomycin [50]	Commercially available from American Type Culture Collection (ATCC) or Dainippon Sumitomo Pharma. Cells need to be kept below 70 % confluence to prevent myotube formation

(continued)

Table 1
(continued)

Cell type	Media	Cell sources
Human myoblasts, bone marrow-derived MSCs, or adipose tissue-derived MSCs	Cultured in myoblast medium, hBMSC medium, or hAMSC medium, respectively, from Lonza	All cells are commercially available from Lonza
Mouse ASMCs	Cultured on 1 % gelatin-coated dishes with IMDM supplemented with 10 % FBS and 1 % penicillin–streptomycin–amphotericin (see **Note 1**) [34]	Isolated from the interscapular adipose tissues of C57BL/6 mice
Human endometrial gland-derived MSCs	Cultured in DMEM supplemented with 10 % FBS and 1 % penicillin–streptomycin [50, 51]	Isolated from human endometrial glands [33]

2.3 Materials for Cell Sheet Transfer

1. Transfer paper (CellShifter™) [52]: Using a transfer paper called CellShifter™ (CellSeed) allows us to adhere/desorb materials, including cells from itself depending on the moisture content. A cell sheet detached from a temperature-responsive culture surface can be adhered to the support paper, and then the adhered cell sheets can be released from the paper and transferred onto tissues under optimal wet conditions.

2. PET membrane: Because polyethylene terephthalate (PET) membrane is nonadhesive to cells, floating cell sheets on culture dishes can be scooped and transferred onto target tissues [32].

3. Cell sheet scooping/transfer device: A cell sheet scooping/ transfer device was recently developed, which allows us to transfer and transplant cell sheets easily and rapidly, similar to micropipetting manipulation [53]. The device is constructed of several parts, (1) a scooping part, which further comprises (2) an inner plate made of polypropylene sheet and (3) an outer cross, made of polytetrafluoroethylene glass (AS ONE, Osaka, Japan), which covers the inner plate, and (4) a handle (Fig. 4). The inner plate is connected to (5) a pushing rod in the handle, thus allowing the inner plate and the outer cross to be extended by pushing the rod in the direction of the tip of the device when the pushing rod is pushed by hand (Fig. 4). The device is available from Furukawakikou Co. Ltd. (Niigata, Japan).

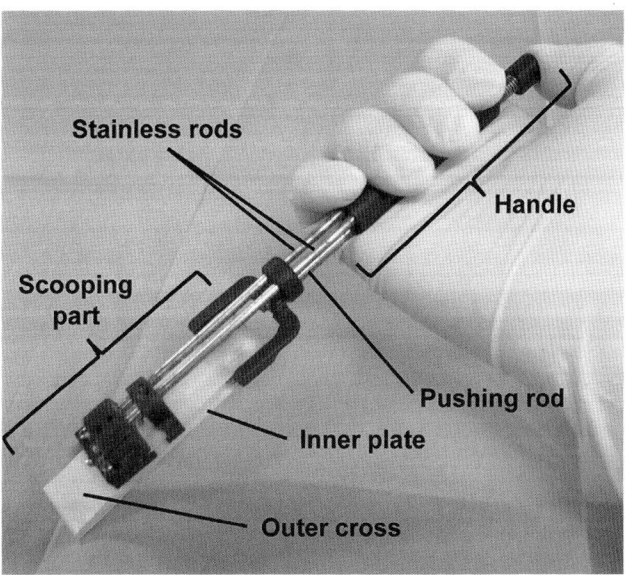

Fig. 4 Cell sheet scooping/transfer device. This photograph shows the device, which has several parts: scooping part, handle, inner plate, outer cross, pushing rod, and stainless rods. The photograph shows the device scooping a red ink-stained paper

3   Methods

3.1   Preparation of Cell Sheets

1. Seed cells onto a temperature-responsive culture surface at an optimal cell concentration, and culture the cells at 37 °C in a CO_2 incubator for several days to reach confluence (*see* **Note 2**).

2. After obtaining confluence, transfer the culture surface into a separate CO_2 incubator set at 20 °C for recovering the cell sheet. As shown in Fig. 2, confluent cells on the culture surface spontaneously detach themselves as an intact cell sheet by simply reducing culture temperature (*see* **Note 3**).

3.2   Layering Cell Sheets for Fabricating 3D Tissue

Three-dimensional tissue can be fabricated by layering detached cell sheets onto a culture dish.

1. After a cell sheet is detached, aspirate the cell sheet with culture medium into the tip of a pipette, and transfer the cell sheet onto another culture dish.

2. After the cell sheet is transferred, smooth the cell sheet by spreading it out using the following three manipulations: rotation of the dish (*see* **Note 4**), slow aspiration, and/or gentle dropping of culture medium (*see* **Note 5**).

3. After removing the remaining culture medium, transfer the dish into a CO_2 incubator, and incubate at 37 °C for 30–60 min to promote adhesion between the cell sheet and the culture surface (*see* **Notes 6** and **7**).

4. After incubation layer the cell sheets by transferring and then spreading the second cell sheet over the first cell sheet by the same manipulation techniques described in **step 2**.

5. After removing the remaining culture medium, incubate the dish in a CO_2 incubator at 37 °C for 30–60 min to allow the two cell sheets to adhere.

6. For reconstructing thicker tissue, repeat **steps 4–5**. Detached cell sheets can be easily layered by using a pipette as described above, resulting in the reconstruction of a cell-dense 3D tissue (Fig. 5).

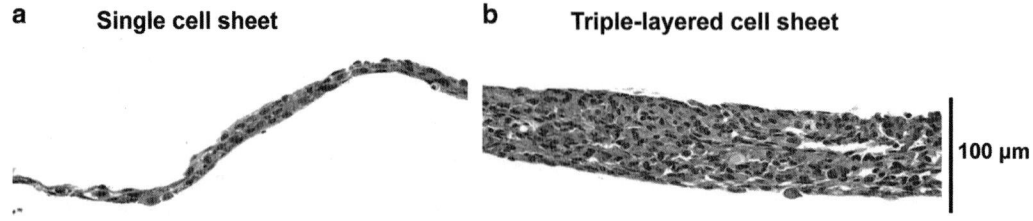

a   Single cell sheet   **b   Triple-layered cell sheet**

100 μm

Fig. 5 Cross-sectional observations of a single C2C12 cell sheet (**a**) and of a triple-layer cell sheet (**b**) treated with hematoxylin–eosin staining

3.3 In Vitro Transfer and In Vivo Transplantation of Cell Sheets

A recovered cell sheet and layered cell sheets can be easily transferred/transplanted onto target surfaces/tissues using several methods [22–61], including a nonadhesive poly(vinylidene difluoride) support membrane (Immobilon-P, DURAPORE; Merck Millipore, Billerica, MA, USA), a wet sheet of woven polyglycolic acid (PGA) (Neoveil, Gunze, Tokyo, Japan), and a plunger-like device containing hydrogel. The detailed manipulation methods for using the PGA sheet and plunger-like device were explained in previous papers [47, 57]. In this chapter, three simple methods using (1) transfer paper (CellShifter™), (2) a PET membrane, and (3) a cell sheet scooping/transfer device are explained.

Transfer of a cell sheet from a culture surface into another surface using CellShifter™

1. Detach a cell sheet or a layered cell sheet prepared on a temperature-responsive culture surface by reducing the culture temperature (*see* **Note 8**).

2. Spread the cell sheet or the layered cell sheet by the same manipulation techniques described in Subheading 3.2, and remove medium.

3. Pick up and immerse CellShifter™ into the medium using forceps, and then remove the extra medium (*see* **Note 9**).

4. Put CellShifter™ onto the cell sheet, and then pick up the paper with the cell sheet attached (*see* **Note 10**).

5. Put the paper with the cell sheet onto a human gloved hand.

6. Pick up the paper, and only the cell sheet will remain on the surface (Fig. 6a) (*see* **Note 11**).

7. Cell sheets can be easily transplanted onto target tissues in rats by a similar method as shown in Fig. 6b and also other tissues in porcine models (*see* **Notes 1** and **12**) [52]. Importantly, after the transplantation, the cell sheet graft will adhere to the host tissues without sutures.

*Transplantation of cell sheets onto a rat heart tissue using a PET membrane (see **Note 1**)*

1. Anesthetize F344 athymic rats with inhaled isoflurane (up to 3.5 %).

2. Sterilize a PET membrane with rubbing alcohol.

3. Expose the heart tissue via a left thoracotomy.

4. Detach a cell sheet or a layered cell sheet prepared on a temperature-responsive culture surface by reducing the culture temperature (*see* **Note 8**).

5. Lift up a floating cell sheet or a layered cell sheet with a nonadhesive PET membrane and transplant over the anterior wall of the heart by sliding it from the film (Fig. 7). Cell sheets can even be easily transplanted onto beating heart tissue without any suturing or cell loss.

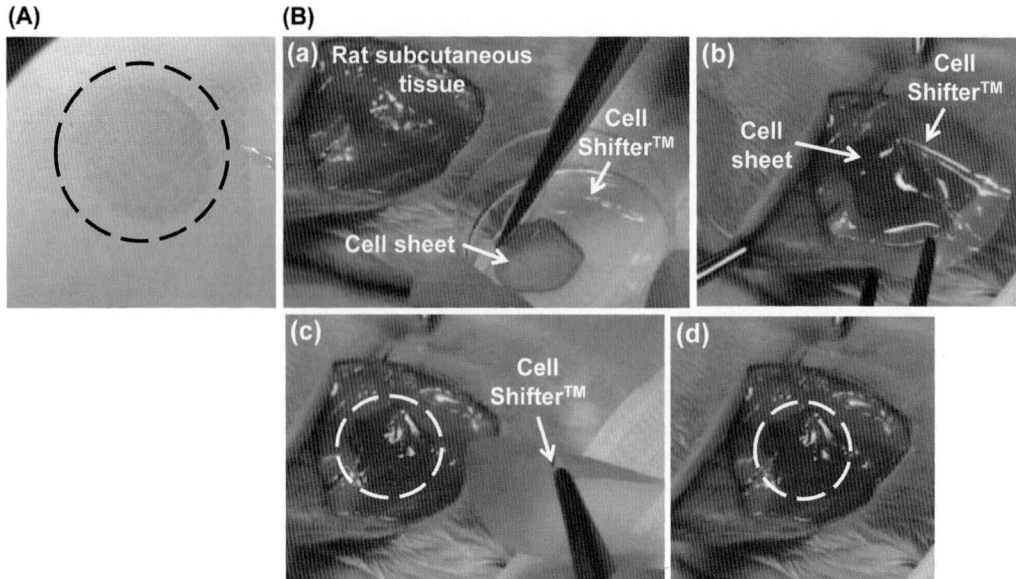

Fig. 6 Transfer/transplantation of a cell sheet using transfer paper, CellShifter™. (**a**) A dashed circle shows the transferred C2C12 cell sheet onto a human gloved hand. (**b**) Transplantation of a C2C12 cell sheet stained with neutral red onto the subcutaneous tissue of a rat. A cell sheet can be picked up (a) and adhered (b) to the subcutaneous tissue of a rat. After the removal of CellShifter™ (c), only the cell sheet will be transplanted onto the tissue (d). White dashed circles show the transplanted cell sheets

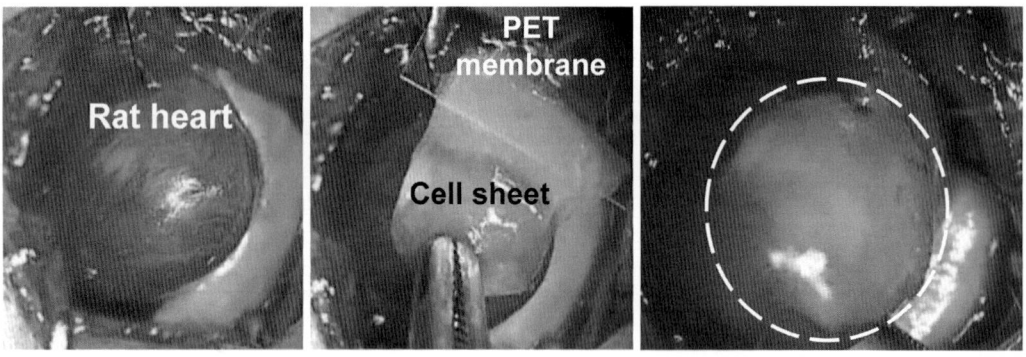

Fig. 7 Transplantation of a triple-layered rat cardiac cell sheet onto rat heart tissue using a PET membrane. A triple-layered rat cardiac cell sheet was lifted onto a polyethylene terephthalate (PET) membrane and transplanted onto rat heart tissue by sliding it from the PET membrane using forceps. The *dashed circle* in the right photograph shows a transplanted layered cell sheet

Transfer of a cell sheet from a culture dish onto another surface using a scooping/transfer device

1. Sterilize the cell sheet scooping/transfer device with rubbing alcohol, an ethylene oxide gas (EOG) sterilizer, or an autoclave instrument.

2. Prepare a cell sheet or a layered cell sheet on a surface.

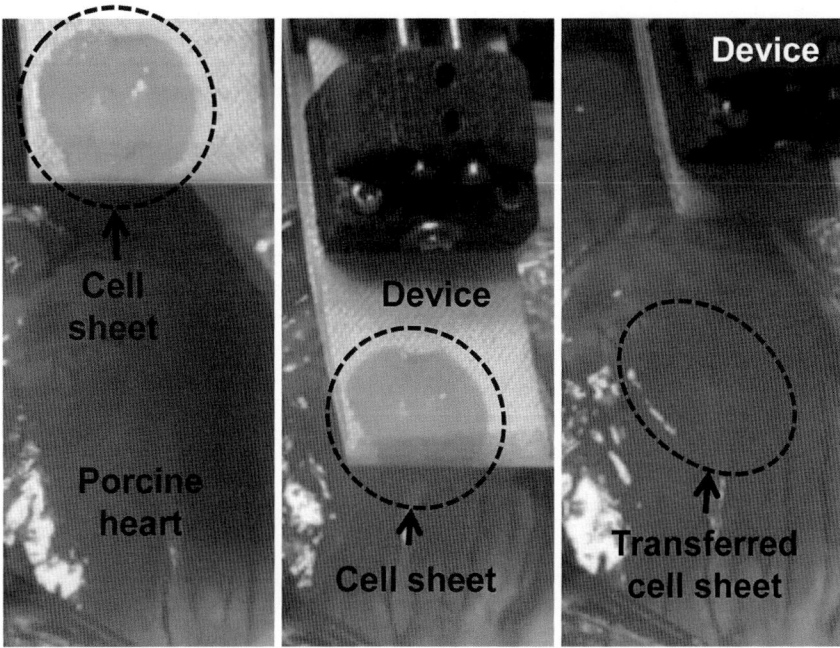

Fig. 8 Transfer of a C2C12 cell sheet onto the heart tissue of a porcine using a scooping/transfer device. The cell sheet stained with neutral red was recovered from a 35-mm temperature-responsive culture surface. The dashed circle in the right photograph shows a transferred cell sheet

3. Extend the tip of the device and tilt toward the near edge of the cell sheet.

4. Push the pushing rod by hand, then slide the inner plate in the direction of the tip, then move the outer cross with the movement of the inner plate, and roll up at the tip simultaneously. The cell sheet can then be scooped up by moving the outer cross.

5. Move the device holding the cell sheet, place the tip of the device onto another surface, and then pull another culture surface into the device to release the cell sheet.

Cell sheets can be transplanted onto the heart tissue of a porcine model by a similar method (Fig. 8). Even an inexperienced technician successfully transferred cell sheets from a culture dish to another surface very rapidly using this device (the transfer time was 3.7 ± 1.6 s, $n = 3$) [53].

The required fabrication conditions for several cell sheets used in cardiac tissue engineering can be found in Table 2. Several types of adult stem/progenitor cells have been used as cell sources for regenerating damaged heart tissues as described above (Fig. 1). The therapeutic effects of most non-pulsatile adult stem/progenitor cells are generally thought to be due to the paracrine effects of

Table 2
Fabrication conditions of several cell sheets for cardiac tissue engineering

Cells	Cell concentration (cells/cm^2)	Cultivation time (days)	References
Neonatal rat cardiac cells	2.5×10^5	4	[24, 47]
Mouse ES cell-derived cardiomyocytes[a]	2.6×10^5	4	[42]
Mouse neonatal cardiac fibroblasts[a]	0.6×10^5		
Human iPS cell-derived cardiomyocytes	2.1×10^5	4	[43, 46]
Mouse cardiac progenitor cells	1.0×10^5	5	[34]
Human myoblasts	6.3×10^4	3	UD[b]
C2C12 mouse myoblast lines	6.3×10^4	3	50]
Human bone marrow-derived MSCs	6.3×10^4	3	UD
Human adipose-derived MSCs	6.3×10^4	3	UD
Mouse adipose-derived MSCs	1.0×10^5	5	[34]
Human endometrial gland-derived MSCs	1.0×10^5	4	[33, 50, 51]

[a]A mouse ES cell-derived cardiac cell sheet can be fabricated by the co-cultivation of mouse ES cell-derived cardiomyocytes (2.6×10^5 cells) and mouse neonatal cardiac fibroblasts ($0.6 \times 10_5$ cells).
[a]UD: Unpublished data

various factors produced from those cells, but not the mechanical support. The transplantation of pulsatile cardiac cells is expected to contribute to the mechanical support of a damaged heart via electrical and functional couplings. In fact, the transplantations of pulsatile grafts fabricated using rat cardiomyocytes induce greater therapeutic effects than those of non-pulsatile grafts in a rat model [62]. However, currently, a suitable cell source of clinically available human cardiomyocytes has not yet been established. ES cells and iPSCs are attractive potential sources of pulsatile cardiomyocytes. The transplantation of human ES-derived cardiomyocytes contributes to a higher recovery of attenuated cardiac functions than that of ES-derived non-cardiomyocytes [63]. Human cardiac cell sheets can be fabricated by using human iPS cell-derived cardiomyocytes. The two cardiac cell sheets couple electrically and beat synchronously after layering [43], resulting in the reconstruction of electrically communicative human 3D cardiac tissue. In near future these spontaneously beating human cardiac cell sheet constructs are expected to contribute to the direct support of mechanically damaged hearts and may soon be used clinically to provide an effective therapy. In addition, electrically communicative human cardiac tissues could contribute to new in vitro tissue models, which offer an alternative to animal experiments.

4   Notes

1. All animal experiments must be performed according to the respective guidelines outlined by governmental and institutional committees for animal experiments.

2. Seeded cell numbers and the timing of cell sheet formation are highly variable depending on the cell type (Tables 1 and 2). The cultivation period before cell sheet formation may be able to be shortened by increasing the initial cell number. In addition, the indexes may be dependent on cell lots. While optimal cell numbers will produce an intact cell sheet that can be recovered, with less than optimal cell numbers, broken cell sheets or cell sheets with holes are commonly recovered.

3. When cell sheets are detached, they tend to shrink horizontally due to the cytoskeletal tensile reorganization. As a result, normal cell sheets consisting of two or three cell layers shrink to become smaller than the culture dish (Figs. 2, 3, and 5).

4. When the dish is rotated, the slow rotation allows us to spread out the cell sheets.

5. Wild medium dropping may induce the destruction of cell sheets.

6. The spreading of a cell sheet and incubation after removing the medium allow the cell sheet to adhere tightly to the culture surface. While short-term incubation may induce insufficient adhesion and detachment of the cell sheet from the dish when medium is added to the dish, long-term incubation may induce cell damage from desiccation. Normally incubation for 30–60 min is optimal. The pretreatment of the culture surface with FBS or cell adhesion proteins for more than 30 min promotes adhesion between the cell sheet and the surface.

7. To recover the second cell sheet during incubation of a dish onto which a first cell sheet was spread at 37 °C, transfer another temperature-responsive culture surface including confluently cultured cells into a CO_2 incubator set at 20 °C.

8. The cell sheets layered on a temperature-responsive culture surface can be recovered from the surface just by reducing the culture temperature to 20 °C.

9. While excess wetness may inhibit adhesion between the paper and a cell sheet, insufficient wetness may induce adhesion that is too strong. Optimal wetness allows us to recover the cell sheet from culture dish and transfer the recovered cell sheet onto the target surfaces.

10. After removing the remaining culture medium, the CellShifter™ will be covered by the cell sheet.

11. When a cell sheet is tightly adhered to the CellShifter™ and cannot be separated from the paper, moisturizing the paper and the cell sheet by dropping medium can weaken the adherence between the cell sheet and the paper.

12. The use of stained cell sheets may be convenient for visually observing the transfer of cell sheets. For example, a cell sheet stained with 0.001 % neutral red, prepared by the dilution of 0.1 % neutral red solution in the culture medium (for 15 min), is shown in Fig. 6b.

Acknowledgements

This work was supported by grants from Formation of Innovation Center for Fusion of Advanced Technologies in the Special Coordination Funds for Promoting Science and Technology "Cell Sheet Tissue Engineering Center (CSTEC)" from the Ministry of Education, Culture, Sports Science, and Technology (MEXT), Japan, and JSPS through the "Funding Program for World-Leading Innovative R&D on Science and Technology (FIRST Program)," initiated by the Council for Science and Technology Policy (CSTP).

References

1. Menasché P (2008) Skeletal myoblasts and cardiac repair. J Mol Cell Cardiol 45:545–553

2. Alaiti MA, Ishikawa M, Costa MA (2010) Bone marrow and circulating stem/progenitor cells for regenerative cardiovascular therapy. Transl Res 156:112–129

3. Bolli R, Chugh AR, D'Amario D, Loughran JH, Stoddard MF, Ikram S, Beache GM, Wagner SG, Leri A, Hosoda T, Sanada F, Elmore JB, Goichberg P, Cappetta D, Solankhi NK, Fahsah I, Rokosh DG, Slaughter MS, Kajstura J, Anversa P (2011) Cardiac stem cells in patients with ischaemic cardiomyopathy (SCIPIO): initial results of a randomised phase 1 trial. Lancet 378:1847–1857

4. Makkar RR, Smith RR, Cheng K, Malliaras K, Thomson LE, Berman D, Czer LS, Marbán L, Mendizabal A, Johnston PV, Russell SD, Schuleri KH, Lardo AC, Gerstenblith G, Marbán E (2012) Intracoronary cardiosphere-derived cells for heart regeneration after myocardial infarction (CADUCEUS): a prospective, randomised phase 1 trial. Lancet 379:895–904

5. Ptaszek LM, Mansour M, Ruskin JN, Chien KR (2012) Towards regenerative therapy for cardiac disease. Lancet 379:933–942

6. Zhang M, Methot D, Poppa V, Fujio Y, Walsh K, Murry CE (2001) Cardiomyocyte grafting for cardiac repair: graft cell death and anti-death strategies. J Mol Cell Cardiol 33:907–921

7. Suzuki K, Murtuza B, Beauchamp JR, Smolenski RT, Varela-Carver A, Fukushima S, Coppen SR, Partridge TA, Yacoub MH (2004) Dynamics and mediators of acute graft attrition after myoblast transplantation to the heart. FASEB J 18:1153–1155

8. Hofmann M, Wollert KC, Meyer GP, Menke A, Arseniev L, Hertenstein B, Ganser A, Knapp WH, Drexler H (2005) Monitoring of bone marrow cell homing into the infarcted human myocardium. Circulation 111:2198–2202

9. Li RK, Jia ZQ, Weisel RD, Mickle DA, Choi A, Yau TM (1999) Survival and Function of Bioengineered Cardiac Grafts. Circulation 100:II63–II69

10. Leor J, Aboulafia-Etzion S, Dar A, Shapiro L, Barbash IM, Battler A, Granot Y, Cohen S (2000) Bioengineered cardiac grafts: A new approach to repair the infarcted myocardium? Circulation 102:III56–III61

11. Zimmermann WH, Didié M, Wasmeier GH, Nixdorff U, Hess A, Melnychenko I, Boy O, Neuhuber WL, Weyand M, Eschenhagen T

(2002) Cardiac grafting of engineered heart tissue in syngenic rats. Circulation 106:I151–I157

12. Ott HC, Matthiesen TS, Goh SK, Black LD, Kren SM, Netoff TI, Taylor DA (2008) Perfusion-decellularized matrix: using nature's platform to engineer a bioartificial heart. Nat Med 14:213–221

13. Chachques JC, Trainini JC, Lago N, Masoli OH, Barisani JL, Cortes-Morichetti M, Schussler O, Carpentier A (2007) Myocardial assistance by grafting a new bioartificial upgraded myocardium (MAGNUM clinical trial): one year follow-up. Cell Transplant 16:927–934

14. Chachques JC, Trainini JC, Lago N, Cortes-Morichetti M, Schussler O, Carpentier A (2008) Myocardial Assistance by Grafting a New Bioartificial Upgraded Myocardium (MAGNUM trial): clinical feasibility study. Ann Thorac Surg 85:901–908

15. Yamada N, Okano T, Sakai H, Karikusa F, Sawasaki Y, Sakurai Y (1990) Thermo-responsive polymeric surface: control of attachment and detachment of cultured cells. Makromol Chem Rapid Commun 11:571–576

16. Okano T, Yamada H, Sakai H, Sakurai Y (1993) A novel recovery system for cultured cells using plasma-treated polystyrene dishes grafted with poly (N-isopropylacrylamide). J Biomed Mater Res 27:1243–1251

17. Matsuda N, Shimizu T, Yamato M, Okano T (2007) Tissue engineering based on cell sheet technology. Adv Mater 19:3089–3099

18. Yang J, Yamato M, Shimizu T, Sekine H, Ohashi K, Kanzaki M, Ohki T, Nishida K, Okano T (2007) Reconstruction of functional tissues with cell sheet engineering. Biomaterials 28:5033–5043

19. Hannachi IE, Yamato M, Okano T (2010) Cell sheet engineering: a unique nanotechnology for scaffold-free tissue reconstruction with clinical applications in regenerative medicine. J Intern Med 267:54–70

20. Haraguchi Y, Shimizu T, Yamato M, Okano T (2011) Regenerative therapies using cell sheet-based tissue engineering for cardiac disease. Cardiol Res Pract 2011:845170

21. Kushida A, Yamato M, Konno C, Kikuchi A, Sakurai Y, Okano T (1999) Decrease in culture temperature releases monolayer endothelial cell sheets together with deposited fibronectin matrix from temperature-responsive culture surface. J Biomed Mater Res 45:355–362

22. Nishida K, Yamato M, Hayashida Y, Watanabe K, Maeda N, Watanabe H, Yamamoto K, Nagai S, Kikuchi A, Tano Y, Okano T (2004) Functional bioengineered corneal epithelial sheet grafts from corneal stem cells expanded ex vivo on a temperature-responsive cell culture surface. Transplantation 77:379–385

23. Ohashi K, Yokoyama T, Yamato M, Kuge H, Kanehiro H, Tsutsumi M, Amanuma T, Iwata H, Yang J, Okano T, Nakajima Y (2007) Engineering functional two- and three-dimensional liver systems in vivo using hepatic tissue sheets. Nat Med 13:880–885

24. Shimizu T, Yamato M, Isoi Y, Akutsu T, Setomaru T, Abe K, Kikuchi A, Umezu M, Okano T (2002) Fabrication of pulsatile cardiac tissue grafts using a novel 3-dimensional cell sheet manipulation technique and temperature-responsive cell culture surfaces. Circ Res 90:e40–e48

25. Shimizu T, Yamato M, Kikuchi A, Okano T (2003) Cell sheet engineering for myocardial tissue reconstruction. Biomaterials 24:2309–2316

26. Shimizu T, Sekine H, Isoi Y, Yamato M, Kikuchi A, Okano T (2006) Long-term survival and growth of pulsatile myocardial tissue grafts engineered by the layering of cardiomyocyte sheets. Tissue Eng 12:499–507

27. Haraguchi Y, Shimizu T, Yamato M, Kikuchi A, Okano T (2006) Electrical coupling of cardiomyocyte sheets occurs rapidly via functional gap junction formation. Biomaterials 27:4765–4774

28. Memon IA, Sawa Y, Fukushima N, Matsumiya G, Miyagawa S, Taketani S, Sakakida SK, Kondoh H, Aleshin AN, Shimizu T, Okano T, Matsuda H (2005) Repair of impaired myocardium by means of implantation of engineered autologous myoblast sheets. J Thorac Cardiovasc Surg 130:1333–1341

29. Hata H, Matsumiya G, Miyagawa S, Kondoh H, Kawaguchi N, Matsuura N, Shimizu T, Okano T, Matsuda H, Sawa Y (2006) Grafted skeletal myoblast sheets attenuate myocardial remodeling in pacing-induced canine heart failure model. J Thorac Cardiovasc Surg 132:918–924

30. Kondoh H, Sawa Y, Miyagawa S, Sakakida-Kitagawa S, Memon IA, Kawaguchi N, Matsuura N, Shimizu T, Okano T, Matsuda H (2006) Longer preservation of cardiac performance by sheet-shaped myoblast implantation in dilated cardiomyopathic hamsters. Cardiovasc Res 69:466–475

31. Miyahara Y, Nagaya N, Kataoka M, Yanagawa B, Tanaka K, Hao H, Ishino K, Ishida H, Shimizu T, Kangawa K, Sano S, Okano T, Kitamura S, Mori H (2006) Monolayered mesenchymal stem cells repair scarred myocardium after myocardial infarction. Nat Med 12:459–465

32. Sekine H, Shimizu T, Hobo K, Sekiya S, Yang J, Yamato M, Kurosawa H, Kobayashi E, Okano T (2008) Endothelial cell coculture

within tissue-engineered cardiomyocyte sheets enhances neovascularization and improves cardiac function of ischemic hearts. Circulation 118:S145–S152

33. Hida N, Nishiyama N, Miyoshi S, Kira S, Segawa K, Uyama T, Mori T, Miyado K, Ikegami Y, Cui C, Kiyono T, Kyo S, Shimizu T, Okano T, Sakamoto M, Ogawa S, Umezawa A (2008) Novel cardiac precursor-like cells from human menstrual blood-derived mesenchymal cells. Stem Cells 26:1695–1704

34. Matsuura K, Honda A, Nagai T, Fukushima N, Iwanaga K, Tokunaga M, Shimizu T, Okano T, Kasanuki H, Hagiwara N, Komuro I (2009) Transplantation of cardiac progenitor cells ameliorates cardiac dysfunction after myocardial infarction in mice. J Clin Invest 119:2204–2217

35. Sekiya N, Matsumiya G, Miyagawa S, Saito A, Shimizu T, Okano T, Kawaguchi N, Matsuura N, Sawa Y (2009) Layered implantation of myoblast sheets attenuates adverse cardiac remodeling of the infarcted heart. J Thorac Cardiovasc Surg 138:985–993

36. Miyagawa S, Saito A, Sakaguchi T, Yoshikawa Y, Yamauchi T, Imanishi Y, Kawaguchi N, Teramoto N, Matsuura N, Iida H, Shimizu T, Okano T, Sawa Y (2010) Impaired myocardium regeneration with skeletal cell sheets—a preclinical trial for tissue-engineered regeneration therapy. Transplantation 90:364–372

37. Sekine H, Shimizu T, Dobashi I, Matsuura K, Hagiwara N, Takahashi M, Kobayashi E, Yamato M, Okano T (2011) Cardiac cell sheet transplantation improves damaged heart function via superior cell survival in comparison with dissociated cell injection. Tissue Eng Part A 17:2973–2980

38. Imanishi Y, Miyagawa S, Maeda N, Fukushima S, Kitagawa-Sakakida S, Daimon T, Hirata A, Shimizu T, Okano T, Shimomura I, Sawa Y (2011) Induced adipocyte cell-sheet ameliorates cardiac dysfunction in a mouse myocardial infarction model: a novel drug delivery system for heart failure. Circulation 124:S10–S17

39. Shudo Y, Miyagawa S, Fukushima S, Saito A, Shimizu T, Okano T, Sawa Y (2011) Novel regenerative therapy using cell-sheet covered with omentum flap delivers a huge number of cells in a porcine myocardial infarction model. J Thorac Cardiovasc Surg 141:1188–1196

40. Kamata S, Miyagawa S, Fukushima S, Nakatani S, Kawamoto A, Saito A, Harada A, Shimizu T, Daimon T, Okano T, Asahara T, Sawa Y. (in press) Improvement of cardiac stem cell-sheet therapy for chronic ischemic injury by adding endothelial progenitor cell transplantation: analysis of layer-specific regional cardiac function. Cell Transplant

41. Sawa Y, Miyagawa S, Sakaguchi T, Fujita T, Matsuyama A, Saito A, Shimizu T, Okano T (2012) Tissue engineered myoblast sheets improved cardiac function sufficiently to discontinue LVAS in a patient with DCM: report of a case. Surg Today 42:181–184

42. Matsuura K, Masuda S, Haraguchi Y, Yasuda N, Shimizu T, Hagiwara N, Zandstra PW, Okano T (2011) Creation of mouse embryonic stem cell-derived cardiac cell sheets. Biomaterials 32:7355–7362

43. Matsuura K, Wada M, Shimizu T, Haraguchi Y, Sato F, Sugiyama K, Konishi K, Shiba Y, Ichikawa H, Tachibana A, Ikeda U, Yamato M, Hagiwara N, Okano T (2012) Creation of human cardiac cell sheets using pluripotent stem cells. Biochem Biophys Res Commun 425:321–327

44. Masumoto H, Matsuo T, Yamamizu K, Uosaki H, Narazaki G, Katayama S, Marui A, Shimizu T, Ikeda T, Okano T, Sakata R, Yamashita JK (2012) Pluripotent stem cell-engineered cell sheets reassembled with defined cardiovascular populations ameliorate reduction in infarct heart function through cardiomyocyte-mediated neovascularization. Stem Cell 30:1196–1205

45. Kawamura M, Miyagawa S, Miki K, Saito A, Fukushima S, Higuchi T, Kawamura T, Kuratani T, Daimon T, Shimizu T, Okano T, Sawa Y (2012) Feasibility, safety, and therapeutic efficacy of human induced pluripotent stem cell-derived cardiomyocyte sheets in a porcine ischemic cardiomyopathy model. Circulation 126:S29–S37

46. Haraguchi Y, Matsuura K, Shimizu T, Yamato M, Okano T (in press) Simple suspension culture system of human iPS cells maintaining their pluripotency for cardiac cell sheet engineering. J Tissue Eng Regen Med

47. Haraguchi Y, Shimizu T, Sasagawa T, Sekine H, Sakaguchi K, Kikuchi T, Sekine W, Sekiya S, Yamato M, Umezu M, Okano T (2012) Fabrication of functional three-dimensional tissues by stacking cell sheets in vitro. Nat Protoc 7:850–858

48. Shioyama T, Haraguchi Y, Muragaki Y, Shimizu T, Okano T (2013) New isolation system for collecting living cells from tissue. J Biosci Bioeng 115:100–103

49. Matsuura K, Wada H, Nagai T, Iijima Y, Minamino T, Sano M, Akazawa H, Molkentin JD, Kasanuki H, Komuro I (2004) Cardiomyocytes fuse with surrounding noncardiomyocytes and reenter the cell cycle. J Cell Biol 167:351–363

50. Haraguchi Y, Sekine W, Shimizu T, Yamato M, Miyoshi S, Umezawa A, Okano T (2010) Development of a new assay system for evalu-

ating the permeability of various substances through three-dimensional tissue. Tissue Eng Part C Methods 16:685–692

51. Sekine W, Haraguchi Y, Shimizu T, Umezawa A, Okano T (2011) Thickness limitation and cell viability of multi-layered cell sheets and overcoming the diffusion limit by a porous-membrane culture insert. J Biochip Tissue Chip S2:001

52. Kanzaki M, Yamato M, Yang J, Sekine H, Takagi R, Isaka T, Okano T, Onuki T (2008) Functional closure of visceral pleural defects by autologous tissue engineered cell sheets. Eur J Cardiothorac Surg 34:864–869

53. Tadakuma K, Tanaka N, Haraguchi Y, Higashimori M, Kaneko M, Shimizu T, Yamato M, Okano T (2013) Development of a simple device for transfer/transplantation of living cell sheets rapidly and completely without cell damage. Biomaterials 34:9018–9025

54. Ohki T, Yamato M, Murakami D, Takagi R, Yang J, Namiki H, Okano T, Takasaki K (2006) Treatment of oesophageal ulcerations using endoscopic transplantation of tissue-engineered autologous oral mucosal epithelial cell sheets in a canine model. Gut 55:1704–1710

55. Kanzaki M, Yamato M, Hatakeyama H, Kohno C, Yang J, Umemoto T, Kikuchi A, Okano T, Onuki T (2006) Tissue engineered epithelial cell sheets for the creation of a bioartificial trachea. Tissue Eng 12:1275–1283

56. Tsuda Y, Shimizu T, Yamato M, Kikuchi A, Sasagawa T, Sekiya S, Kobayashi J, Chen G, Okano T (2007) Cellular control of tissue architectures using a three-dimensional tissue fabrication technique. Biomaterials 28:4939–4946

57. Iwata T, Yamato M, Tsuchioka H, Takagi R, Mukobata S, Washio K, Okano T, Ishikawa I (2009) Periodontal regeneration with multi-layered periodontal ligament-derived cell sheets in a canine model. Biomaterials 30:2716–2723

58. Shimizu H, Ohashi K, Utoh R, Ise K, Gotoh M, Yamato M, Okano T (2009) Bioengineering of a functional sheet of islet cells for the treatment of diabetes mellitus. Biomaterials 30:5943–5949

59. Elloumi HI, Itoga K, Kumashiro Y, Kobayashi J, Yamato M, Okano T (2009) Fabrication of transferable micropatterned-co-cultured cell sheets with microcontact printing. Biomaterials 30:5427–5432

60. Asakawa N, Shimizu T, Tsuda Y, Sekiya S, Sasagawa T, Yamato M, Fukai F, Okano T (2010) Pre-vascularization of in vitro three-dimensional tissues created by cell sheet engineering. Biomaterials 31:3903–3909

61. Sasagawa T, Shimizu T, Sekiya S, Haraguchi Y, Yamato M, Sawa Y, Okano T (2010) Design of prevascularized three-dimensional cell-dense tissues using a cell sheet stacking manipulation technology. Biomaterials 31:1646–1654

62. Zimmermann WH, Melnychenko I, Wasmeier G, Didié M, Naito H, Nixdorff U, Hess A, Budinsky L, Brune K, Michaelis B, Dhein S, Schwoerer A, Ehmke H, Eschenhagen T (2006) Engineered heart tissue grafts improve systolic and diastolic function in infarcted rat hearts. Nat Med 12:452–458

63. Shiba Y, Fernandes S, Zhu WZ, Filice D, Muskheli V, Kim J, Palpant NJ, Gantz J, Moyes KW, Reinecke H, Van Biber B, Dardas T, Mignone JL, Izawa A, Hanna R, Viswanathan M, Gold JD, Kotlikoff MI, Sarvazyan N, Kay MW, Murry CE, Laflamme MA (2012) Human ES-cell-derived cardiomyocytes electrically couple and suppress arrhythmias in injured hearts. Nature 489:322–325

Chapter 14

Design and Fabrication of Biological Wires

Jason W. Miklas, Sara S. Nunes, Boyang Zhang, and Milica Radisic

Abstract

Cardiac tissue engineering using human pluripotent stem cell-derived cardiomyocytes (hPSC-CMs) has facilitated the creation of in vitro diagnostic platforms to study novel small molecules and cardiac disease at the tissue level. Yet, due to the immaturity of hPSC-CMs, there is a low fidelity between tissue-engineered cardiac tissues and adult cardiac tissues. To address this challenge, we have developed a platform that combines both physical and electrical cues to guide hPSC-CMs towards a more mature state in vitro.

Key words Cardiac tissue engineering, Stem cells, Cardiomyocyte, Electrical stimulation, Microfabrication

1 Introduction

In order to create high-fidelity in vitro cardiac tissues, bioengineering techniques have been employed to create microenvironments that are conducive to cardiomyocyte survival and function. This includes customizing the appropriate matrix for the cells to attach to [1–3], customizing media/growth factors [4], and using mechanical stimulation [5–7] and electrical stimulation [8, 9]. Previously, it has been shown that electrical stimulation [8] or cyclic mechanical stimulation [7] of neonatal rat cardiomyocyte-based cardiac tissues promoted maturation of cardiomyocytes to an adult-like phenotype.

With the advent of readily accessible cardiac stem cell differentiation protocols [10–12], many in vitro human cardiac tissues have been created to study various aspects of human cardiac diseases and responses to novel drugs. However, the major limitation with human pluripotent stem cell-derived cardiomyocytes (hPSC-CMs) has been their immaturity [13, 14]. Since differentiation protocols produce cardiomyocytes that are usually associated with fetal-like maturity, it becomes difficult to model the behavior of the adult myocardium using hPSC-CMs.

Milica Radisic and Lauren D. Black III (eds.), *Cardiac Tissue Engineering: Methods and Protocols*, Methods in Molecular Biology, vol. 1181, DOI 10.1007/978-1-4939-1047-2_14, © Springer Science+Business Media New York 2014

Capitalizing on our prior knowledge of cardiac tissue engineering, many efforts have been made to try and mature hPSC-CMs using gel compaction and cyclic mechanical stretch stimulation [5, 15]. However, in both cases a moderate maturation response was seen and minimal electrophysiological and calcium handling enhancements were found. To address these shortcomings, we have developed a platform named biological wire (biowire) [16] that provides both structural cues and electrical field stimulation to mature hPSC-CMs. Using a microfabricated platform, we have been able to guide cardiomyocyte alignment along a rigid suture while concurrently providing a custom electrical field stimulation protocol to promote cardiomyocyte maturation, both structurally and electrophysiologically. Here, we describe in detail the design and fabrication of the biowire technology.

2   Materials

All solutions are to be prepared with ultrapure water (sensitivity of 18.2 MΩ cm at 25 °C) and analytical grade reagents.

2.1   Soft Lithography

1. Master components: SU-8 50 (MicroChem), SU-8 2050, 4-inch silicon wafer, propylene glycol monomethyl ether acetate.

2.2   Electrical Stimulation

1. Electrical stimulation chamber components: 3 mm Diameter carbon rods, 0.2 mm platinum wire.

2. Tools required: Dremel, drill bit 1, drill bit size 2.

3. Poly(dimethylsiloxane) (PDMS): Ratio of 1:10 for cross-linker to base, mix, let air bubbles dissipate.

2.3   Biowire Preparation

1. PDMS: Ratio of 1:10 for cross-linker to base, mix, let air bubbles dissipate.

2. Collagenase type I: Weigh 1 g of collagenase type I, add 100 mL of fetal bovine serum (FBS) and 400 mL of phosphate-buffered saline (PBS) with Ca^{2+} and Mg^{2+}. Mix, sterile filter through 0.22 μm filter, prepare 12 mL aliquots, and store at –20 °C. Upon thawing a 12 mL aliquot, re-aliquot into 2 mL aliquots and refreeze extra.

3. DNase I: Make a stock solution of 1 mg/mL DNase I in water. Filter sterilize, prepare 0.5 mL aliquots, and store at –20 °C. Use aliquots once and discard excess.

4. Trypsin/ethylenediaminetetraacetic acid (EDTA): Weigh 2.5 g of trypsin and add 2 mL of EDTA (0.5 M pH 8) and 1,000 mL of PBS without Ca^{2+} and Mg^{2+}. Mix, sterile filter, prepare 10 mL aliquots, and store at –20 °C. Upon thawing, store excess at 4 °C for 2 weeks.

5. Wash medium: Iscove's modified Dulbecco's medium (IMDM) and 1 % penicillin/streptomycin.

6. Stop medium: 50 % Wash medium, 50 % FBS.

7. Culture medium: Per 50 mL of StemPro-34 add 1.3 mL of supplement, 500 µL of 100× glutamine, 250 µL of 30 mg/mL transferrin, 500 µL of 5 mg/mL ascorbic acid, 150 µL of 26λ/2 mL monothioglycerol (MTG), and 500 µL (1 %) penicillin/streptomycin.

8. Collagen gel preparation (final concentrations): 2.1 mg/mL of rat tail collagen type I in 24.9 mM glucose, 23.8 mM $NaHCO_3$, 14.3 mM NaOH, 10 mM HEPES, in 1× M199 media with 10 % of growth factor-reduced Matrigel.

3   Methods

3.1   Master Design and Creation

Fabricate device using soft lithography techniques. A two-layer SU-8 master is used to mold PDMS.

1. Design the device in AutoCAD. Draw each layer of the master separately to create two corresponding film masks.

2. Print designs on two film masks (CADART) setting the device features as transparent and the surrounding regions dark (Fig. 1a).

3. Spin SU-8 50 onto a 4-inch silicon wafer at 2000 RPM. for 30 s.

4. Bake at 95 °C for 10 min.

5. Expose to UV light at a dose of 200 mJ/cm^2 to create the base layer.

6. Spin SU-8 2050 onto the wafer at 2500 RPM for 30 s.

7. Bake at 95 °C for 15 min, and then let cool to room temperature.

8. Repeat **steps 6–7** two more times.

9. Expose the coated wafer to UV light under the first-layer mask at a dose of 270 mJ/cm^2 to create the first layer including the suture channel and the chamber with thickness of 185 µm.

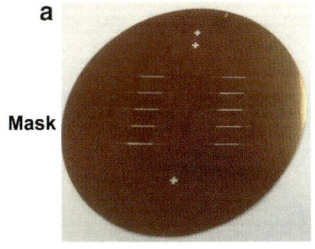
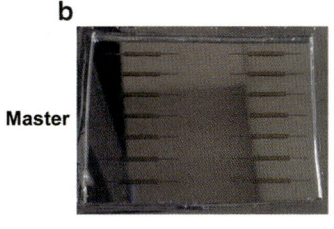
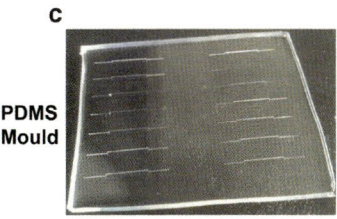

a Mask   **b** Master   **c** PDMS Mould

Fig. 1 Images of the biowire template fabrication. (**a**) Mask used to expose silicon wafer. (**b**) Master mould used to imprint biowire dimensions on PDMS. (**c**) PDMS mould of biowire

10. Bake at 95 °C for 15 min.

11. Create the second layer of SU-8 2050 by repeating **steps 6–7** twice.

12. Align the second-layer mask using a mask aligner, with manual xyz stage, to the features on the first layer.

13. UV expose at a total dose of 240 mJ/cm^2. The exposure set time was derived based on the required total UV dose and the UV light output intensity measured on that specific day prior to use.

14. Bake at 95 °C for 15 min.

15. Develop the wafer for 30 min using propylene glycol mono-methyl ether by immersing the wafer within the solution and stirring using an orbital shaker (Fig. 1b).

3.2 PDMS Biowire Template

1. Pour PDMS onto the master ensuring that all features are covered.

2. Place the PDMS-covered master into an oven for 2 h at 70 °C. After 2 h, remove the master from the oven and cut out the PDMS biowire templates (Fig. 1c, *see* **Note 1**).

3. Under sterile conditions, place the PDMS microwells into a Petri dish using sterile forceps.

4. Under sterile conditions, cut 6-0 silk suture to a desired length. Place a piece of sterile surgical silk suture (6-0) centrally in the channel of the PDMS microwells by placing it in the grooves located at both ends of the channel. Handle the sutures with sterile tweezers (Fig. 2a, *see* **Note 2**).

3.3 Electrical Stimulation Chamber

1. From a sheet of polycarbonate, cut rectangular pieces with dimensions of 2 cm length, 0.35 cm thickness, and 0.85 cm width. A pair of rectangular pieces will constitute the frame of the electrical stimulation chamber.

2. Using a dremel with a 3 mm diameter drill bit attached, create two holes that are 2 cm apart along the center line of each polycarbonate frame piece.

3. Cut a pair of carbon rods to a length of 1.5 cm. Using a dremel tool with a 1 mm diameter drill bit attached, create a hole through the carbon rod about 3 mm from the end of the carbon rod. Through this hole, string platinum wire through and wrap the wire tightly against the carbon rod. Secure a clip to the free end of the platinum wire, which will serve as the attachment point for the wire from the cardiac stimulator to the electrical stimulation chamber (Fig. 2b).

4. Construct the electrical stimulation chamber by placing the carbon rods into the polycarbonate frame pieces. Place the assembled frame with carbon rods into a well of a six-well plate.

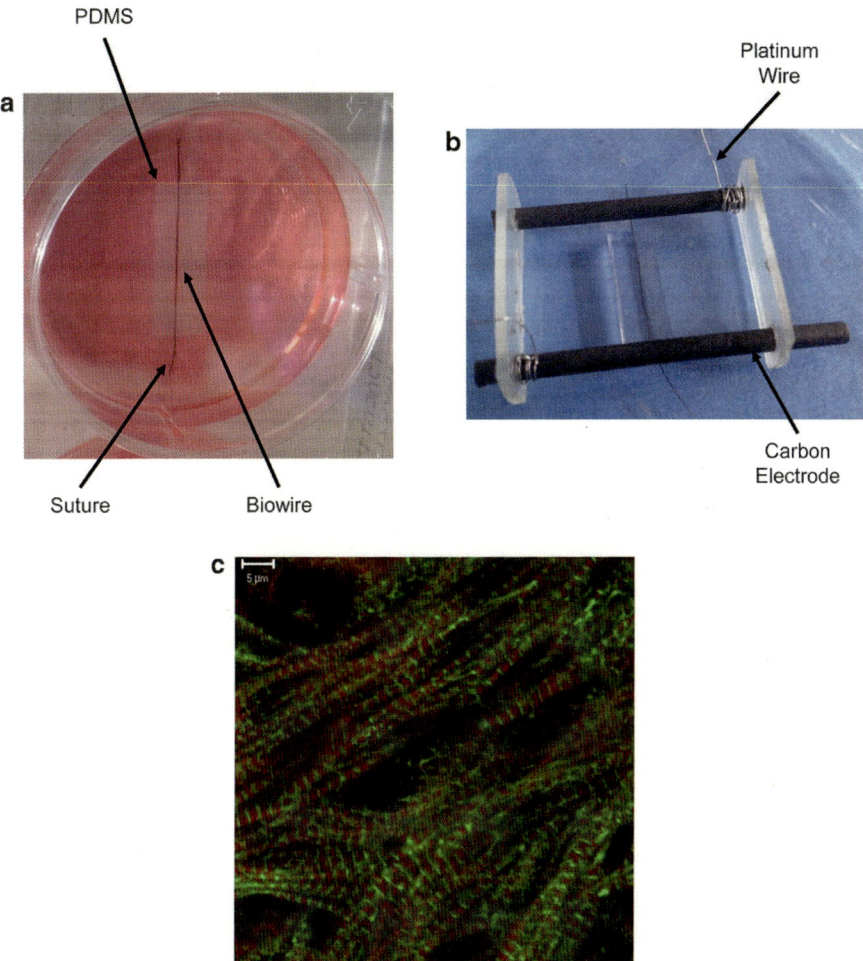

Fig. 2 Biowire culture setup. (**a**) Biowire PDMS well after 1 week of culture showing gel compaction of the cardiac tissue. After 1 week of culture, the biowire should be transferred to the electrical stimulation chamber. (**b**) Electrical stimulation chamber used during the second week of biowire cultivation with the biowire placed perpendicular to the electrodes. (**c**) Confocal microscopy showing well-aligned and mature sarcomeres after culture in the biowire platform. *Green*—α-actinin, *red*—actin staining using phalloidin

5. Prepare PDMS and pour 2 mL into the well ensuring that the bottom of the polycarbonate frame pieces is submersed in PDMS while the PDMS level approaches but does not reach the bottom of the carbon rods. Place the six-well plate into an oven for 2 h at 70 °C to cure the PDMS. This will ensure that the carbon rods remain securely in place during biowire cultivation.

6. Remove the finished electrical stimulation chamber from the six-well plate and autoclave (*see* **Notes 3** and **4**).

**3.4 hPSC-CM
Single-Cell Isolation**

1. Warm collagenase type I to 37 °C, and add 10 µL of DNAse per 1 mL of collagenase type I.

2. Directed differentiation of cardiomyocytes in embryoid bodies (EBs) is described elsewhere in this book. We use EBs from day 20 to day 34 of differentiation to make biowires. Remove and transfer EBs from the low cluster plate to a conical tube. Spin EBs at 125 r.c.f for 5 min. Aspirate media, and then suspend the EBs in collagenase and place in the incubator for 2 h. Agitate tube every 30 min during incubation (*see* **Notes 5** and **6**).

3. Add 5 mL of wash medium, and spin EBs at 125 r.c.f for 5 min.

4. Aspirate supernatant, resuspend in 2 mL of trypsin/EDTA solution, and incubate at 37 °C for 5 min. Agitate tube twice during the 5-min incubation.

5. Quench trypsin with 1 mL of stop medium and 30 µL DNase.

6. Disrupt EBs into single-cell suspension by drawing and plunging the EB suspension from a 5 mL syringe three to four times using a 20-gauge needle (*see* **Notes 7** and **8**).

7. Add 7 mL of wash medium, and spin the cells at 280 r.c.f for 5 min.

8. Aspirate media, resuspend cells in culture medium, and place cells on ice.

**3.5 Collagen
Gel Preparation**

All gel components should be at 4 °C and placed on ice when ready and waiting to be used.

1. Into a sterile 1.5 mL Eppendorf tube add all components of the collagen gel except the collagen and Matrigel.

2. Add collagen type I, and pipette up and down to ensure that the solution is well mixed. Add reduced growth factor Matrigel, and mix.

3.6 Biowire Seeding

1. Count cells, remove desired number of cells in suspension, and spin at 280 r.c.f for 5 min. Aspirate the media, and add 3.5 µL of collagen gel mixture for every 500,000 cells. Mix the gel cell slurry well by gently pipetting up and down, ensuring that minimal air bubbles are formed.

2. Using a P10 pipette tip, pipette 4 µL of the cardiac cell suspension (8 µL/cm of channel length) into the channel so that it surrounds the suture. If necessary, adjust the position of the suture with sterile forceps before the gel polymerizes.

3. Prior to placing the microwells with seeded cells into a 37 °C incubator for polymerization, add enough sterile PBS to coat the bottom of the Petri dish (*see* **Note 9**).

4. After polymerization for 30 min, aspirate the PBS and add 12 mL of culture medium to cover the cells in the gel-seeded microwells. Culture the biowires for 1 week to allow gel compaction around the suture, changing media every other day (*see* **Notes 10–12**).

3.7 Electrical Stimulation of Biowire Setup

1. Take a six-well plate, open in sterile conditions, and add 2 mL of sterile PBS to each well that will contain an electrical stimulation chamber.

2. Open the autoclaved electrical stimulation device, and using blunt-tipped tweezers, place the device into one of the PBS-primed wells of a six-well plate. Gently insert the electrical stimulation chamber into the well until it is flush with the bottom of the well. Aspirate the PBS from the well.

3. Add 5 mL of warmed culture medium to each well ensuring that the carbon rods are fully submersed in the media.

4. Using sterile tweezers, transfer the biowires to the electrical stimulation chamber placing them perpendicular to the carbon rods (*see* **Note 13**).

5. Place the six-well plate lid on top of the plate ensuring that platinum wires are sufficiently outside the plate (*see* **Notes 14** and **15**).

6. Place the six-well plate into the incubator, and attach wires from the cardiac stimulator to the clips on the end of the platinum wire (*see* **Note 16**).

7. Turn on the cardiac stimulator. Select biphasic repeating pulse, 1-ms pulse duration, 6 V per phase (3 V/cm), and 1 pulse per second (PPS) for the pacing frequency. Every 24 h, the PPS should be increased to the following values: 1.83, 2.66, 3.49, 4.82, 5.15, and 6. Change culture medium every other day.

8. After a week of electrical stimulation, if further culture is desired all further electrical stimulation should be performed at 1 PPS. Continue to change culture medium every other day.

4   Notes

1. Trim each PDMS biowire template to a desired length and width. Autoclave the PDMS biowire templates.

2. Expose the Petri dish containing the PDMS biowire templates with sutures to UV light overnight in a laminar flow hood to ensure that the templates are sterilized after handling.

3. It is easiest to remove the stimulation chamber by only gripping the polycarbonate frame and gently twisting and pulling the chamber out of the well.

4. The first time the stimulation chamber is created it should be tested to confirm that the connections between the platinum wire and the carbon rod are sound. To do this, hook up the stimulation chamber to the cardiac stimulator and set a desired voltage. Using a volt meter, place a volt meter lead onto each of the two carbon rods. The outputted voltage from the stimulation chamber should be the same voltage being read on the volt meter.

If this is not the case, PDMS may have coated the platinum wire that is wrapped around the carbon rod or there is some other faulty connection in the system.

5. Collagenase incubation time depends on the age of the hPSC-CMs and the type of differentiation protocol used (monolayer versus EB format). For day-21 cells, 2 h of collagenase is recommended for the EB format. Day-21 cells using the monolayer format for differentiation require only 30 min of collagenase digestion. For cells cultured longer or shorter, a longer or a shorter collagenase digestion will be required and should be determined by the user based on the viability and contractile function of the resulting cells.

6. One hour before digestion is finished start making the collagen gel (Subheading 3.5).

7. It is very important to carefully attach the syringe to the needle tip by only handling the very end of the syringe. Since the entire syringe will be going into the 15 mL conical tube any contamination on the outside of the syringe will easily transfer to the cells.

8. Depending on the time point of the cells, type of differentiation, and how forcefully one disrupts the cells, a different number of plunges will be required to dissociate the cells. It is best to use as few plunges as possible and to minimize air bubble formation during each plunge.

9. Avoid touching the seeded cells and gel in the PDMS microwells. PBS is added in order to prevent the gel from drying out and for the biowires to be formed properly. As PBS evaporates in the incubator it will create a moist environment.

10. If microwells used are longer than 1 cm, adjust polymerization time accordingly; that is, allow for longer gelation.

11. Longer biowires may require more frequent media changes due to the increased number of cells. Adjust as necessary.

12. Biowire PDMS wells should be attached to the bottom of the Petri dish and should not be floating.

13. Once all biowires are placed in the electrical stimulation chamber and submersed with media, ensure that no biowires have tipped over or are floating. All biowires should be perpendicular to the carbon rods and be pushed down if floating.

14. Gently close the lid so that the platinum wire does not snap during the process.

15. If lid does not close fully, tape lid down to ensure that the lid sits properly on the plate.

16. When closing the incubator doors, try to place the thin portion of the wires at door closing points while leaving the thicker portions of wires outside of the incubator.

References

1. Chiu LL, Iyer RK, Reis LA, Nunes SS, Radisic M (2012) Cardiac tissue engineering: current state and perspectives. Front Biosci (Landmark Ed) 17:1533–1550

2. Bronshtein T, Au-Yeung GC, Sarig U, Nguyen EB, Mhaisalkar PS, Boey FY, Venkatraman SS, Machluf M (2013) A mathematical model for analyzing the elasticity, viscosity, and failure of soft tissue: comparison of native and decellularized porcine cardiac extracellular matrix for tissue engineering. Tissue Eng Part C Methods 19(8):620–630. doi:10.1089/ten.TEC.2012.0387

3. Ye Z, Zhou Y, Cai H, Tan W (2011) Myocardial regeneration: Roles of stem cells and hydrogels. Adv Drug Deliv Rev 63(8):688–697. doi:10.1016/j.addr.2011.02.007

4. Naito H, Melnychenko I, Didie M, Schneiderbanger K, Schubert P, Rosenkranz S, Eschenhagen T, Zimmermann WH (2006) Optimizing engineered heart tissue for therapeutic applications as surrogate heart muscle. Circulation 114(1 Suppl):I72–I78. doi:10.1161/CIRCULATIONAHA.105.001560

5. Schaaf S, Shibamiya A, Mewe M, Eder A, Stohr A, Hirt MN, Rau T, Zimmermann WH, Conradi L, Eschenhagen T, Hansen A (2011) Human engineered heart tissue as a versatile tool in basic research and preclinical toxicology. PLoS One 6(10):e26397. doi:10.1371/journal.pone.0026397

6. Tiburcy M, Didie M, Boy O, Christalla P, Doeker S, Naito H, Karikkineth BC, El-Armouche A, Grimm M, Nose M, Eschenhagen T, Zieseniss A, Katschinksi D, Hamdani N, Linke WA, Yin X, Mayr M, Zimmermann WH (2011) Terminal Differentiation, Advanced Organotypic Maturation, and Modeling of Hypertrophic Growth in Engineered Heart Tissue. Circ Res. doi:10.1161/CIRCRESAHA.111.251843

7. Zimmermann WH, Schneiderbanger K, Schubert P, Didie M, Munzel F, Heubach JF, Kostin S, Neuhuber WL, Eschenhagen T (2002) Tissue engineering of a differentiated cardiac muscle construct. Circ Res 90(2):223–230

8. Radisic M, Park H, Shing H, Consi T, Schoen FJ, Langer R, Freed LE, Vunjak-Novakovic G (2004) Functional assembly of engineered myocardium by electrical stimulation of cardiac myocytes cultured on scaffolds. Proc Natl Acad Sci U S A 101(52):18129–18134. doi:10.1073/pnas.0407817101

9. Thavandiran N, Nunes SS, Xiao Y, Radisic M (2013) Topological and electrical control of cardiac differentiation and assembly. Stem Cell Res Ther 4(1):14. doi:10.1186/scrt162

10. Yang L, Soonpaa MH, Adler ED, Roepke TK, Kattman SJ, Kennedy M, Henckaerts E, Bonham K, Abbott GW, Linden RM, Field LJ, Keller GM (2008) Human cardiovascular progenitor cells develop from a KDR + embryonic-stem-cell-derived population. Nature 453(7194): 524–528. doi:10.1038/nature06894

11. Zhang J, Klos M, Wilson GF, Herman AM, Lian X, Raval KK, Barron MR, Hou L, Soerens AG, Yu J, Palecek SP, Lyons GE, Thomson JA, Herron TJ, Jalife J, Kamp TJ (2012) Extracellular matrix promotes highly efficient cardiac differentiation of human pluripotent stem cells: the matrix sandwich method. Circ Res 111(9):1125–1136. doi:10.1161/CIRCRESAHA.112.273144

12. Lian X, Hsiao C, Wilson G, Zhu K, Hazeltine LB, Azarin SM, Raval KK, Zhang J, Kamp TJ, Palecek SP (2012) Robust cardiomyocyte differentiation from human pluripotent stem cells via temporal modulation of canonical Wnt signaling. Proc Natl Acad Sci U S A 109(27):E1848–E1857. doi:10.1073/pnas.1200250109

13. Snir M, Kehat I, Gepstein A, Coleman R, Itskovitz-Eldor J, Livne E, Gepstein L (2003) Assessment of the ultrastructural and proliferative properties of human embryonic stem cell-derived cardiomyocytes. Am J Physiol Heart Circ Physiol 285(6):H2355–H2363. doi:10.1152/ajpheart.00020.2003

14. Dolnikov K, Shilkrut M, Zeevi-Levin N, Gerecht-Nir S, Amit M, Danon A, Itskovitz-Eldor J, Binah O (2006) Functional properties of human embryonic stem cell-derived cardiomyocytes: intracellular Ca2+ handling and the role of sarcoplasmic reticulum in the contraction. Stem Cells 24(2):236–245. doi:10.1634/stemcells.2005-0036

15. Tulloch NL, Muskheli V, Razumova MV, Korte FS, Regnier M, Hauch KD, Pabon L, Reinecke H, Murry CE (2011) Growth of engineered human myocardium with mechanical loading and vascular coculture. Circ Res 109(1):47–59. doi:10.1161/CIRCRESAHA.110.237206

16. Nunes SS, Miklas JW, Liu J, Aschar-Sobbi R, Xiao Y, Zhang B, Jiang J, Masse S, Gagliardi M, Hsieh A, Thavandiran N, Laflamme MA, Nanthakumar K, Gross GJ, Backx PH, Keller G, Radisic M (2013) Biowire: a platform for maturation of human pluripotent stem cell-derived cardiomyocytes. Nat Methods 10(8):781–787. doi:10.1038/nmeth.2524

Chapter 15

Collagen-Based Engineered Heart Muscle

Malte Tiburcy, Tim Meyer, Poh Loong Soong, and Wolfram-Hubertus Zimmermann

Abstract

Cardiac muscle engineering has evolved over nearly 20 years from a scientific oddity to a mainstream technology with a wide range of applications. Of the many published methods it appears that hydrogels constitute the preferred scaffolds for myocardial tissue engineering and support of organotypic development. Here we describe a simple and highly robust protocol for the generation of engineered heart muscle using a collagen-based hydrogel method.

Key words Tissue engineering, Heart, Stem cells

1 Introduction

Cardiac tissue engineering technologies have advanced considerably in recent years with great potential for in vitro studies of cardiomyocyte maturation and hypertrophy [1], drug screening [2], and regenerative therapies [3]. Various methods of cardiac tissue assembly have been developed [4], including (1) seeding heart cells onto natural or synthetic preformed matrices [5–7], (2) scaffold-free approaches [8, 9], and (3) the hydrogel method [10–13]. All of these tissue engineering methods generate heart muscle with features of native myocardium.

Until recently, the most comprehensive evidence for organotypic function has been derived from cardiac tissue generated by hydrogel-based methods [14]. The most prominent hydrogel-based tissue engineering technologies use fibrin or collagen [2, 10, 12, 15]. Fibrin can be isolated from blood, and its polymerization is facilitated by the addition of thrombin to mixtures of heart cells and fibrinogen. Collagen type I is the most abundant extracellular matrix (ECM) protein in the body; it is widely used in plastic and reconstructive surgery and can be easily prepared from connective tissue. Moreover, it is an integral part for the structural and functional integrity of the heart, providing tensile strength, orientation, and

Milica Radisic and Lauren D. Black III (eds.), *Cardiac Tissue Engineering: Methods and Protocols*, Methods in Molecular Biology, vol. 1181, DOI 10.1007/978-1-4939-1047-2_15, © Springer Science+Business Media New York 2014

mechanical support and aiding nutrient supply [16]. It is produced by fibroblasts in the heart and connects with cardiomyocytes via integrins [17]. Importantly, the formation of an interstitial collagen network is closely linked with heart development [18] and a failure of collagen–cardiomyocyte interaction leads to cardiac malformations [19]. These data collectively highlight the cardio-instructive role of collagen type I. Accordingly, we prefer the collagen-hydrogel method for cardiac muscle engineering. A further rationale for this approach stems from experimental evidence demonstrating near-physiological development of the collagen-engulfed heart cells into a complex three-dimensional functional syncytium with robust tissue biomechanics [1, 11].

The collagen-hydrogel method entraps cell mixtures reminiscent of the cell populations found in the native myocardium in a defined space and facilitates their self-assembly into cardiac organoids. In our original tissue engineering studies we defined Matrigel™ as an essential component [12], but more recently identified conditions that do not require this ill-defined ECM supplement (*see* **Note 5**). We refer to the collagen-hydrogel technology as engineered heart muscle (EHM) as these constructs contain the most abundant cellular and extracellular components of native heart muscle and to emphasize their functionality, which largely resembles that of native myocardium. Achieving proper contractile function is of fundamental importance to myocardial tissue engineering. This includes a positive force-preload behavior (Frank–Starling mechanism); a negative or a positive force-frequency behavior (Bowditch phenomenon) in immature and mature myocardium, respectively; and cardiotypic inotropic, chronotropic, as well as lusitropic responses to pharmacological agents, for example, mediators of the autonomic nervous system (e.g., catecholamines and acetylcholine).

Here, we describe a simple and versatile protocol for collagen I-based EHM generation and outline the principles of quality control by isometric force measurements.

2   Materials

2.1   Cell Culture Components

1. Iscove's MEM.
2. Fetal bovine serum (FBS).
3. 10,000 U/ml Penicillin–10,000 µg/ml streptomycin (P/S).
4. 200 mM L-Glutamine.
5. 100× MEM nonessential amino acids.
6. 2-Mercaptoethanol.
7. L-Ascorbic acid 2-phosphate sesquimagnesium salt hydrate.

8. EHM medium: Add 120 ml FBS, 6.2 ml P/S, 6.2 ml 200 mM L-glutamine, 6.2 ml 100× MEM nonessential amino acids, 4.34 µl mercaptoethanol, and 54 mg L-ascorbic acid 2-phosphate sesquimagnesium salt hydrate to 500 ml Iscove's medium (*see* **Note 1**).

2.2 EHM Generation Components

All protocol steps must be performed under sterile conditions. Casting molds and stretchers (see below) can be autoclaved and reused.

1. Collagen type I (typically 3–7 mg/ml): Make sure to use acid-solubilized collagen type I. We typically use rat or bovine collagen type I, but anticipate that collagen from many other species can be applied. Note that commercial collagen solutions exhibit a considerable batch-to-batch variation with respect to their gelling properties. We typically prefer to produce collagen type 1 from rat tails in-house according to a standard acid-solubilization protocol [20].

2. 10× Concentrated medium (10× DMEM): Solubilize 1.38 g DMEM powder in 10 ml deionized water. Filter through a 0.22 µm syringe filter.

3. 2× Concentrated medium (2× DMEM): Mix 2 ml 10× DMEM, 2 ml FBS, 0.2 ml of 200 mM L-glutamine, 0.2 ml of P/S, and 5.6 ml deionized water. Filter through a 0.22 µm syringe filter.

4. NaOH 0.1 N.

5. 50 ml Centrifuge tubes, pre-chilled to 4 °C.

6. 2 and 5 ml serological pipettes, pre-chilled to 4 °C.

7. Casting molds (Fig. 1).

8. Mechanical stretchers (Fig. 2).

3 Methods

3.1 Preparation of Cells for EHM Generation

1. A single-cell suspension of heart cells needs to be prepared. This can be obtained either from neonatal rat or mouse hearts [12] or cardiomyogenically differentiated pluripotent stem cells [21]. Resuspend cells in EHM medium and keep on ice.

2. Determine the cell number to calculate the volume of master mix (Table 1) needed. Typically, we prepare EHM with an individual volume of 450 or 900 µl in custom-made casting molds (Fig. 1). Down- and up-scaling is possible, but requires careful adaptation of mold dimensions.

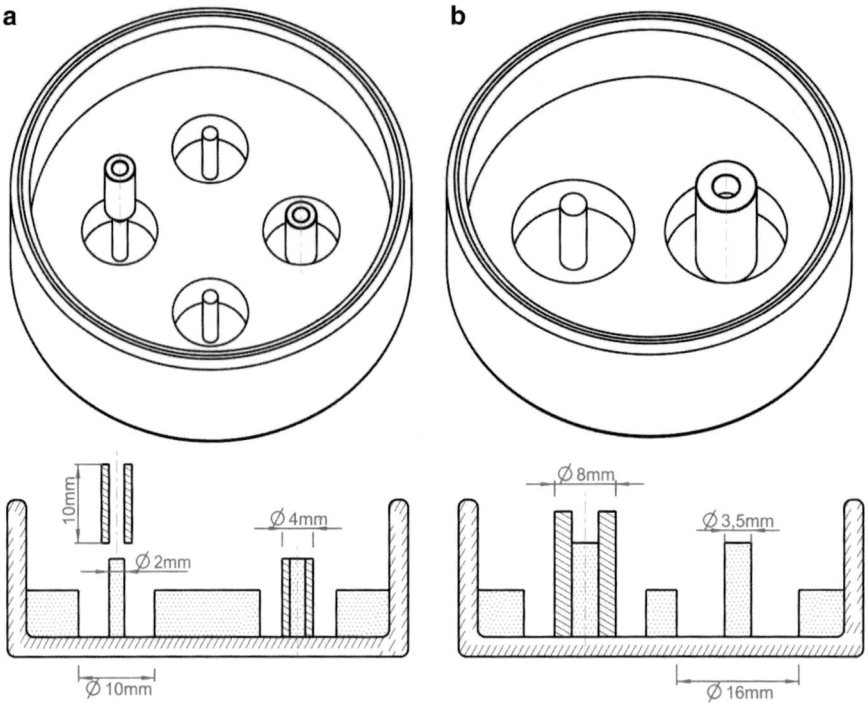

Fig. 1 Technical drawing of circular molds for EHM casting. (**a**) Example of circular molds for generation of 4 EHMs with 450 μl volume. (**b**) Example of circular molds for generation of 2 EHMs with 900 μl volume. These molds are made from polydimethylsiloxane (PDMS, also known as silicone) and removable PDMS or polytetra-fluoroethylene (Teflon®) tubes in glass dishes (diameter 60 mm)

3.2 EHM Generation

All materials should be pre-chilled to 4 °C. All the following steps should be performed on ice.

1. The amount of master mix needed depends on the required amount of EHM. We typically prepare master mixes for 4–8 EHMs with a collagen content of 0.8–1.2 mg/ml and a cell density of $2.5–3 \times 10^6$ cells per ml.

2. With a serological pipette, transfer the required volume of collagen solution into a pre-chilled 50 ml centrifuge tube (Table 1). The master mix volumes are adjusted to account for volume lost (~15 %) during pipetting of the viscous master mix.

3. Add the same volume of 2× DMEM, and mix well by shaking the tube. Do not use the pipette to mix as there will be too much hydrogel lost in the pipette.

4. Neutralize the pH by adding 0.1 N NaOH. NaOH should be added dropwise while shaking the tube until the phenol red indicator goes from yellow to red/pink. Volume may have to be adjusted according to the volume of the collagen solution.

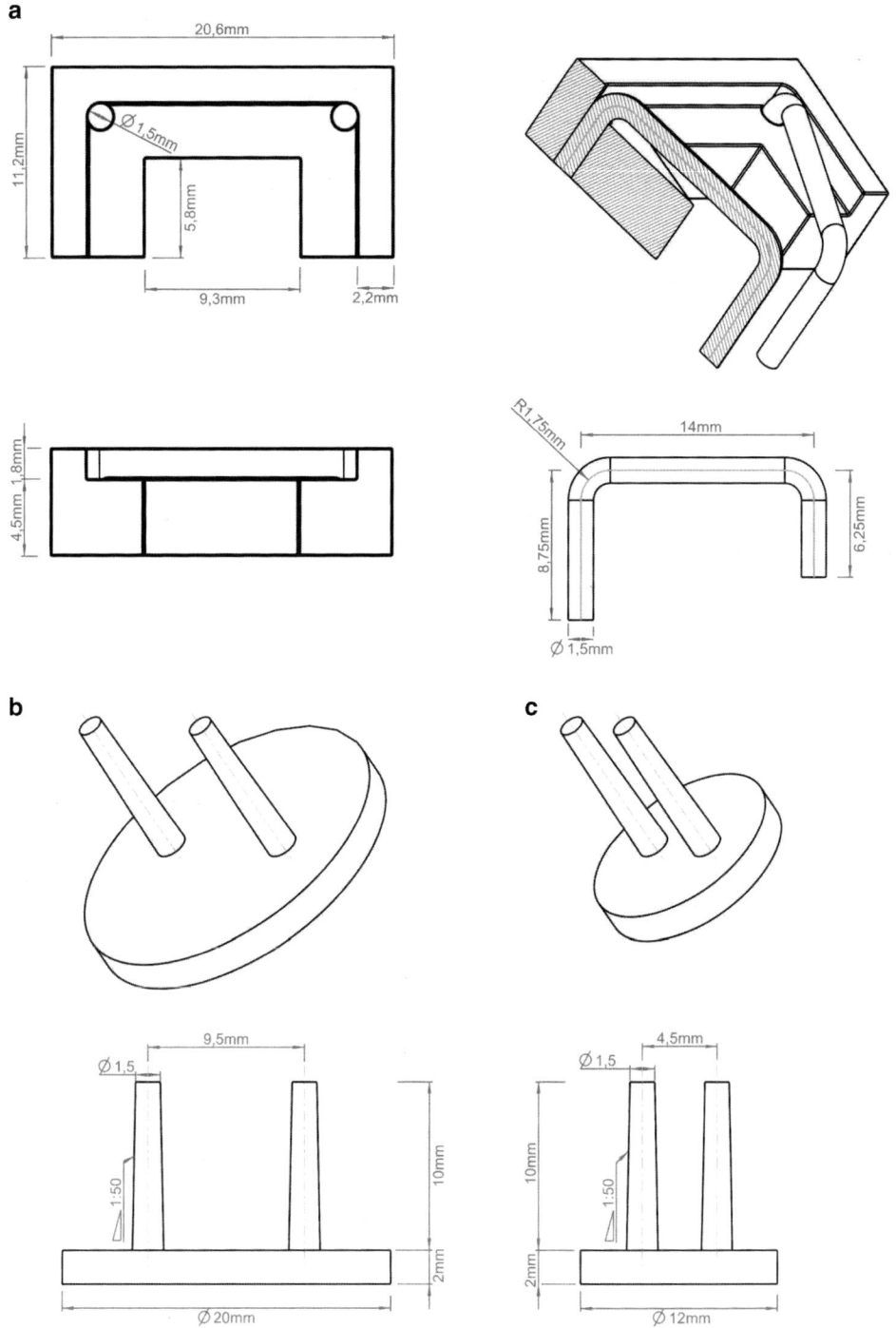

Fig. 2 Technical drawing of mechanical stretchers. (**a**) Static stretcher made from a Teflon® base and stainless steel holders. (**b**) Dynamic mechanical stretcher made from silicone for 900 μl EHM. (**c**) Dynamic mechanical stretcher made from silicone for 450 μl EHM

Table 1
Composition of EHM master mix

Master mix components	Volume (μl)
Collagen (7.25 mg/ml)[a]	515
2× DMEM	515
0.1 N NaOH	100
Total cell number (11.6×10^6)[b,c]	3,070
Total volume of master mix[d]	4,200
Number of EHM (900 μl EHM)	4
Number of EHM (450 μl EHM)	8

[a]We typically use 0.8–1.2 mg/ml final collagen concentration
[b]Resuspend cells in EHM medium
[c]EHMs comprise either 2.5 or 1.25×10^6 cells for 900 μl and 450 μl EHM, respectively
[d]Extra volume (~15 %) facilitates pipetting exact volumes of the viscous EHM mixture

5. Resuspend the cells with a serological pipette and add to the master mix. Mix well to ensure homogenous distribution of cells.

6. Use the same serological pipette to distribute hydrogel into appropriate casting molds (Fig. 3a).

7. Allow the hydrogel to solidify for 1 h at 37 °C in a humidified incubator with 5 % CO_2. The solidified gel will turn opaque (Fig. 3b, *see* **Note 2**).

8. After gelation, carefully add pre-warmed EHM culture medium to the casting mold so that the hydrogel is completely immersed, but not disrupted.

9. Change medium the next day and then every other day.

10. During tissue formation the hydrogel will "condense" around the central pole (Figs. 1 and 3a–d, *see* **Note 3**). When fully condensed the tissue can be transferred to a stretch device for mechanical loading. Mechanical load is essential for the formation of functional EHM. The simplest way of mechanical loading is by putting the EHM onto static or dynamic mechanical stretchers (Fig. 3e, f, *see* **Note 4**).

11. Typically, analysis of EHM function is performed following a minimum of 10 days in culture.

3.3 Quality Control by Isometric Force Measurements

1. EHM should be contracting visibly and synchronously (i.e., the whole tissue beating in unison) by day 10 at 37 °C. Note that in contrast to rodents, human EHMs are very temperature sensitive and stop beating quickly if the temperature drops below 37 °C.

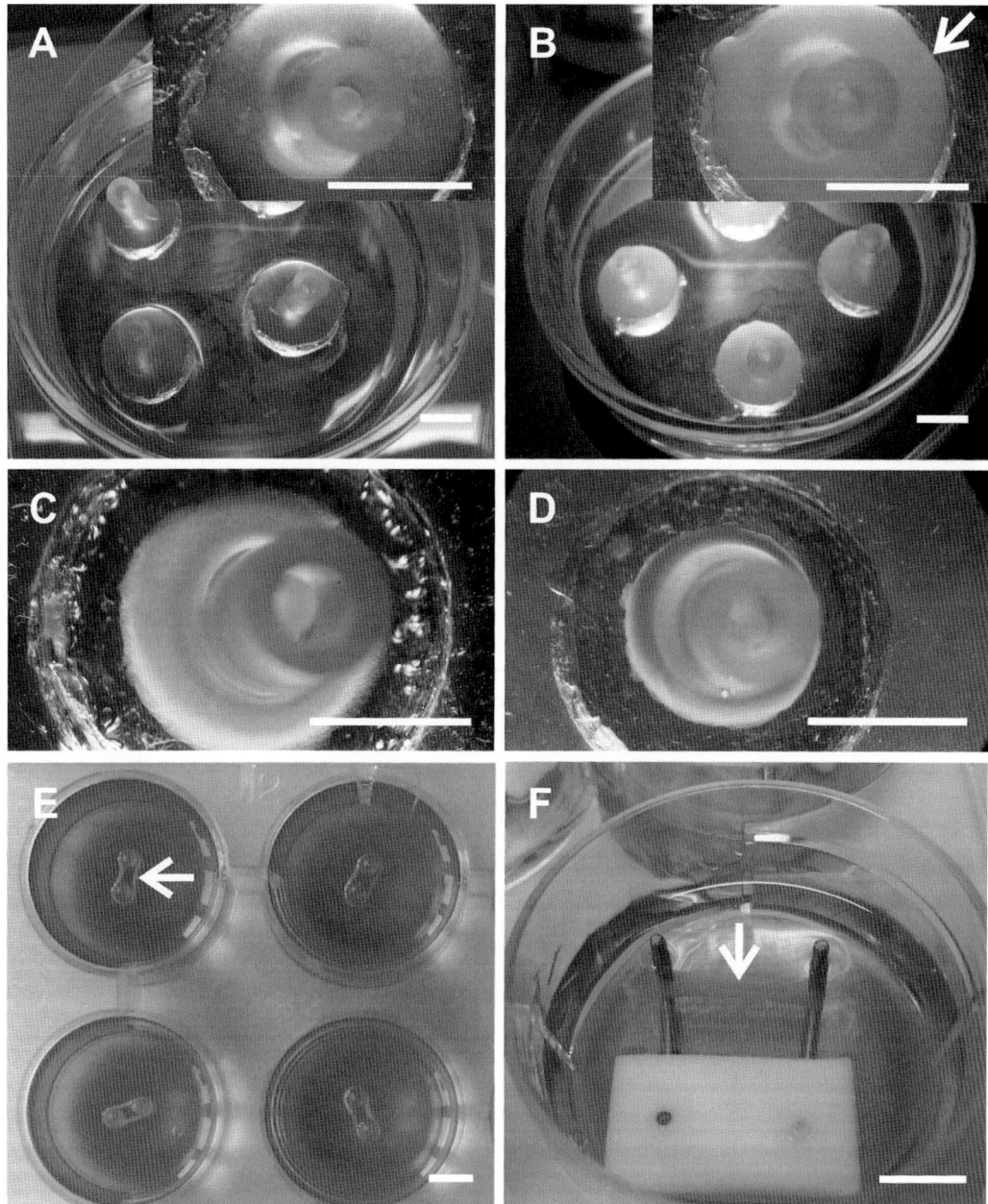

Fig. 3 EHM formation. (**a**) Mold with circular EHM (translucent) directly after casting. *Inset*: Magnified view of single mold with cell-containing hydrogel. (**b**) Mold with EHM (*opaque*) 1 h after casting and incubation at 37 °C before addition of culture medium. *Inset*: Magnified view of single mold. Note the space (*arrow*) appearing between hydrogel and mold indicating beginning condensation. (**c**) Advanced condensation on culture day 1. (**d**) Full condensation on culture day 3 (*see* **Note 3**). (**e**) 4 EHMs on dynamic mechanical stretchers on culture day 10. Note that these are 450 μl volume EHMs (*arrow*: single EHM suspended between silicone poles). (**f**) EHM on static stretcher on culture day 10. Note that this is a 900 μl volume EHM (*arrow*: EHM suspended between stainless steel poles). Bars: 5 mm

2. Measurement of contractile force under isometric conditions is, in our view, the "gold standard" quality test for formation of a functional syncytium in EHM as it depends on many critical factors, including cardiomyocyte maturation, cardiomyocyte coupling, cardiomyocyte/non-myocyte ratio, ECM composition, and tissue viscoelastic properties.

3. Isometric force is typically measured under field stimulation at 0.5–3 Hz in modified Tyrode's solution as described before [12]. Isometric force should be >0.2 mN, and active: passive force ratio should be >1 [22]. We prefer to report absolute force values, because twitch tension would require relating the force amplitude to the specific muscle cross section. Each individual EHM would then have to be analyzed histologically for individual muscle cross section. Under standard conditions, muscle cross section of a regular circular EHM (diameter: 800 μm) with a total cross-sectional area of 1 mm^2 [$2\times(\pi\times0.4$ mm$^2)$] is ~10 %. Increasing and decreasing the total EHM volume do not typically change absolute force, but total cross-sectional area. However, specific muscle cross-sectional area remains typically constant resulting in similar absolute force despite differences in total volume (refer to [23] for a detailed discussion).

4   Notes

1. The culture medium may be adapted depending on the species and source of the starting cardiac cell population. We use the medium described here mainly for human pluripotent stem cell-derived EHM. For EHM comprising primary rat or mouse cardiomyocytes, a different medium containing DMEM containing 1 g glucose/L without L-glutamine, 10 % horse serum, 2 % chick embryo extract (Egg Tech, Salisbury, UK), and P/S may be used. Batch testing of serum is critical for the reproducible generation of EHM. All media are passed through 0.2 μm sterile filters before use.

2. Collagen gelation is an essential prerequisite for proper cell entrapment in EHM. If the collagen is not nicely gelled and opaque after 1 h at 37 °C (Fig. 3b), it is either degraded or too diluted. We typically use 0.8–1.2 mg/ml final concentration of collagen in our master mix. Be aware that acid-solubilized collagen type 1 is slowly degraded at 4 °C and should not be kept for more than 6 months.

3. Condensation of the collagen-hydrogel during tissue formation is essential to obtain functional EHM (Fig. 3a–d). Condensation depends crucially on the cell content and composition. If condensation does not occur, the following should be considered:

- Are the entrapped cells alive? Lack of cell spreading indicates cell death.

- Are the entrapped cells capable of condensing the collagen-hydrogel? Cell mixtures with low fibroblast content fail to condense properly. Collagen-hydrogels containing only fibroblasts condense maximally within 6 h. Increase fibroblast content if condensation of EHM is not completed within 5 days.

4. The transfer to stretchers (Fig. 2) has been described before [21]. We prefer dynamic mechanical stretchers (Figs. 2b and 3e) over static stretchers (Figs. 2a and 3f) to facilitate auxotonic contractions [3].

5. Here, we describe a protocol that does not require supplementation with Matrigel®. Also, in rodent EHM, Matrigel® is dispensable if cell survival is enhanced by supplementation of the culture medium with 10 µg/ml insulin and 0.1 nmol/L triiodothyronine for at least 24 h after casting [24].

Acknowledgements

M.T. is supported by the University Medical Center Göttingen (Startförderung I). W.H.Z. is supported by the DZHK (German Center for Cardiovascular Research), German Federal Ministry for Science and Education (BMBF FKZ 01GN0827, FKZ 01GN0957), the German Research Foundation (DFG ZI 708/7-1, 8-1, 10-1, SFB 1002), European Union FP7 CARE-MI, CIRM/BMBF FKZ 13GW0007A, and NIH U01 HL099997.

References

1. Tiburcy M, Didie M, Boy O, Christalla P, Doker S, Naito H, Karikkineth BC, El-Armouche A, Grimm M, Nose M, Eschenhagen T, Zieseniss A, Katschinksi DM, Hamdani N, Linke WA, Yin X, Mayr M, Zimmermann WH (2011) Terminal differentiation, advanced organotypic maturation, and modeling of hypertrophic growth in engineered heart tissue. Circ Res 109(10):1105–1114, PMCID: PMID: 21921264

2. Hansen A, Eder A, Bonstrup M, Flato M, Mewe M, Schaaf S, Aksehirlioglu B, Schwoerer AP, Uebeler J, Eschenhagen T (2010) Development of a drug screening platform based on engineered heart tissue. Circ Res 107(1):35–44, PMCID: PMID: 20448218

3. Zimmermann WH, Melnychenko I, Wasmeier G, Didie M, Naito H, Nixdorff U, Hess A, Budinsky L, Brune K, Michaelis B, Dhein S, Schwoerer A, Ehmke H, Eschenhagen T (2006) Engineered heart tissue grafts improve systolic and diastolic function in infarcted rat hearts. Nat Med 12(4):452–458, PMCID: PMID: 16582915

4. Ye L, Zimmermann WH, Garry DJ, Zhang J (2013) Patching the heart: cardiac repair from within and outside. Circ Res 113(7):922–932, PMCID: PMID: 24030022

5. Radisic M, Park H, Shing H, Consi T, Schoen FJ, Langer R, Freed LE, Vunjak-Novakovic G (2004) Functional assembly of engineered myocardium by electrical stimulation of cardiac myocytes cultured on scaffolds. Proc Natl Acad Sci U S A 101(52):18129–18134, PMCID: 539727, PMID: 15604141

6. Papadaki M, Bursac N, Langer R, Merok J, Vunjak-Novakovic G, Freed LE (2001) Tissue engineering of functional cardiac muscle: molecular, structural, and electrophysiological studies. Am J Physiol Heart Circ

Physiol 280(1):H168–H178, PMCID: PMID: 11123231

7. Ott HC, Matthiesen TS, Goh SK, Black LD, Kren SM, Netoff TI, Taylor DA (2008) Perfusion-decellularized matrix: using nature's platform to engineer a bioartificial heart. Nat Med 14(2):213–221, PMCID: PMID: 18193059

8. Shimizu T, Yamato M, Isoi Y, Akutsu T, Setomaru T, Abe K, Kikuchi A, Umezu M, Okano T (2002) Fabrication of pulsatile cardiac tissue grafts using a novel 3-dimensional cell sheet manipulation technique and temperature-responsive cell culture surfaces. Circ Res 90(3):e40, PMCID: PMID: 11861428

9. Kelm JM, Ehler E, Nielsen LK, Schlatter S, Perriard JC, Fussenegger M (2004) Design of artificial myocardial microtissues. Tissue Eng 10(1–2):201–214, PMCID: PMID: 15009946

10. Eschenhagen T, Fink C, Remmers U, Scholz H, Wattchow J, Weil J, Zimmermann W, Dohmen HH, Schafer H, Bishopric N, Wakatsuki T, Elson EL (1997) Three-dimensional reconstitution of embryonic cardiomyocytes in a collagen matrix: a new heart muscle model system. FASEB J 11(8): 683–694, PMCID: PMID: 9240969

11. Zimmermann WH, Schneiderbanger K, Schubert P, Didie M, Munzel F, Heubach JF, Kostin S, Neuhuber WL, Eschenhagen T (2002) Tissue engineering of a differentiated cardiac muscle construct. Circ Res 90(2):223–230, PMCID: PMID: 11834716

12. Zimmermann WH, Fink C, Kralisch D, Remmers U, Weil J, Eschenhagen T (2000) Three-dimensional engineered heart tissue from neonatal rat cardiac myocytes. Biotechnol Bioeng 68(1):106–114, PMCID: PMID: 10699878

13. Bian W, Liau B, Badie N, Bursac N (2009) Mesoscopic hydrogel molding to control the 3D geometry of bioartificial muscle tissues. Nat Protoc 4(10):1522–1534, PMCID: 2924624, PMID: 19798085

14. Eschenhagen T, Eder A, Vollert I, Hansen A (2012) Physiological aspects of cardiac tissue engineering. Am J Physiol Heart Circ Physiol 303(2):H133–H143, PMCID: PMID: 22582087

15. Baar K, Birla R, Boluyt MO, Borschel GH, Arruda EM, Dennis RG (2005) Self-organization of rat cardiac cells into contractile 3-D cardiac tissue. FASEB J 19(2):275–277, PMCID: PMID: 15574489

16. Weber KT (1989) Cardiac interstitium in health and disease: the fibrillar collagen network. J Am Coll Cardiol 13(7):1637–1652, PMCID: PMID: 2656824

17. Terracio L, Rubin K, Gullberg D, Balog E, Carver W, Jyring R, Borg TK (1991) Expression of collagen binding integrins during cardiac development and hypertrophy. Circ Res 68(3): 734–744, PMCID: PMID: 1835909

18. Carver W, Terracio L, Borg TK (1993) Expression and accumulation of interstitial collagen in the neonatal rat heart. Anat Rec 236(3):511–520, PMCID: PMID: 8363055

19. Ieda M, Tsuchihashi T, Ivey KN, Ross RS, Hong TT, Shaw RM, Srivastava D (2009) Cardiac fibroblasts regulate myocardial proliferation through beta1 integrin signaling. Dev Cell 16(2):233–244, PMCID: 2664087, PMID: 19217425

20. Eschenhagen T, Didie M, Munzel F, Schubert P, Schneiderbanger K, Zimmermann WH (2002) 3D engineered heart tissue for replacement therapy. Basic Res Cardiol 97(Suppl 1):I146–I152, PMCID: PMID: 12479248

21. Soong PL, Tiburcy M, Zimmermann WH (2012) Cardiac differentiation of human embryonic stem cells and their assembly into engineered heart muscle. Curr Protoc Cell Biol, Chapter 23: p. Unit23 8, PMCID: PMID: 23129117

22. Zimmermann WH, Eschenhagen T (2003) Cardiac tissue engineering for replacement therapy. Heart Fail Rev 8(3):259–269, PMCID: PMID: 12878835

23. Tiburcy M, Zimmermann WH (2014) Modeling myocardial growth and hypertrophy in engineered heart muscle. Trends Cardiovasc Med 24(1):7–13, PMCID: PMID: 23953977

24. Naito H, Melnychenko I, Didie M, Schneiderbanger K, Schubert P, Rosenkranz S, Eschenhagen T, Zimmermann WH (2006) Optimizing engineered heart tissue for therapeutic applications as surrogate heart muscle. Circulation 114(1 Suppl):I72–I78, PMCID: PMID: 16820649

Creation of a Bioreactor for the Application of Variable Amplitude Mechanical Stimulation of Fibrin Gel-Based Engineered Cardiac Tissue

Kathy Y. Morgan and Lauren D. Black III

Abstract

This chapter details the creation of three-dimensional fibrin hydrogels as an engineered myocardial tissue and introduces a mechanical stretch bioreactor system that allows for the cycle-to-cycle variable amplitude mechanical stretch of the constructs as a method of conditioning the constructs to be more similar to native tissue. Though mechanical stimulation has been established as a standard method of improving construct development, most studies have been performed under constant frequency and constant amplitude, even though variability is a critical aspect of healthy cardiac physiology. The introduction of variability in other organ systems has demonstrated beneficial effects to cell function in vitro. We hypothesize that the introduction of variability in engineered cardiac tissue could have a similar effect.

Key words Cardiac tissue engineering, Mechanical stimulation, Variable amplitude stimulation, Bioreactor

1 Introduction

Engineered cardiac tissues have been developed as an innovative method of repairing or replacing damaged myocardium after an infarction [1]. Mechanical distension stretch mimics the mechanical stretch observed by the cells in the heart as the chambers of the heart fill with blood and is often used in the creation process to engineer tissues with similar functional properties to native tissue. Zimmermann et al. initially demonstrated that mechanical stimulation of tissue-engineered constructs improved the organization and contraction force [2]. Since this study, many groups have used stretch as a standard method of conditioning engineered myocardium in vitro [3–5]. Furthermore, these studies have shown that cyclic stretch also activates FAK and RhoA pathways, which aids in ECM formation [6] and initiates myotube formation [7].

Milica Radisic and Lauren D. Black III (eds.), *Cardiac Tissue Engineering: Methods and Protocols*, Methods in Molecular Biology, vol. 1181, DOI 10.1007/978-1-4939-1047-2_16, © Springer Science+Business Media New York 2014

However, mechanical stretch in most of the previous studies has been performed under constant frequency and constant amplitude mechanical stretch, even though it is known that variability is a critical aspect of healthy cardiac physiology. Heart rate variability is an indication of the physiological factors that modulate normal heart rhythm [8] and R–R intervals in a healthy patient form a pseudo-Gaussian distribution over a 24-h period [9]. Furthermore, the loss of beat-to-beat variability in R–R intervals is an indicator of progressing heart failure [10].

Variable stretch patterns similar to the ones observed in normal physiology have been shown to be beneficial to cell function in vitro in other organ systems. For example, the introduction of variable stretch to mimic the variations in tidal breathing in the lung enhances surfactant secretion in monolayers of alveolar type II epithelial cells [11], differing levels of variability cause a modified mRNA expression of key cell–matrix interaction molecules in lung fibroblasts [12], and the accretion of stretch amplitudes has been showed to improve mechanical properties and collagen deposition in porcine valve interstitial cells in fibrin gel constructs [13]. The response mechanism is postulated to be due to the prevention of cellular adaptation to the mechanical stimuli. However, variable stretch has not yet been investigated in engineered cardiac constructs.

In this chapter, we detail the process of creating three-dimensional (3D) fibrin-engineered myocardial tissue as well as introduce a bioreactor system that allows for us to mechanically stimulate the constructs with variable amplitude stretch on a beat-to-beat basis. This bioreactor system is based on a previously published construct [14, 15] and bioreactor system [16] (*see* Fig. 1). We can utilize the construct's ringlike structure by creating a bioreactor system that allows us to uniformly stretch our constructs in a circumferential direction. Furthermore, one advantage of our system is the ability to multiplex and increase sample sizes by adding bioreactors in parallel to the same air regulator line.

2   Materials

2.1   Bioreactor

1. A bioreactor system: Use a wide-mouth 2 oz glass jar w/autoclavable lid as the basis for the bioreactor system. A spacer inside of the bioreactor can be added by taking a ½″ OD, ¼″ ID Teflon rod, cut to be ½″ in length. To allow for sterile airflow, a leur-lock male adaptor is attached to the lid of the glass jar to allow for a 0.22 μm 13 mm sterile syringe filter to be attached (*see* Fig. 1a).

2. Bioreactor tubing: Create by taking latex tubing (Kent Elastomer, 1/4″ ID, 1/32″ wall thickness) and attaching a ¼″ Teflon rod as a stopper to one end with a nontoxic, autoclavable epoxy.

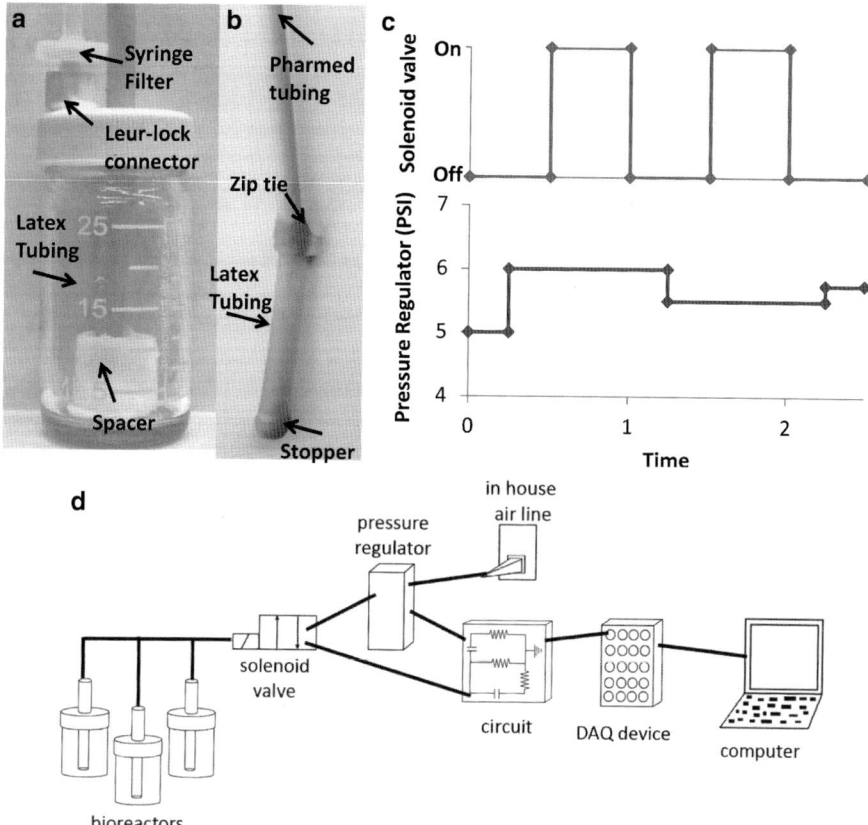

Fig. 1 (**a**) Representative custom distention bioreactor with spacer, latex tubing, and leur-lock connector for a syringe filter. (**b**) Representative latex tubing for construct distention with stopper and Pharmed tubing. (**c**) Representative outputs of the solenoid valve and the pressure regulator over time. (**d**) Schematic of bioreactor system, which shows the bioreactors connected to the solenoid valve, which is connected to a pressure regulator which regulates the in-house air line. The regulator and solenoid valves are controlled by a LabVIEW program running on a computer and relayed through the DAQ device

Tygon Pharmed ¼″ tubing (#57318) is attached to the other side of the latex tubing with a zip tie. Add epoxy between the zip tie and the Pharmed tubing to prevent leakage (*see* Fig. 1b).

2.2 Variable Pressure System

1. A variable pressure system is created through two separate pressure regulation systems that work in parallel. One pressure system controls the amplitude of pressure, while the other controls the periodic stretch (*see* Fig. 1c). A NI-DAQ (National Instruments USB-6221) with a customized LabVIEW code is used to control both systems concurrently.

2. The pressure amplitude is controlled through a current-to-pressure regulator. A voltage-to-current converter is used to run the solenoid valve (*see* Fig. 1d for schematic).

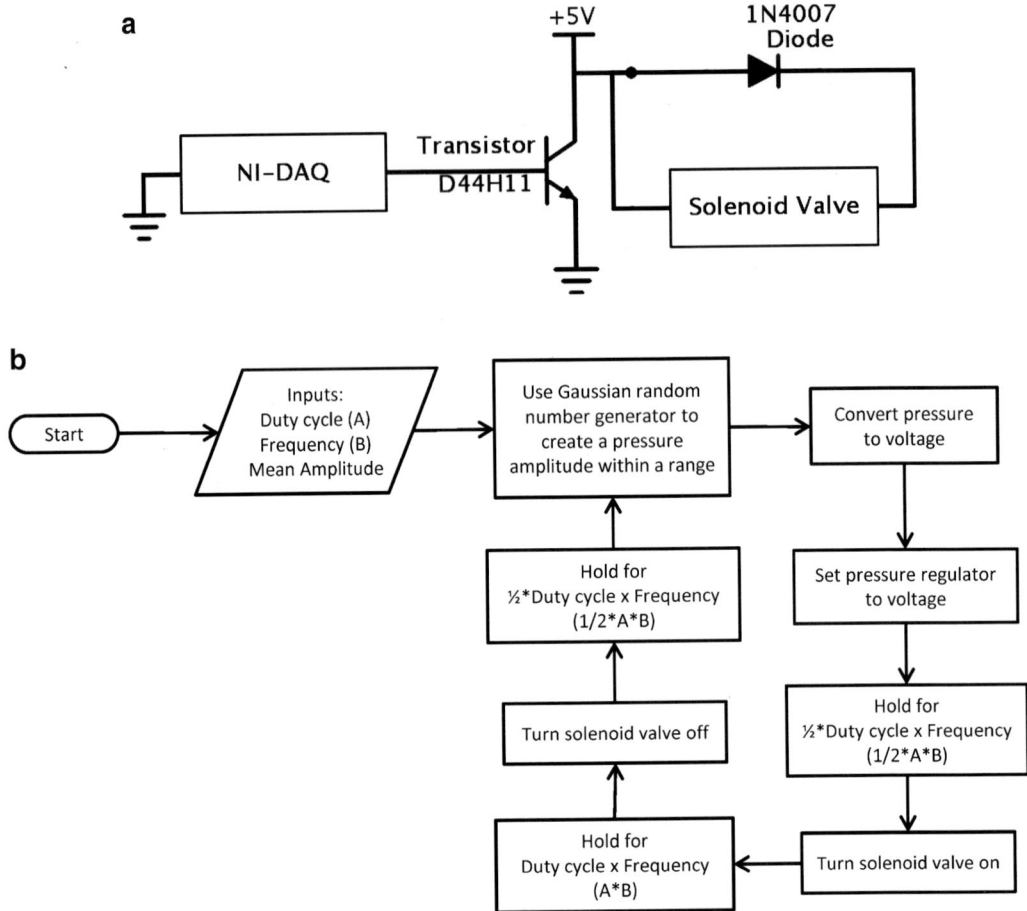

Fig. 2 (a) Electrical circuit used to power the solenoid valve. **(b)** Flow chart describing the LabVIEW program

3. The periodic stretch is controlled through a solenoid valve, and a custom electrical circuit (*see* Fig. 2a) is used to turn the solenoid valve on and off.

4. Customized LabVIEW program is used to control the pressure regulation systems (*see* Fig. 2b for schematic). Briefly the voltage output is set, and the solenoid valve is opened. The program then waits for a set amount of time before the solenoid valve is closed. The program then waits for another set amount of time before the voltage output is adjusted. Then the cycle is allowed to repeat.

2.3 Construct Mold

1. A construct mold is created by first creating a 3–4 cm ¼ in. diameter long Teflon rod. A Teflon tube with an ¼″ ID, 5/16″ OD, cut to be 1 cm long is placed around the Teflon rod. The ends are each capped with a Teflon ring with an ¼″ ID, ½″ OD, and a silicone O-ring, 1/3″ ID, ½″ OD (*see* **Note 1** and Fig. 3).

2. The construct mold is placed into a cut and beveled 6 cc syringe and pushed into place with a cut 3 cc syringe.

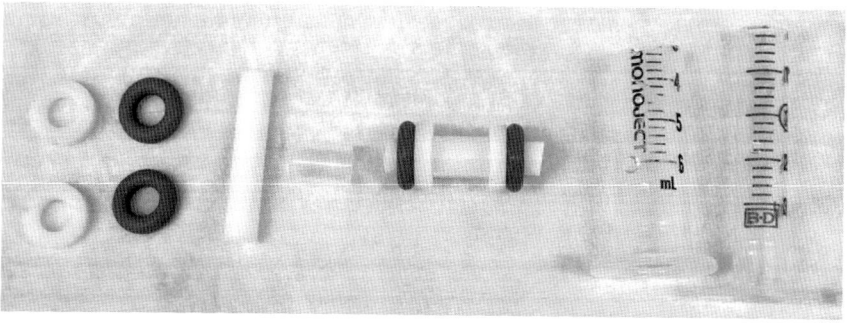

Fig. 3 Separate and combined mold parts for casting constructs. From *left* to *right*: Two Teflon washers with notches, two silicone O-rings, a Teflon rod, Teflon tubing, a completed mandrel, the outer casing, and plunger

2.4 Contraction Force Device

1. Create a post, and attach it to the end of an isometric transducer (for example, a Harvard Apparatus 724490). Attach the transducer to a stand (*see* Fig. 5a).

2. Create a water bath with spacing for two carbon rods and the construct to be stretched. Wrap silver wire around each carbon rod and epoxy into place. In between the two carbon rods, on one side, place a non-conductive rod to hold the construct in place (*see* **Note 2**).

3. Connect to a nerve and muscle electrical pulse stimulator.

2.5 Solutions

1. 5 % Pluronic F-127 in water: Dissolve 50 g of Pluronic F-127 into 700 mL of deionized water. Bring solution volume up to 1 L with additional deionized water. Sterile filter before use.

2. Myocyte construct media: 10 % Horse serum, 2 % FBS, 2 μM ε-aminocaproic acid, 2 μg/mL insulin, 50 μg/mL ascorbic acid (*see* **Note 3**) in DMEM.

3. Fibrinogen stock: 33 mg/mL Stock solution in 20 mM HEPES buffer in 0.9 % saline. Mix fibrinogen into the HEPES-buffered saline at 37 °C over several hours. After allowing the solution to settle overnight at 2–8 °C, warm the solution to 37 °C and filter the solution through a series of consecutive filters with prefilters until the solution is sterile. Aliquot solution and store at −80 °C.

4. Thrombin stock: 25 U/mL of Thrombin. Add 500 U of thrombin into 18 mL of 0.9 % saline and 2 mL of deionized water. Sterile filter, aliquot, and freeze at −80 °C.

5. 2 N Calcium chloride solution: Dissolve 11.01 g of calcium chloride in 80 mL of deionized water. Bring the solution volume up to 100 mL with additional deionized water.

6. F solution: 3.3 mg/mL Final fibrinogen concentration in 1 mL of fibrin gel. Add 112 μl of the fibrinogen stock to 558 μl of 20 mM HEPES buffer in 0.9 % saline solution. Create F solution immediately before use.

7. T solution: 0.425 U/mL Final thrombin concentration in 1 mL of fibrin gel. Add 17 µl of the thrombin stock and 1.3 µl of 2 N calcium chloride solution to 135 µl of DMEM. Create T solution immediately before use.

8. Cell solution: 6× Desired final concentration of cells in 1 mL of fibrin gel. Prepare the cell solution by spinning down the cells and resuspending in 170 µl of DMEM.

3 Methods

3.1 Creation of the Bioreactor

1. Drill a ¼ in. hole into the middle of the jar lid, and then drill a smaller 2/32 in. hole between the first hole and the edge of the jar.

2. Epoxy a leur-lock male adaptor into the smaller hole to allow for air circulation in the bioreactor.

3. Create a spacer by taking a Teflon tube and cutting it into ¾″ lengths. The spacer helps the tubing to stay in the middle of the jar as well as reduces the volume of media necessary to cover the tube.

4. Before culturing constructs in bioreactor, screw in a sterile 0.22 µm syringe filter into the leur-lock adaptor to allow for sterile airflow into the bioreactor.

3.2 Creation of the Bioreactor Tubing

1. Cut the latex tubing to be 2/3 to 3/4 of the height of the bioreactor jar.

2. Soak the latex tubing in 70 % isopropyl alcohol for a week, changing it once halfway through. This step allows for the impurities in the latex tube to leach out and preps the latex tubing for cell culture.

3. Cut 1 cm long stoppers for the latex tube out of a ¼ in. Teflon stock. Epoxy the stoppers into one end of each latex tube, sealing the stopper completely in epoxy, and allow to dry overnight.

4. Ziptie the other end of the latex tubing onto some Pharmed tubing. Then epoxy the edge between the latex tubing and the Pharmed tubing to prevent air leaks.

5. Push air into the latex tubing to check for leaks, and then autoclave before using (*see* **Note 4**).

3.3 Creation of the Variable Pressure Solenoid Valve System

1. Connect the variable pressure regulator into the NI-DAQ using a voltage to current adaptor to an analog output circuit.

2. Connect the solenoid valve to the NI-DAQ using a voltage amplifier to a digital output circuit.

3.4 Creation of the Custom LabVIEW Program

1. Create a program that controls both the variable pressure regulator and the solenoid valve.

2. Set the voltage output range and the rate of stimulation.

3. The solenoid valve opens when a voltage is applied to the solenoid valve. Thus, the program applies a voltage to the solenoid valve and then waits for a set amount of time (as previously set). Then the solenoid valve closes for a period of time. The cycle repeats until the program is stopped.

4. While the solenoid valve is closed and no pressure is being applied to the tubing, the pressure regulator can be modified to be set at different pressures, allowing for a change in the amplitude of stretch observed by the construct.

3.5 Preparation of the Construct Mold

1. Mandrel assembly: Put together one Teflon rod, one Teflon tube, two Teflon washers, and two O-rings. Autoclave before use.

2. Mold assembly: Cut off the leur-lock ends of both the 6 and 3 cc syringes. In the 6 cc syringe, bevel the edge to allow for easy insertion and removal of the autoclave before use.

3. Soak the mold parts in sterile Pluronic solution for 2–3 h.

4. After 2–3 h, sterile filter the Pluronic solution back into the bottle. The Pluronic solution can be reused up to three times.

5. Place the mold parts on top of sterile drapes. Push the mandrel inside the 6 cc syringe casing. Use the 3 cc syringe as a plunger to insure a tight fit. Take care to keep the mold parts sterile.

3.6 Casting of the Construct

1. Prepare the neonatal rat cardiac cells, keeping them on ice. Instructions for this step are not provided in this protocol (*see* **Note 5**).

2. Create F and T solution immediately before use.

3. Mix the cell solution and the T solution together.

4. Incorporate the cells and T solution into the F solution, taking care to mix gently and thoroughly to prevent bubbles. After the F and T solutions are mixed together, the fibrin solidifies in approximately 15 min.

5. Pick up the solution with a 18G needle, tapping to remove air pockets in the syringe.

6. Taking care to keep the needle upright, remove the 18G needle and replace with a 21G needle.

7. Inject the solution into the molds.

8. Allow the construct to solidify for 15 min in a 37 °C incubator.

9. Take the 3 cc syringe plunger and carefully push the construct out into a beaker of DMEM.

10. Take sterile forceps and a sterile pick and push the constructs out of the edges of the mandrel and move the construct into a new jar with myocardial construct media.

4   Moving the Construct into the Bioreactor

1. After 2 days of culture, one Teflon ring and silicone O-ring were removed and the construct was slid off of the mandrel into a shallow petri dish containing warm DMEM.

2. The sterile latex tube is placed through the hole in the lid of the jar.

3. With a pair of sterile forceps, the construct was stretched around the sterile latex tubing and placed in the middle of the tubing (*see* **Notes 6** and **7**).

4. The lid of the bioreactor is screwed shut, a sterile filter is placed in the leur-lock fitting on the top, and the edge is sealed with parafilm.

5. Connect the bioreactor to the solenoid/regulator system, and turn on the pressure.

4.1 Culturing Conditions

1. Place constructs in a 5 % CO_2 incubator.

2. Each construct requires 21 mL of myocyte construct media, taking care to completely cover the construct with media. Change the media every other day. When changing the media, take care not to disturb the construct, but push the construct towards the middle of the latex tube and keep the construct submerged.

3. Culture for 12 days in the bioreactor, with feedings every other day (*see* **Note 8**), before removing it for analysis. Note that the construct will compact over culture time (*see* Fig. 4). Visible contractions can be observed after 2 weeks.

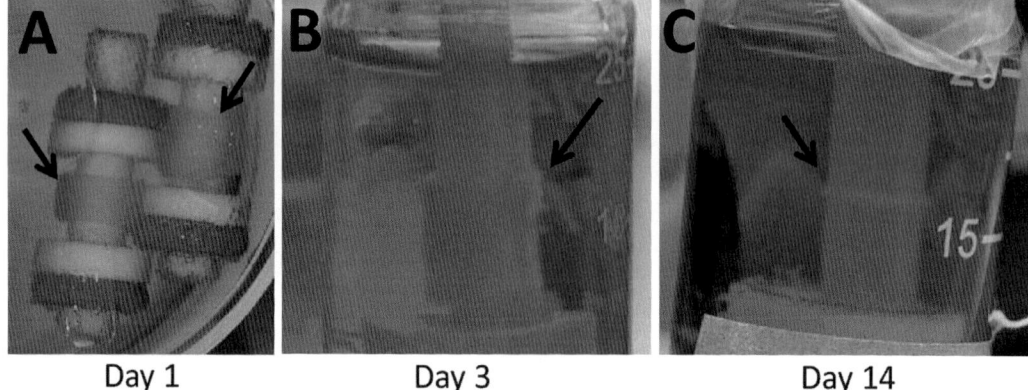

Fig. 4 Cultured constructs on day 1 (**a**), immediately after casting; day 3 (**b**), immediately after transfer; and day 14 (**c**), immediately before contraction analysis. *Arrows* point to the engineered construct

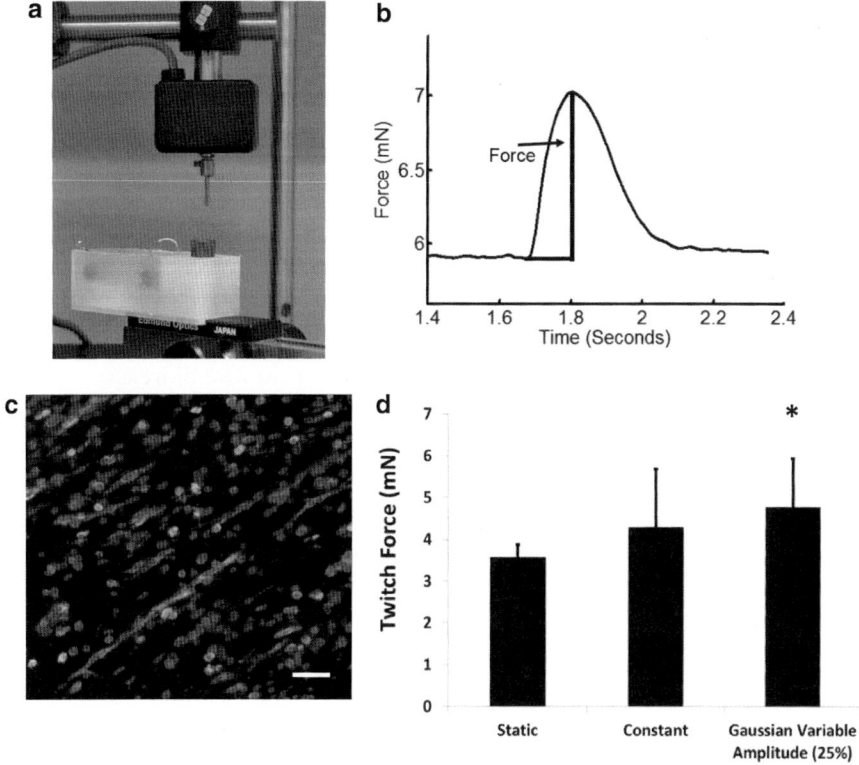

Fig. 5 (**a**) Custom-designed contraction force device for measuring twitch force of constructs. (**b**) Sample twitch force profile indicating the definition of twitch force. (**c**) Representative histological section of construct with Cx43 (*red*) and cardiac α-actin (*green*) shown. Cell nuclei are stained with DAPI (bar = 100 μm). (**d**) Twitch force data for constructs cultured statically under constant amplitude mechanical stimulation and or under variable amplitude mechanical stimulation (Gaussian distribution with a width of ±25 % of the mean). The mean strain for both mechanical stimulation groups was 5 %. Note that *asterisk* represents $p < 0.05$ compared to static constructs

4.2 Analysis
Methods

1. After culturing the constructs for 2 weeks, analyze the constructs via contraction force with a custom contraction force device, and immunohistochemistry (*see* Fig. 5).

5 Notes

1. The Teflon tubes were cut so that each construct was approximately 1 mL in initial volume.

2. The construct will be wrapped around the two posts (one connected to the force transducer, and the other in the water bath) and stretched to approximately 1 mN in tension before being paced.

3. The ascorbic acid and insulin need to be added immediately before each feeding to prevent degradation.

4. Latex tubing is very sensitive to heat. Thus, the latex tubing can only be autoclaved once before the latex melts.

5. This method refers to cells from neonatal Sprague-Dawley rats, isolated via a previously published method [14]. We created each construct with approximately five million cells per construct, though cell numbers can vary due to cell type. We find that only 80 % of the cells initially seeded were maintained after 2 h in culture.

6. When moving the construct onto the bioreactor, take care that you do not pinch the tissue with the forceps, but rather lift the outer layer of the construct that stays on the outside—because of the diffusion limitations into the construct, the outer layer of the construct has more live cells than the inner layer.

7. Constructs will look more compact and have increased elasticity even after 2 days in culture. After 2 weeks, the construct will have compacted to approximately ¼ of the initial width (*see* Fig. 4).

8. At each feeding, check the media color for signs of possible leakage. The media will turn purple indicating a pH change to a more basic pH if there is an air leak. Change the tubing by carefully removing the construct from the tubing and replacing with new tubing.

Acknowledgements

This work was supported by a grant from the National Institutes of Health—National Heart, Lung and Blood Institute (R00 HL093358 to LDB). We also acknowledge assistance of Professor Robert Tranquillo and members of his laboratory at the University of Minnesota during the initial development of the bioreactor.

References

1. Ye KY, Black LD (2011) Strategies for tissue engineering cardiac constructs to affect functional repair following myocardial infarction. J Cardiovasc Transl Res 4(5):575, Available from: http://www.pubmedcentral.nih.gov/articlerender.fcgi?artid=3182851&tool=pmcentrez&rendertype=abstract

2. Zimmermann W-H (2001) Tissue engineering of a differentiated cardiac muscle construct. Circ Res 90(2):223, Available from: http://circres.ahajournals.org/cgi/doi/10.1161/hh0202.103644

3. Shimko VF, Claycomb WC (2008) Effect of mechanical loading on three-dimensional cultures of embryonic stem cell-derived cardiomyocytes. Tissue Eng Part A 14(1):49, Available from: http://www.pubmedcentral.nih.gov/articlerender.fcgi?artid=2562769&tool=pmcentrez&rendertype=abstract

4. Birla RK, Huang YC, Dennis RG (2007) Development of a novel bioreactor for the mechanical loading of tissue-engineered heart muscle. Tissue Eng 13(9):2239, Available from: http://www.ncbi.nlm.nih.gov/pubmed/17590151

5. Kensah G, Gruh I, Ph D, Schumann H, Dahlmann J, Meyer H et al (2011) A novel miniaturized multimodal bioreactor for continuous

in situ assessment of bioartificial cardiac tissue. Tissue Eng Part C Methods 17(4): 463–473

6. Akhyari P, Fedak PWM, Weisel RD, Lee TJ, Mickle DAG, Li R (2002) Mechanical stretch regimen enhances the formation of bioengineered autologous cardiac muscle grafts. Circulation 106:I137

7. Zhang J, Montañez SI, Jewell CM, Lynn DM (2007) Multilayered films fabricated from plasmid DNA and a side-chain functionalized poly(beta-amino ester): surface-type erosion and sequential release of multiple plasmid constructs from surfaces. Langmuir 23(22):11139, Available from: http://www.ncbi.nlm.nih.gov/pubmed/17887783

8. Acharya UR, Joseph KP, Kannathal N, Lim CM, Suri JS (2006) Heart rate variability: a review. Med Biol Eng Comput 44(12):1031, Available from: http://www.ncbi.nlm.nih.gov/pubmed/17111118

9. Goldberger AL, Amaral LAN, Glass L, Hausdorff JM, Ivanov PC, Mark RG et al (2000) {PhysioBank, PhysioToolkit, and PhysioNet}: components of a new research resource for complex physiologic signals. Circulation 101(23):e215, Available from: http: //circ.ahajournals.org/cgi/content/full/ 101/23/e215

10. Cysarz D, Lange S, Matthiessen PF, Leeuwen PV (2007) Regular heartbeat dynamics are associated with cardiac health. Am J Physiol Regul Integr Comp Physiol 292(1):R368, Available from: http://www.ncbi.nlm.nih.gov/pubmed/16973939

11. Arold SP, Bartolák-Suki E, Suki B (2010) Variable stretch pattern enhances surfactant secretion in alveolar type II cells in culture. Am J Physiol Lung Cell Mol Physiol 296(4): L574–L581

12. Imsirovic J, Derricks K, Buczek-Thomas JA, Rich CB, Nugent MA, Suki B (2013) A novel device to stretch multiple tissue samples with variable patterns: application for mRNA regulation in tissue-engineered constructs. Biomatter 3(2):1, Available from: http://www.ncbi.nlm.nih.gov/pubmed/23628870

13. Syedain ZH, Weinberg JS, Tranquillo RT (2008) Cyclic distension of fibrin-based tissue constructs: evidence of adaptation during growth of engineered connective tissue. Proc Natl Acad Sci U S A 105(18):6537, Available from: http://www.pubmedcentral.nih.gov/articlerender.fcgi?artid=2373356&tool=pmcentrez&rendertype=abstract

14. Ye KY, Sullivan KE, Black LD (2011) Encapsulation of cardiomyocytes in a fibrin hydrogel for cardiac tissue engineering. J Vis Exp (55):e3251, Available from: http://www.jove.com/details.php?id=3251

15. Black LD, Meyers JD, Weinbaum JS, Shvelidze YA, Tranquillo RT (2009) Cell-induced alignment augments twitch force in fibrin gel-based engineered myocardium via gap junction modification. Tissue Eng Part A 15(10):3099, Available from: http://www.pubmedcentral.nih.gov/articlerender.fcgi?artid=2792050&tool=pmcentrez&rendertype=abstract

16. Isenberg BC, Tranquillo RT (2003) Long-term cyclic distention enhances the mechanical properties of collagen-based media-equivalents. Ann Biomed Eng 31(8):937

Chapter 17

Preparation of Acellular Myocardial Scaffolds with Well-Preserved Cardiomyocyte Lacunae, and Method for Applying Mechanical and Electrical Simulation to Tissue Construct

Bo Wang, Lakiesha N. Williams, Amy L. de Jongh Curry, and Jun Liao

Abstract

Cardiac tissue engineering/regeneration using decellularized myocardium has attracted great research attention due to its potential benefit for myocardial infarction (MI) treatment. Here we describe an optimal decellularization protocol to generate 3D porcine myocardial scaffolds with well-preserved cardiomyocyte lacunae and a multi-stimulation bioreactor that is able to provide coordinated mechanical and electrical stimulation for facilitating cardiac construct development.

Key words Cardiac tissue engineering/regeneration, Acellular myocardial scaffolds, Decellularization, Mechanical simulation, Electrical simulation, Bioreactor

1 Introduction

Myocardial infarction (MI) and heart failure are major causes of mortality worldwide [1]. Currently, the only successful treatment for end-stage heart failure is whole-heart transplantation, which is unfortunately limited by the persistent shortage of suitable heart donors. Newer strategies, including cellular transplantation, intramyocardial gene transfer, and cardiac tissue engineering (TE), have come to the forefront as alternative therapeutic approaches [2–4].

The purpose of cardiac tissue engineering is to develop functional cardiac tissue through integrating cellular components within scaffolds that serve as a structural guide [5–8]. Two major types of scaffold materials have been commonly used for cardiac tissue engineering: synthetic biodegradable material and tissue-derived acellular scaffolds [9–14]. The use of synthetic biodegradable polymers still faces challenges, including inflammatory response, mismatched material properties, nonpliability, and difficulty in controlling the degradation rate [13, 15, 16].

Milica Radisic and Lauren D. Black III (eds.), *Cardiac Tissue Engineering: Methods and Protocols*, Methods in Molecular Biology, vol. 1181, DOI 10.1007/978-1-4939-1047-2_17, © Springer Science+Business Media New York 2014

Acellular scaffolds, which are derived from native tissues or organs via decellularization, are able to preserve the extracellular matrix (ECM) composition, overall ultrastructure, shape compatibility, ECM mechanical integrity, and bioactive molecules that benefit cell–ECM adhesion, cell–cell interaction, and de novo ECM formation [17–23].

In practice, it is important to determine the optimal decellularization protocol that can mostly remove cells, cell debris, chromosome fragments, and xenogeneic antigens in order to diminish immunogenicity while at the same time preserving the needed structural and mechanical integrity of the native tissue ECM, which is important for target tissue functionalities [23–26]. For the applications of myocardial ECM scaffolds, Ott et al. decellularized whole rat heart and were able to keep the intact chamber geometry, perfusable vasculature, and competent acellular valves [27]. Badylak et al. and Taylor et al. attempted scaled-up research on whole porcine heart [21, 28]. Yet, many challenges still exist in whole-heart regeneration, such as preservation of myocardial ECM structure, reseeding homogeneity and thoroughness across the ventricle wall thickness, feasibility of reviving the existing vasculature network, and functional integration [21, 22, 29, 30]. Thus, our group has undertaken an effort to harness the potential of decellularized porcine myocardium as a TE scaffold material and focus on tissue-level application [31–33].

Heart walls are constructed of cardiac muscles that consist of cardiomyocytes, which are connected via gap junctions and structurally organized by highly vascularized ECM [34, 35]. As visually demonstrated by our previous diffusion tension MRI study [36], the heart muscles have a highly organized, multilayered helical structure (Fig. 1a, b). The compelling structural beauty of heart muscle fibers hints at the uniqueness and importance of heart muscle ECM. Indeed, the intriguing myocardial ECM network does play key roles in maintaining structural integrity, tethering cardiomyocytes, mediating contraction/relaxation of muscle fibers, and preventing excessive stretching [37–39]. As shown in Fig. 1c, d, removal of the heart muscle fibers (red staining) from the myocardial collagenous network (blue staining) will leave an ECM network that possesses a three-dimensional (3D) morphology and structural anisotropy. Hence, it is understandable that determining how to preserve the 3D ultrastructure of myocardial ECM represents a real challenge in myocardium decellularization via current available decellularization means, which are often beneficial in certain aspects but disruptive at some levels or to certain components [21–23].

In this chapter, we introduce an optimal decellularization protocol to generate 3D porcine acellular myocardial scaffolds in which 3D cardiomyocyte lacunae, ECM networks, vasculature templates, and mechanical anisotropy can be well preserved [31, 32]. To further improve the effectiveness and efficiency of cell

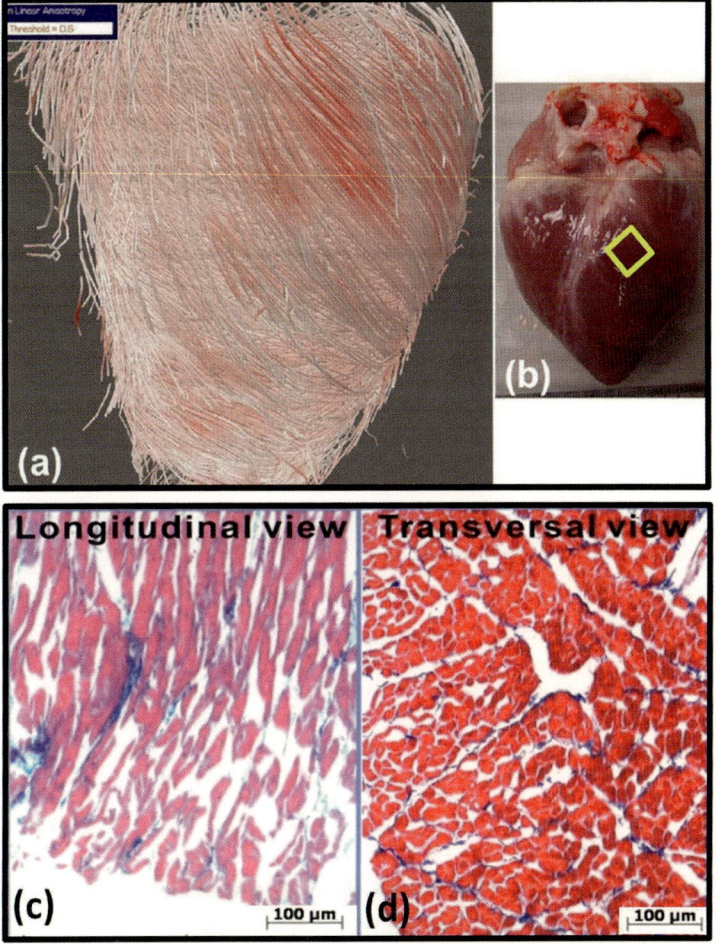

Fig. 1 (**a**) Heart muscles have well-organized multilayered helical architecture, which is mediated by 3D myocardial ECM (Diffusion Tensor MRI image by Zhang and Liao, 2010) [36]; (**b**) porcine heart used for DT-MRI imaging. Mason's trichrome staining of longitudinal section (**c**) and cross section (**d**) of the native myocardium (*red*: cardiomyocytes; *blue*: collagen). Figures reproduced with permission [31, 36]

differentiation and tissue remodeling of the reseeded acellular myocardial scaffolds, we also describe a bioreactor conditioning protocol that is able to apply combined mechanical and electrical stimulations to tissue constructs fabricated with the acellular myocardial scaffolds [33].

2 Materials

2.1 Decellularization Stock Solutions and Working Solution

1. 0.1 M Phenylmethylsulfonyl fluoride (PMSF): 0.174 g PMSF dissolved in 10 ml of 1-propanol.

2. DNase (5 mg/ml): 50 mg of DNase dissolved in 10 ml 1× PBS.

3. RNase A (5 mg/ml): 50 mg of Ribonuclease A from bovine pancreas (RNase) dissolved in 10 ml 1× PBS.

4. 100× antibiotic-antimycotic solution.

5. 1× Trypsin solution.

Note that all the above solutions were stored at –20 °C.

6. 1× Phosphate-buffered saline (PBS) pH 7.4 stored at 4 °C.

7. 1 % Sodium dodecyl sulfate (SDS) solution: 1 g SDS dissolved in 100 ml 1× PBS stored at room temperature.

8. Decellularization working solution: For 50 ml working solution, add 41.8 ml 1× PBS, 2 ml DNase, 200 µl RNase, 5 ml 1 % SDS, 500 µl 0.1 M PMSF, 500 µl ABAM, and 5 µl trypsin solution to make a final solution of 0.1 % (SDS), 0.01 % trypsin, 1 mM PMSF, 0.2 mg/ml DNase, and 20 µg/ml RNase A.

2.2 Preparation of Porcine Myocardium

1. Fresh porcine hearts were obtained from juvenile pigs (~6 months old).

2. The porcine hearts were transported to the laboratory in 1× PBS on ice.

3. A myocardium square (20×20×~3 mm) was dissected from the middle region of the anterior left ventricular wall of the porcine heart (Fig. 1b) (*see* **Note 1**).

4. All the heart samples were kept in 1× PBS solution at –80 °C for preservation (*see* **Note 2**).

2.3 Cell Culture

1. Mesenchymal stem cell (MSC) medium: Low-glucose Dulbecco's modified Eagle's medium (L-DMEM) with 10 % fetal bovine serum (FBS), 1 % MSC growth supplement (Sciencell), and 100 U/ml penicillin and 100 µg/ml streptomycin.

2. Differentiation medium: L-DMEM, 10 % FBS, 3 µmol/L 5-azacytidine (MP Biomedicals), and 100 U/ml penicillin and 100 µg/ml streptomycin (*see* **Note 3**).

3. Complete medium: L-DMEM, 10 % FBS, 1 % cardiac myocyte growth supplement (Sciencell), 100 U/ml penicillin, and 100 µg/ml streptomycin.

4. All the above media were stored at 4 °C.

2.4 Preparation of the Multi-stimulation Bioreactor

1. The bioreactor used in this study consists of one tissue culture chamber in which all the structural elements were machined from polysulfone, which provided excellent thermal and chemical stability.

2. Inside the tissue culture chamber, a maximum of four pieces of tissue construct (20 mm×20 mm×~3 mm) can be mounted between a fixed clamp and a movable clamp (Fig. 2c).

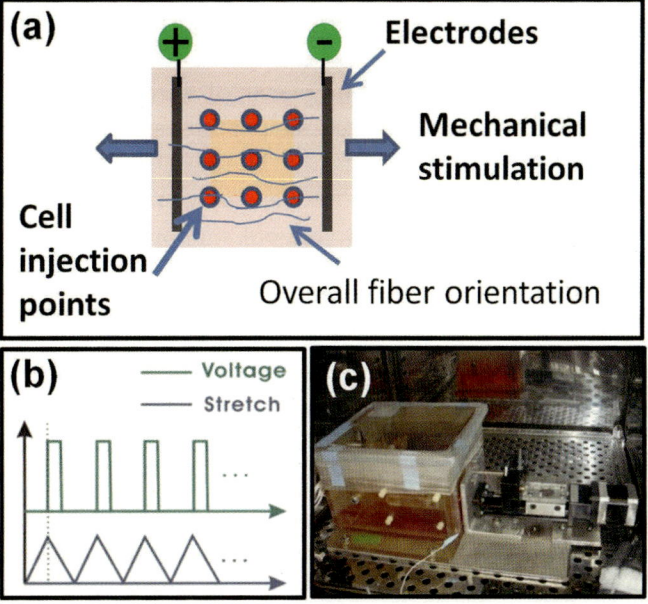

Fig. 2 (**a**) Schematic illustration of the acellular myocardial scaffold subjected to cell injection, mechanical, and electrical stimulations; (**b**) wave forms of the applied mechanical stretch and electrical pulses; (**c**) the multi-stimulation bioreactor placed in the incubator. Figures reproduced with permission [33]

3. Sutures (# 0) were used for connecting the sample with the two clamps.

4. The cover of the tissue culture chamber is fabricated with ¼ in. thick, clear polycarbonate. A hole (2 cm in diameter) was cut on the cover and sealed with a PIFE membrane with a 0.2 µm pore size to enable air exchange (Fig. 2c).

5. Linear movement was applied by the movable clamp that was driven by an Xslide assembly and a stepper motor (Velmex) (Fig. 2c) (*see* **Notes 4** and **5**).

6. Electrodes were made from Teflon-coated silver wire (75 µm diameter, A-M Systems). The end part (2 cm) of the Teflon insulation was stripped off, and the naked wires were inserted into the two opposite edges of the tissue construct.

7. The frequency and amplitude of the cyclic stretch were controlled by a customized LabView program (version 2010, National Instruments). To simulate the in vivo stimulatory environment of myocardial tissue, electrical pulses were applied at the initial stage of each unloading cycle (Fig. 2b). The frequency and amplitude of the electrical pulses were controlled by the LabView program, which was capable of delivering multiple protocols of mechanical stretching and various waveforms of electrical stimulation.

3 Methods

3.1 Myocardium Decellularization, Bioreactor Setup, and Sterilization

1. Four pieces of native myocardium samples were thawed at room temperature and washed four times with 100 ml distilled water for 10 min in a 100 ml glass media storage bottle on an orbital shaker.

2. A frame-pin supporting system was designed to better maintain tissue macrogeometry during decellularization (Fig. 3). Briefly, four corners of the myocardium sample were perforated with four $27G \times 3.5''$ needles and then mounted onto customized rectangular plastic frames (*see* **Note 6**).

3. After being mounted onto the frame-pin system, the myocardium samples were immersed inside 100 ml decellularization working solution in a 100 ml glass media storage bottle sealed with cap.

4. The myocardium patches were then decellularized with agitated decellularization working solution on an orbital shaker at 30 revolutions per minute at room temperature.

5. A 10-min ultrasonic treatment (50 HZ) was applied each day, and the decellularization solution was changed every day to avoid contamination and tissue deterioration.

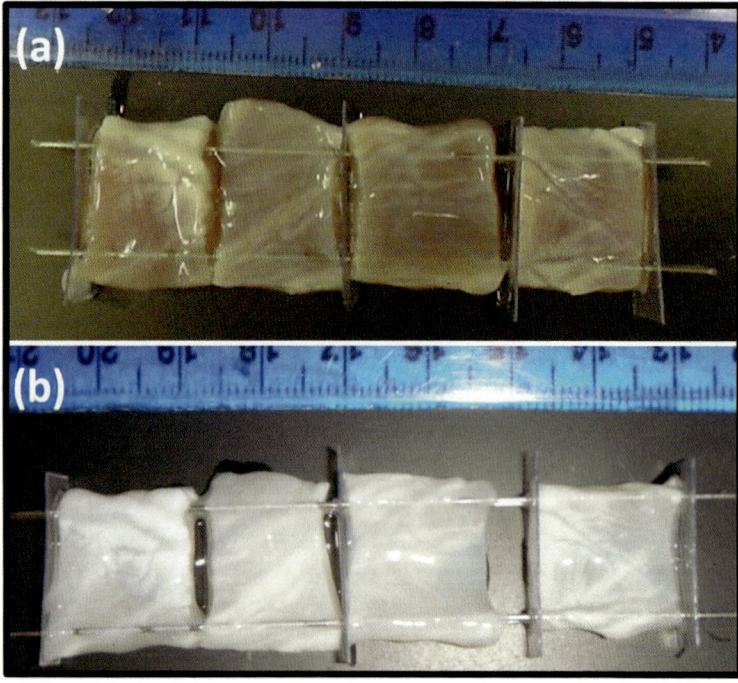

Fig. 3 The frame-pin supporting system: (**a**) Sample morphology after 3-day decellularization; (**b**) sample morphology after 2.5-week decellularization. Figures reproduced with permission [32]

6. The completeness of myocardium decellularization could be determined when the myocardium patches became a bright white color, a typical color of collagenous materials; the whole procedure lasted for approximately 2.5 weeks (*see* **Notes 7** and **8**).

7. After decellularization, all the myocardial scaffolds were removed from the frame-pin system and washed four times with 100 ml distilled water for 10 min and then washed four times with 100 ml PBS for 10 min, each time in a 100 ml glass media storage bottle on an orbital shaker.

8. After sample washing, the naked ends of the positive and negative electrodes were inserted into the two opposite edges of the acellular myocardial scaffolds (20 mm × 20 mm × ~3 mm), and the other ends of the Teflon-coated silver wires were dangled for later bioreactor connection.

9. The acellular myocardial scaffolds (with electrodes mounted) were transferred into the culture chamber of the bioreactor and mounted between a fixed clamp and a movable clamp using surgical sutures (# 0). Two to four pieces of the acellular scaffold samples could be placed in the bioreactor at one time.

10. The other ends of the Teflon-coated silver wires were then connected with the electrical control module of the bioreactor.

11. The acellular myocardial scaffolds were then sterilized in 70 % ethanol in the tissue culture chamber for 2 h (*see* **Note 9**).

12. After ethanol sterilization, all the samples were rinsed thoroughly four times with sterilized PBS while still sitting in the culture chamber (*see* **Note 9**).

13. For sterilizing the other parts of the bioreactor that could not be immersed in ethanol, the whole tissue culture chamber with the mounted acellular scaffold samples was further treated with UV light for 20 min (*see* **Note 9**).

3.2 MSC Culture, Reseeding, Differentiation, and Bioreactor Conditioning

1. Well-characterized Lewis rat MSCs (fourth passage) were obtained from the stem/progenitor cell standardization core (SPCS) at the Texas A&M Health Science Center (NIH/NCRR grant) (*see* **Note 10**).

2. After receiving the cells, MSCs were resuspended in MSC medium and seeded onto 175 mm flasks at a density of 2×10^3 cells/cm^2. The medium was changed twice a week.

3. The fifth to eighth passages of MSCs were resuspended after HyQtase (Thermo Scientific) treatment and two washes with Tyrode's balanced salt solution.

4. Next, the MSCs were used for scaffold recellularization. The density of MSCs used for scaffold recellularization was 10^6 cells/ml in MSC medium.

5. After the completion of the sterilization protocols for the bioreactor and tissue constructs, each scaffold sample was injected with totally 10^6 MSCs (1 ml MSC solution used; cell density of MSC solution: 10^6 cells/ml).

6. A 1 ml syringe with 26G permanent needle was used for cell injection, and 1 ml MSCs were injected evenly at nine injection points (~0.1 ml/point) located as a 3×3 array within the middle region of the square sample (Fig. 2a) (*see* **Note 11**).

7. The movable clamp was adjusted to make sure that no stretch is applied on the reseeded scaffold sample, and this clamp-to-clamp distance was set as the reference distance to calculate the applied strain (*see* **Note 12**).

8. The tissue culture chamber was filled with differentiation medium until the medium fully immersed all the reseeded scaffold samples.

9. The tissue culture chamber was covered and the top edges sealed with sterilized parafilm.

10. The bioreactor chamber was carefully moved into the incubator with incubation conditions set at 37 °C in a humid atmosphere with 5 % CO_2 (*see* **Note 13**).

11. After wires were connected to the bioreactor control modules, the LabView program initiated the application of mechanical stretch (20 % strain) and electrical pulses (5 V).

12. Both the triangular strain waveform and square wave electrical pulse were set at a frequency of 1 Hz (Fig. 2b), which simulates the physiological frequency experienced by the heart muscles (*see* **Note 14**).

13. Note that the differentiation medium was added in the bioreactor chamber to facilitate the cardiomyocyte differentiation during the first 24 h of tissue culture.

14. After the 24-h differentiation medium treatment, the medium was changed to the complete medium for the remaining bioreactor conditioning protocol, and the medium was changed every 3 days.

4   Notes

1. For the square myocardium patch dissection, one edge was aligned along the muscle fiber preferred direction (PD) and the other edge aligned along the cross-fiber preferred direction (XD); note that the PD direction was determined based on the overall muscle fiber texture and heart anatomy (Fig. 1b).

2. Snap freezing can disrupt cellular membranes by forming intracellular ice crystals and cause cell lysis [40, 41]; therefore, we deep-froze the myocardium samples before decellularization treatment in order to increase the efficiency of decellularization.

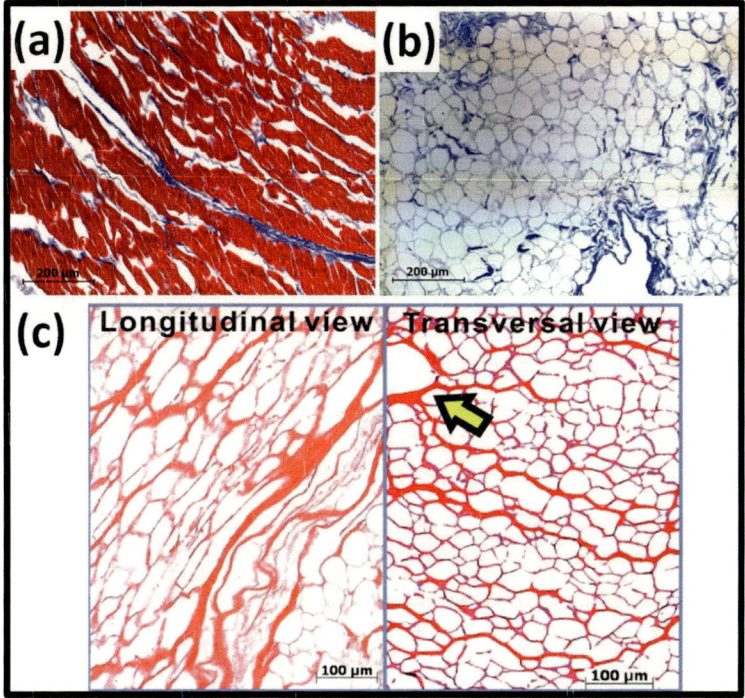

Fig. 4 Mason's trichrome staining of (**a**) the native myocardium and (**b**) the acellular myocardial scaffold (*red*: cardiomyocytes; *blue*: collagen). (**c**) H&E staining of the longitudinal and transversal views of the acellular myocardial scaffolds showed structural anisotropy (*red*: collagen); *arrow* indicates that vasculature channel was preserved after decellularization. Figures reproduced with permission [31, 32]

3. 5-Azacytidine is a member of the cytosine analogue, which had been reported to induce uncontrolled myogenic specification at random [42–44]. It was reported that treating MSCs with 5-azacytidine could generate a cardiomyocyte differentiation rate of ~30 % [43, 44].

4. For cyclic mechanical stretching, a stepper motor was chosen because its motion could be precisely controlled and easily programmed.

5. To monitor the real-time tension level in the tissue construct, load cells can be applied in the bioreactor design to oversee the mechanical forces experienced by the construct during the tissue remodeling process.

6. Frame-pin supporting system was applied for preventing scaffold contraction; this design resulted in well-preserved 3D cardiomyocyte lacunae during decellularization procedures (Fig. 5b).

7. The efficiency of cardiomyocyte removal could be verified by histology (Figs. 4b, c and 5a) and quantitative DNA analysis [31, 32]. Xenogeneic antigens, such as porcine α-Gal, were

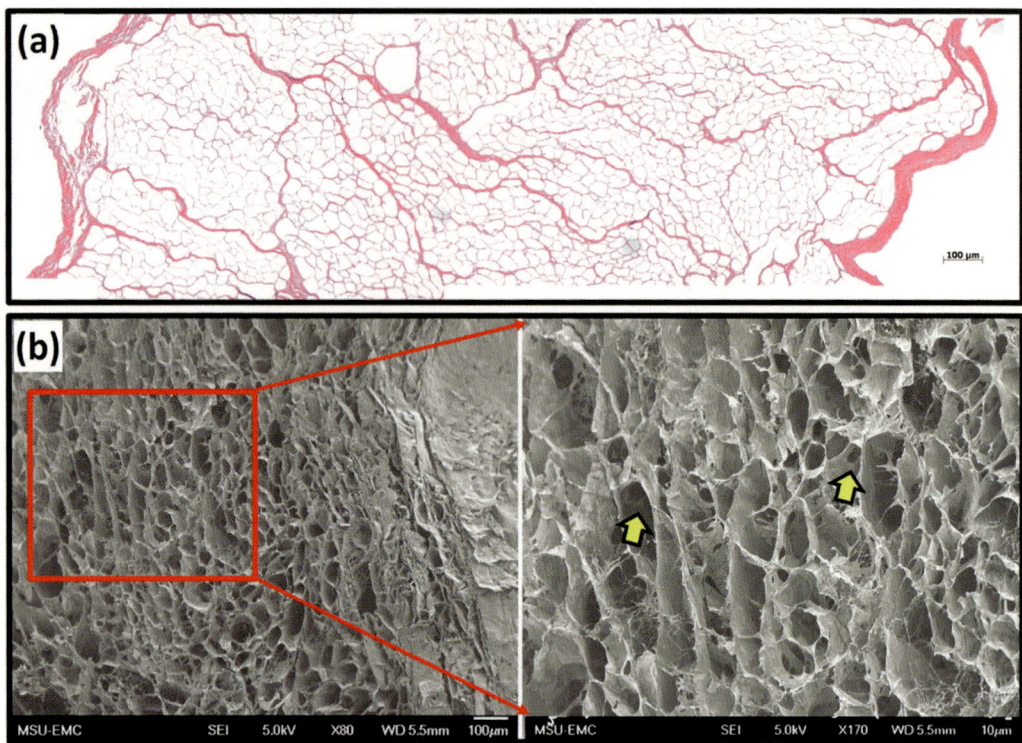

Fig. 5 (**a**) Edge-to-edge view of the acellular myocardial scaffold revealed by H&E staining; thorough decellularization and preservation of cardiomyocyte lacunae (porous structures) were verified. (**b**) 3D topography of the acellular myocardial scaffold revealed by SEM; enlarged view showed more details of the aligned 3D cardiomyocyte lacunae; note that *arrows* highlight the interconnecting openings inside the cardiomyocyte lacunae. Figures reproduced with permission [31, 32]

found to be completely removed from the acellular myocardial scaffolds [32]. We also showed that the vasculature templates (acellular blood vessel structures) were preserved in the acellular myocardial scaffolds [31, 32].

8. The decellularization protocol described here generated acellular myocardial scaffolds with thorough decellularization and ECM preservation; however, it took a relatively long treatment time (~2.5 weeks). To achieve effective decellularization within a shorter time period, the concentration of SDS can be increased to 0.5 %.

9. Effective sterilization is an essential step for tissue culture, especially for a bioreactor that has many complicated components and was constructed for repeated use. For our application, a combined method that used both 70 % ethanol sterilization and UV light sterilization was adopted. This sterilization protocol can be further optimized to reduce sterilizing treatments to a minimum [33]. Moreover, for thinner

samples, the duration of the 70 % ethanol treatment should be largely reduced, and a thorough rinse must be performed to remove residual ethanol.

10. For better cardiomyocyte differentiation, other cell sources can be utilized including cardiac stem cells, embryonic stem cells (ESCs), or induced pluripotent stem (iPS) cells [45–49].

11. To obtain a tissue-engineered cardiac construct with a high cell density, we employed a needle injection method for cell implantation with a total cell amount of 10^6 cells/scaffold. However, due to the porous structure of the acellular myocardial scaffold, a small amount of leakage may happen during the process of the cell injection. When this occurs, we inject the leaked medium back into the same region of the acellular myocardial scaffold.

12. The applied strain was estimated by normalizing the displacement of the movable clamp to the reference distance of two sample mounting clamps. The use of the clamp-to-clamp displacement in strain estimation was not an ideal method to accurately measure the tissue construct strain. To achieve more accurate measurement of the tissue construct strain, a camera can be used for real-time tracking of markers placed on the tissue construct.

13. Temperature is another important parameter in bioreactor conditioning. Heat exchange has to be carefully designed to maintain the incubator/bioreactor at a constant temperature (37 °C). In our application, we placed both the culture chamber and the stepper motor inside the incubator. We noticed that the heat generated by the stepper motor after long hours of operation greatly affected the motor performance. This problem was solved by designing a water refrigeration system in which cold copper coils were wrapped around the step motor, and cooling water was circulated by a rotation pump outside of the incubator. The water refrigeration system dissipated the heat generated by the step motor effectively, and the temperature was maintained in a reasonable range without causing any motor malfunction.

14. Previous studies have shown that cyclic mechanical stimulation can assist in cell alignment, stimulate ECM formation [50, 51], and improve cardiomyocyte development and function [52, 53]. Electrical stimulation is believed to be able to induce transient calcium levels and facilitate cell proliferation and promote the formation and localization of electric gap junctions [54, 55]. The benefit of combined mechanical and electrical stimulation was demonstrated by good cell viability, repopulation, differentiation, and positive tissue remodeling within a short period of time (2–4 days) [33].

Acknowledgments

This study is supported by NIH National Heart, Lung, and Blood Institute grant HL097321. The authors also would like to acknowledge the support from American Heart Association (13GRNT17150041) and MAFES Strategic Research Initiative (CRESS MIS-361020).

References

1. Rosamond W, Flegal K, Friday G, Furie K, Go A, Greenlund K et al (2007) Heart disease and stroke statistics—2007 update: a report from the American Heart Association Statistics Committee and Stroke Statistics Subcommittee. Circulation 115:e69–e171

2. Sharma R, Raghubir R (2007) Stem cell therapy: a hope for dying hearts. Stem Cells Dev 16:517–536

3. Grauss RW, Winter EM, van Tuyn J, Pijnappels DA, Steijn RV, Hogers B et al (2007) Mesenchymal stem cells from ischemic heart disease patients improve left ventricular function after acute myocardial infarction. Am J Physiol Heart Circ Physiol 293:H2438–H2447

4. Strauer BE, Kornowski R (2003) Stem cell therapy in perspective. Circulation 107:929–934

5. Langer R, Vacanti J (1993) Tissue engineering. Science 260:920–926

6. Zimmermann WH, Melnychenko I, Eschenhagen T (2004) Engineered heart tissue for regeneration of diseased hearts. Biomaterials 25: 1639–1647

7. Thompson RB, Emani SM, Davis BH, van den Bos EJ, Morimoto Y, Craig D et al (2003) Comparison of intracardiac cell transplantation: autologous skeletal myoblasts versus bone marrow cells. Circulation 108(Suppl 1): II264–II271

8. Bursac N, Papadaki M, Cohen RJ, Schoen FJ, Eisenberg SR, Carrier R et al (1999) Cardiac muscle tissue engineering: toward an in vitro model for electrophysiological studies. Am J Physiol 277:H433–H444

9. Fujimoto KL, Tobita K, Merryman WD, Guan J, Momoi N, Stolz DB et al (2007) An elastic, biodegradable cardiac patch induces contractile smooth muscle and improves cardiac remodeling and function in subacute myocardial infarction. J Am Coll Cardiol 49:2292–2300

10. Fujimoto KL, Guan J, Oshima H, Sakai T, Wagner WR (2007) In vivo evaluation of a porous, elastic, biodegradable patch for reconstructive cardiac procedures. Ann Thorac Surg 83:648–654

11. Li W-J, Laurencin CT, Caterson EJ, Tuan RS, Ko FK (2002) Electrospun nanofibrous structure: a novel scaffold for tissue engineering. J Biomed Mater Res 60:613–621

12. Smith IO, Liu XH, Smith LA, Ma PX (2009) Nanostructured polymer scaffolds for tissue engineering and regenerative medicine. Wiley Interdisc Rev Nanomed Nanobiotechnol 1:226–236

13. Ozawa T, Mickle DA, Weisel RD, Koyama N, Wong H, Ozawa S et al (2002) Histologic changes of nonbiodegradable and biodegradable biomaterials used to repair right ventricular heart defects in rats. J Thorac Cardiovasc Surg 124:1157–1164

14. Ozawa T, Mickle DA, Weisel RD, Koyama N, Ozawa S, Li RK (2002) Optimal biomaterial for creation of autologous cardiac grafts. Circulation 106:I176–I182

15. Grad S, Zhou L, Gogolewski S, Alini M (2003) Chondrocytes seeded onto poly (L/DL-lactide) 80 %/20 % porous scaffolds: a biochemical evaluation. J Biomed Mater Res A 66:571–579

16. Weber B, Emmert MY, Schoenauer R, Brokopp C, Baumgartner L, Hoerstrup SP (2011) Tissue engineering on matrix: future of autologous tissue replacement. Semin Immunopathol 33:307–315

17. Badylak SF, Record R, Lindberg K, Hodde J, Park K (1998) Small intestinal submucosa: a substrate for in vitro cell growth. J Biomater Sci Polym Ed 9:863–878

18. Badylak SF, Tullius R, Kokini K, Shelbourne KD, Klootwyk T, Voytik SL et al (1995) The use of xenogeneic small intestinal submucosa as a biomaterial for Achilles tendon repair in a dog model. J Biomed Mater Res 29:977–985

19. Voytik-Harbin SL, Brightman AO, Kraine MR, Waisner B, Badylak SF (1997) Identification of extractable growth factors from small intestinal submucosa. J Cell Biochem 67:478–491

20. Baptista PM, Vyas D, Soker S (2012) Liver regeneration and bioengineering—the emergence of whole organ scaffolds. In: Baptista PM (ed) Liver regeneration. InTech, Rijeka

21. Badylak SF, Taylor D, Uygun K (2011) Whole-organ tissue engineering: decellularization and recellularization of three-dimensional matrix scaffolds. Annu Rev Biomed Eng 13: 27–53

22. Crapo PM, Gilbert TW, Badylak SF (2011) An overview of tissue and whole organ decellularization processes. Biomaterials 32:3233–3243

23. Gilbert TW, Sellaro TL, Badylak SF (2006) Decellularization of tissues and organs. Biomaterials 27:3675–3683

24. Chan BP, Leong KW (2008) Scaffolding in tissue engineering: general approaches and tissue-specific considerations. Eur Spine J 17 (Suppl 4):467–479

25. Badylak SF (2004) Xenogeneic extracellular matrix as a scaffold for tissue reconstruction. Transpl Immunol 12:367–377

26. Badylak SF, Freytes DO, Gilbert TW (2009) Extracellular matrix as a biological scaffold material: structure and function. Acta Biomater 5:1–13

27. Ott HC, Matthiesen TS, Goh SK, Black LD, Kren SM, Netoff TI et al (2008) Perfusion-decellularized matrix: using nature's platform to engineer a bioartificial heart. Nat Med 14: 213–221

28. Wainwright JM, Czajka CA, Patel UB, Freytes DO, Tobita K, Gilbert TW et al (2010) Preparation of cardiac extracellular matrix from an intact porcine heart. Tissue Eng Part C Methods 16:525–532

29. Vunjak-Novakovic G, Tandon N, Godier A, Maidhof R, Marsano A, Martens TP et al (2010) Challenges in cardiac tissue engineering. Tissue Eng Part B Rev 16:169–187

30. Steinhauser ML, Lee RT (2011) Regeneration of the heart. EMBO Mol Med 3:701–712

31. Wang B, Borazjani A, Tahai M, Curry AL, Simionescu DT, Guan J et al (2010) Fabrication of cardiac patch with decellularized porcine myocardial scaffold and bone marrow mononuclear cells. J Biomed Mater Res A 94: 1100–1110

32. Wang B, Tedder ME, Perez CE, Wang G, de Jongh Curry AL, To F et al (2012) Structural and biomechanical characterizations of porcine myocardial extracellular matrix. J Mater Sci Mater Med 23:1835–1847

33. Wang B, Wang G, To F, Butler JR, Claude A, McLaughlin RM et al (2013) Myocardial scaffold-based cardiac tissue engineering: application of coordinated mechanical and electrical stimulations. Langmuir 29(35):11109–11117

34. Rakusan K, Flanagan MF, Geva T, Southern J, Van Praagh R (1992) Morphometry of human coronary capillaries during normal growth and the effect of age in left ventricular pressure-overload hypertrophy. Circulation 86:38–46

35. Korecky B, Hai CM, Rakusan K (1982) Functional capillary density in normal and transplanted rat hearts. Can J Physiol Pharmacol 60:23–32

36. Zhang S, Crow JA, Yang X, Chen J, Borazjani A, Mullins KB et al (2010) The correlation of 3D DT-MRI fiber disruption with structural and mechanical degeneration in porcine myocardium. Ann Biomed Eng 38:3084–3095

37. Weber KT (1989) Cardiac interstitium in health and disease: the fibrillar collagen network. J Am Coll Cardiol 13:1637–1652

38. Holmes JW, Borg TK, Covell JW (2005) Structure and mechanics of healing myocardial infarcts. Annu Rev Biomed Eng 7:223–253

39. Humphery JD (2002) Cardiovascular solid mechanics. Springer, New York

40. Jackson DW, Grood ES, Cohn BT, Arnoczky SP, Simon TM, Cummings JF (1991) The effects of in situ freezing on the anterior cruciate ligament. An experimental study in goats. J Bone Joint Surg Am 73:201–213

41. Roberts TS, Drez D Jr, McCarthy W, Paine R (1991) Anterior cruciate ligament reconstruction using freeze-dried, ethylene oxide-sterilized, bone-patellar tendon-bone allografts. Two year results in thirty-six patients. Am J Sports Med 19:35–41

42. Chiu CP, Blau HM (1985) 5-Azacytidine permits gene activation in a previously noninducible cell type. Cell 40:417–424

43. Tomita Y, Makino S, Hakuno D, Hattan N, Kimura K, Miyoshi S et al (2007) Application of mesenchymal stem cell-derived cardiomyocytes as bio-pacemakers: current status and problems to be solved. Med Biol Eng Comput 45:209–220

44. Fukuda K (2003) Regeneration of cardiomyocytes from bone marrow: use of mesenchymal stem cell for cardiovascular tissue engineering. Cytotechnology 41:165–175

45. Carrier RL, Papadaki M, Rupnick M, Schoen FJ, Bursac N, Langer R et al (1999) Cardiac tissue engineering: cell seeding, cultivation parameters, and tissue construct characterization. Biotechnol Bioeng 64:580–589

46. Birla RK, Borschel GH, Dennis RG (2005) In vivo conditioning of tissue-engineered heart muscle improves contractile performance. Artif Organs 29:866–875

47. Birla RK, Borschel GH, Dennis RG, Brown DL (2005) Myocardial engineering in vivo: formation and characterization of contractile, vascularized three-dimensional cardiac tissue. Tissue Eng 11:803–813

48. Borschel GH, Dow DE, Dennis RG, Brown DL (2006) Tissue-engineered axially vascularized contractile skeletal muscle. Plast Reconstr Surg 117:2235–2242

49. Vouyouka AG, Powell RJ, Ricotta J, Chen H, Dudrick DJ, Sawmiller CJ et al (1998) Ambient pulsatile pressure modulates endothelial cell proliferation. J Mol Cell Cardiol 30: 609–615

50. Rubbens MP, Driessen-Mol A, Boerboom RA, Koppert MM, van Assen HC, TerHaar Romeny BM et al (2009) Quantification of the temporal evolution of collagen orientation in mechanically conditioned engineered cardiovascular tissues. Ann Biomed Eng 37:1263–1272

51. Akhyari P, Fedak PW, Weisel RD, Lee TY, Verma S, Mickle DA et al (2002) Mechanical stretch regimen enhances the formation of bioengineered autologous cardiac muscle grafts. Circulation 106:I137–I142

52. Zimmermann WH, Melnychenko I, Wasmeier G, Didie M, Naito H, Nixdorff U et al (2006) Engineered heart tissue grafts improve systolic and diastolic function in infarcted rat hearts. Nat Med 12:452–458

53. Fink C, Ergun S, Kralisch D, Remmers U, Weil J, Eschenhagen T (2000) Chronic stretch of engineered heart tissue induces hypertrophy and functional improvement. FASEB J 14: 669–679

54. Wikswo JP Jr, Lin SF, Abbas RA (1995) Virtual electrodes in cardiac tissue: a common mechanism for anodal and cathodal stimulation. Biophys J 69:2195–2210

55. McDonough PM, Glembotski CC (1992) Induction of atrial natriuretic factor and myosin light chain-2 gene expression in cultured ventricular myocytes by electrical stimulation of contraction. J Biol Chem 267: 11665–11668

Chapter 18

Patch-Clamp Technique in ESC-Derived Cardiomyocytes

Jie Liu and Peter H. Backx

Abstract

ESC-derived cardiomyocytes are excitable cells that express many of the ion channels also found in adult cardiomyocytes. The patch-clamp technique is a powerful technique to characterize both the electrophysiological properties of excitable cells as well as the underlying ion channel currents responsible for these electrophysiological properties. The technique also allows rapid and accurate screening of the pharmacological actions of agents used to modulate cardiomyocyte properties. In this chapter we illustrate the use of the whole-cell patch clamp technique by recording hERG (human ether-à-go-go-related gene) currents and action potentials in ESC-derived cardiomyocytes and also examine the effects of hERG channel blocker, dofetilide.

Key words Electrophysiology, Patch-clamp technique, Cardiomyocyte, Embryonic stem cells, hERG channel, Action potential, Dofetilide, Whole-cell recording

1 Introduction

Cardiomyocytes derived from ESCs hold considerable promise for regenerative therapies in the treatment of heart disease. Because ESC-derived cardiomyocytes express ion channels that are present in adult cardiomyocyte [1], these cells also provide new opportunities for drug discovery. The patch-clamp technique is an established technique for characterizing the electrophysiological properties of excitable cells like ESC-derived cardiomyocytes as well as their responses to various pharmacological agents. The patch-clamp technique was developed by Erwin Neher and Bert Sakmann in 1970s [2]. The technique allows the measurement of either the transmembrane voltage or the currents flowing across membranes using glass micropipette [3] which are capable of forming tight electrical seals to lipid bilayers via strong chemical interactions between the microelectrode glass and the membrane. The tips of the glass microelectrodes are typically configured into characteristic shapes by using a microelectrode "puller" which is designed to temporarily melt the glass by heating while pulling on

Milica Radisic and Lauren D. Black III (eds.), *Cardiac Tissue Engineering: Methods and Protocols*, Methods in Molecular Biology, vol. 1181, DOI 10.1007/978-1-4939-1047-2_18, © Springer Science+Business Media New York 2014

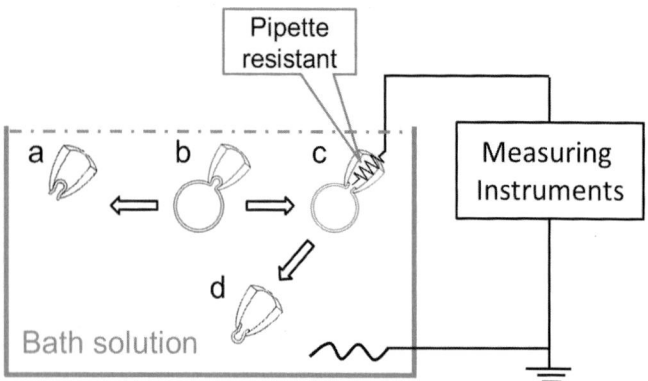

Fig. 1 Patch-clamp configurations and electrical connection of pipette, biological membrane, and patch-clamp measuring instruments. (*a*) Inside-out configuration obtained by rupturing the membrane from cell-attached configuration. (*b*) Cell-attached configuration. (*c*) Whole-cell configuration established after breakthrough of the membrane from cell-attached configuration. (*d*) Outside-out configuration formed by pulling pipette away from whole-cell configuration

the ends of the microelectrode, thereby creating microelectrode tips with outer diameters ranging from 1 to 3 μm. The tips are then filled with solutions, carefully selected on the basis of the primary objectives of the planned recordings. After forming an electrical seal between the forged fluid-filled microelectrode, the membrane found under the glass electrode can be left intact (called an attached patch configuration, Fig. 1b), ripped away from the cell (an "inside-out" excised patch configuration, Fig. 1a), or ruptured, thereby allowing the microelectrode solution to have direct access inside of the cell (a whole-cell configuration, Fig. 1c). It is also possible to tear off membrane after forming the whole-cell configuration (an outside-out excised patch configuration, Fig. 1d). Finally, the pipette solution can include molecules that are capable of forming relatively large nonselective pores in the lipid bilayers (the perforated-patch technique), thereby allowing electrical access into the cell while preventing the mobile internal cellular contents from diffusing into the large volume of the pipette. The whole-cell configuration is probably the most popular method since it offers opportunities to rapidly evaluate the physiological and electrical responses of excitable cells while also controlling the cellular content by choosing different solutions to place in the pipette. In the whole-cell configuration, the patch-clamp amplifiers are operated in one of the two modes: voltage clamp and current clamp. In the voltage-clamp mode, the amplifier attempts to control the membrane potential while simultaneously recording the current flowing across the membrane, typically in response to various voltage waveforms applied to the cell. In the current-clamp mode, it is possible to inject charge (or current) into the cell and simultaneously record

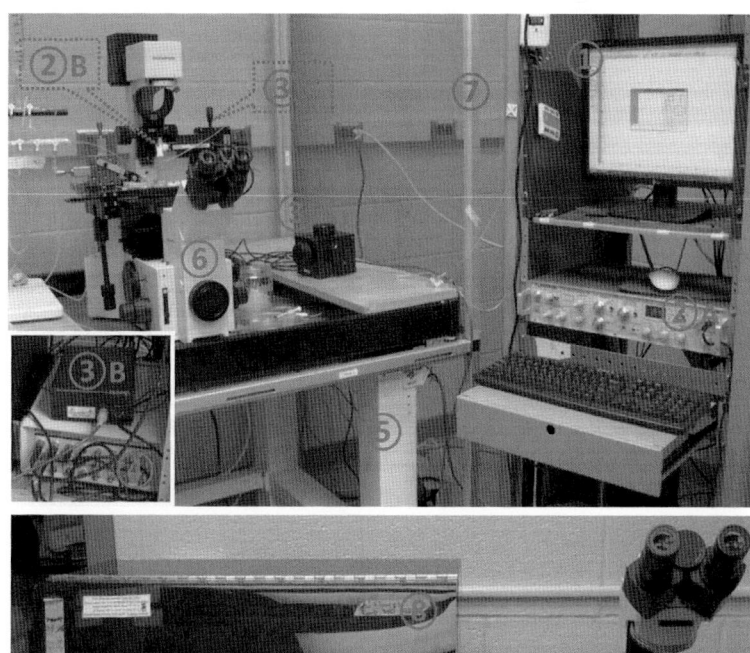

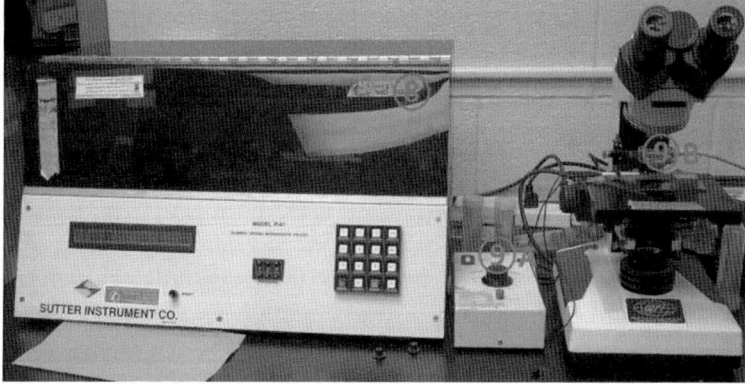

Fig. 2 Basic setup for patch-clamp experiments. ① Computer, ② A and B amplifier, pipette holder, and headstage, ③ A, B, and C micromanipulator, ④ digitizer, ⑤ vibration-isolated table, ⑥ converted microscope, ⑦ Faraday cage, ⑧ pipette puller, ⑨ A and B microforge

the electrical potential across the membrane. The equipment required to make the patch-clamp recordings possible is shown in Fig. 2. The equipment consists of an amplifier and a digitizer interfaced with a computer. These are connected to a micromanipulator to allow positioning and movement of the microelectrodes which is typically mounted on the stage of a microscope which is in turn placed on a "vibration-free" isolation table which is surrounded by a grounded metal encasement (called a Faraday cage). Until recently, the patch-clamp measurements were tedious and the standard amplifiers allowed one or two recordings to be made simultaneously. Although these traditional approaches have yielded a wealth of insights into the properties of excitable cells, the process of data generation is slow and painstaking. More recently, new technologies allow up to 64 patch-clamp recordings to be made simultaneously (Qpatch from Sophion Bioscience, Denmark;

Autopatch from Xention Ltd, UK; and IONWORK from MDS, USA), thereby immensely increasing the throughput using this technique.

While the mechanics of making recordings are relatively straightforward, the ability to record meaningful high-quality measurements require thorough training in electrophysiological principles and appreciation of several inherent assumptions that are invariably made when making electrophysiological measurements. Some factors that require a clear understanding in order to ensure meaningful interpretations of measurements include the following: (1) the need for Ag/AgCl electrodes in the pipette and external bathing solutions; (2) liquid junction potentials arising from solution interfaces; (3) surface electrode potentials arising from the differences in the chemical activity of Cl⁻ ions in the bath and pipette solutions; (4) the existence, quantification, and elimination/minimization of a series resistance of the pipette (i.e., the resistance between the Ag/AgCl electrode in the pipette and the cell membrane); (5) the need for, and effects of, signal filtering; and (6) effects of sampling frequency of recordings to satisfy the Nyquist sampling theorem (preventing complicating due to aliasing). Although various commercial suppliers provide instructions related to these matters, a thorough understanding remains essential to ensure reliable recordings.

Below we use the whole-cell recording technique to measure hERG currents using the voltage-clamp mode and action potentials using the current-clamp mode in ESC-derived cardiomyocytes. The software interfaces are Clampex 8.2 and Clampfit 10. For further reading, we suggest these books: The Axon Guide for Electrophysiology and Biophysics Laboratory Techniques (Axon Instruments, Inc.), Axopatch 200B Patch Clamp Theory and Operation (Axon Instruments, Inc.), and ref. 3.

2   Materials

All solutions are prepared using ultrapure water (18.4 MΩ at 24 °C) and analytical grade reagents. Note that this ultrapure water typically still contains measurable amounts of ions. Of most relevance is the observation that this water typically contains 10–30 μM Ca^{2+} which mandates the need for some Ca^{2+} buffering in all pipette solutions that are used for whole-cell recordings.

2.1   Cell Isolation

1. Ca^{2+}-free modified Tyrode's solution: 120 mM NaCl, 5.4 mM KCl, 5 mM $MgSO_4$, 5 mM Na-pyruvate, 20 mM glucose, 20 mM taurine, 10 mM HEPES, pH 6.9 with NaOH.

2. Digestion solution: Ca^{2+}-free modified Tyrode's solution with 1 mg/mL type II collagenase (352 IU/mg).

3. KB solution: 85 mM KCl, 20 mM K_2HPO_4, 5 mM $MgSO_4$, 1 mM EGTA, 5 mM Na-pyruvate, 5 mM creatine, 20 mM taurine, 20 mM glucose, 2 mM Na_2ATP, pH 7.4 with KOH.

4. 50 µg/ml Laminin solution

5. Glass cover slip

2.2 Patch-Clamp Experiment

1. External solution: 140 mM NaCl, 4 mM KCl, 1 mM $MgCl_2$, 1.2 mM $CaCl_2$, 10 mM HEPES, 10 mM d-glucose, pH 7.35 with NaOH (*see* **Note 1**).

2. Internal pipette solution: 115 mM potassium aspartate, 15 mM KCl, 4 mM NaCl, 1 mM $MgCl_2$, 5 mM MgATP, 5 mM HEPES, 5 mM EGTA, pH 7.2 with KOH (*see* **Note 1**).

3. Dofetilide was added to external solution in some experiments to block hERG current [4], and final concentration was 100 µM (*see* **Note 2**).

4. Preparation of pipettes:
   A pipette puller (Sutter instrument, Model P-87, parameter value setting was based on ramp test) is used to forge a pipette from glass capillary. The ends of the electrodes (i.e., the tips) are polished with a microforge which heats the ends after pulling to smooth the glass in the rim of the microelectrode tip. The pipettes used in these experiments are typically 4–6 MΩ (tip diameter ~1.3 µm) after they are filled with internal solution (*see* **Note 3**).

5. Use of a 3 M KCl agar bridge:
   We routinely cover our external bath Ag/AgCl electrodes with a polyethylene capillary tube that has been filled with 2 % agar dissolved in 3 M KCl. These are made by heating a 3 M KCl solution with 2 % (by weight) agarose. Once the agarose is dissolved, place one end of a polyethylene capillary tube (diameter 0.5 mm) with a length of about 6 in. into the molten agar. Apply gentle suction onto the free-end of the capillary tube, thereby slowing drawing the heated agar into the tube. Ensure that no air bubbles are formed. The agar-filled tubes are then cut into 1 in. lengths and placed over the Ag/AgCl electrodes located in the bath solution.

3 Methods

3.1 Cell Isolation

1. To prepare laminin-coated glass cover slips (dimensions 4×10 mm²) to which isolated cardiomyocytes are attached, place glass cover slips into 50 µg/ml laminin solution for 1 min. Allow cover slips to dry at room temperature.

2. To remove culture medium and prepare for digestion, place 15–20 beating embryoid bodies (EBs) from cardiac differentiation in 1 ml Eppendorf tubes containing StemPro-34

(Invitrogen) culture media and centrifuge at $21 \times g$ for 3 min. Protocols for cardiac differentiation can be found in other chapters of this book. After aspirating the culture media, the EBs are resuspended in Ca^{2+}-free modified Tyrode's solution and then transferred to 10 mL culture tubes containing Ca^{2+}-free modified Tyrode's solution. Leave the EBs in Ca^{2+}-free solution at room temperature for 20 min.

3. To digest the EBs, spin down the EBs at $21 \times g$ for 3 min, carefully remove the supernatant as much as possible (*see* **Note 4**), add 3 ml digestion solution pre-warmed to 37 °C, and incubate the EBs at 37 °C for 20 min with gentle periodic shaking by hand.

4. To judge the yield of digestion, take 50 μl aliquots from the culture tube and check for the presence of dissociated cardiomyocytes every 5 min after 20 min of digestion.

5. When large numbers of cardiomyocytes appear, the digestion is terminated (*see* **Note 5**) by centrifuging the suspension at $233 \times g$ for 3 min and aspirating the supernatant. The pellet is then resuspended in 3 ml KB solution pre-warmed to 37 °C.

6. Next, a small aliquot of cells is placed into culture dishes (35 mm) containing 20 laminin-coated cover slips in serum-free culture medium (*see* **Note 6**) and incubated at 37 °C for 24 h before patch-clamp experiments. Ensure that the cells are transferred with a minimum of KB solution (minimum dilution of 1:30).

7. Patch-clamp recordings are performed within 48 h after dissociation and culturing.

3.2 Whole-Cell Recording

1. Before starting electrophysiological experiments, ensure that the recording equipment are collectively grounded to minimize electrical noise. Each day, Ag/AgCl electrodes should be re-chloridized and fresh agar bridges placed over the bath electrode (*see* **Note 7**). Check that the amplifier and digitizer are well calibrated following the manufacturer's instructions. All electrical components need to be turned on for at least 1 h before beginning recordings. We typically placed our pipette and ground electrodes in their corresponding solutions for at least 1 h before recording in order to minimize electrical drift associated with slow equilibration times.

2. Amplifier settings on the front panel of the Axopatch 200B use the following settings to begin:

   PIPETTE OFFSET at about 4.0.

   ZAP at 0.5 ms.

   PIPETTE CAPACITANCE COMPENSATION at the off position (i.e. at full counterclockwise position).

SERIES RESISTANCE COMP. % PREDICTION at 0 %.

SERIES RESISTANCE COMP. % CORRECTION at 0 %.

SERIES RESISTANCE COMP. LAG at 100 μs (fully clockwise).

WHOLE-CELL CAP. at 0 pF, OFF.

SERIES RESISTANCE at 0 MΩ.

HOLDING COMMAND at 0 mV, toggle ×1, OFF.

SEAL TEST at ON.

METER at I.

MODE at V-CLAMP.

CONFIG. at WHOLE CELL.

OUTPUT GAIN at $\alpha = 20$ or as desired.

LOW-PASS BESSEL FILTER at 1 kHz.

LEAK SUBTRACTION at 0 MΩ.

3. Recording protocols:

Recording protocols are created using the Clampex 8.2 software, which is used to control the actions of the amplifier: setting cell membrane potential, injecting current, etc. The Clampex program is also used to record output generated by the amplifier. Note that the sampling frequency should be at least four times faster than the cutoff frequency used to filter the output signals from the amplifier.

Action potential recording protocols are achieved by accessing the Acquisition menu in Clampex 8.2. In this menu select Edit Protocol. In Mode/Rate page, check Episodic Stimulation under Acquisition Mode part. In Trial Hierarchy, set Run/trial at 1, sweep/run at 10, and samples/sweep/signal at 5,160. In Sample interval per signal part, select First interval (μs) at 50; in Start-to-Start interval part, set Sweep (s) at 1 s. In Experiment Type select Current-clamp; this means we will use current-clamp mode recording of the action potential. In Outputs page, set Analog OUT holding levels (mV) part at 0; this means there will be no current inject to cell when the cell is held for next stimulation. Then in Wave 0 page, check Epochs in Analog Waveform part and set Inter-sweep Holding level at Use holding; this means no current injection will occur during the sweep interval. In Epoch description part, set the following in column A: Type at Step, First level at 1.2 threshold value; this means amplifier will inject 1.2-fold of the threshold current into cell to depolarize cell membrane (*see* **Note 8**). Set First duration (samples) at 40, and First duration (ms) will automatically show "2"; this means amplifier will output a 2-ms current pulse. Save this protocol for later use.

To set the protocols required to make hERG current recordings, return to the Acquisition menu in Clampex 8.2. Select Edit Protocol. In the Mode/Rate page, select Episodic Stimulation under Acquisition Mode; in Trial Hierarchy part, set Run/trial at 1, sweep/run at 1, and samples/sweep/signal at 15,000. In Sample interval per signal part, set First interval (μs) at 200. In Experiment Type set at voltage clamp; this means we will use voltage-clamp mode, recording hERG tail current. In Output Page, set Analog OUT holding levels (mV) part at –80; this means cell membrane potential will be held at –80 mV when waiting for stimulation. Then in Wave 0 page, check Epochs in Analog Waveform part and set Inter-sweep Holding level at Use holding; this means membrane potential will be held at –80 mV between recording sweep intervals. In Epoch description part set column A to the following: Type at Step, First level (mV) at 60; First duration (samples) at 2,500, and First Duration (ms) will automatically show "500"; this means amplifier will clamp the membrane potential at 60 mV for 500 ms. Then set column B to the following: Type at Step, First level (mV) at –50; First duration (samples) at 11,000, and First Duration (ms) will automatically show "2,200"; this means the amplifier will clamp the membrane potential at –50 mV for 2,200 ms. Save this protocol for later use.

4. Mount pipette to the pipette holder, apply small positive air pressure inside the pipette (*see* **Note 9**), lower the pipette into bath solution, and run seal test in Clampex 8.2. Before doing these, touch your hand to grounded metal to discharge static on your body (*see* **Note 10**).

5. Pipette offset adjustment:
Set holding potential at 0 mV, and adjust the PIPETTE OFFSET until the current on the meter reads zero.

6. Pipette resistance measurement:
Apply a 2 mV pulse for 10 ms at a frequency of 50 Hz (the values are adjustable in Seal test setting). This is achieved in the Clampex software by using the seal test protocol. With this protocol, you will observe repeated step increases in current with an amplitude of about 1 nA passing through the electrode tip into the bath. The current magnitude is used to determine an estimate of the series resistance (R_S) of the pipette.

7. Giga seal formation:
With the seal test protocol activated, move the pipette towards the cell. As contact is made with the cell, the current level decreases indicating an increase in resistance as a result of the opening of the pipette being occluded by the cell membrane. When the current is reduced by about 50–70 %, the positive

pressure on the pipette is relieved and suction is applied to the pipette to promote a strong seal formation between the glass electrode and the membrane (a giga seal).

8. Gaining access into the cell:
   Set holding potential at −50 mV (*see* **Note 11**), and apply voltage steps of +5 mV for 10 ms at a rate of 50 Hz. This is typically achieved by using a second test seal function in Clampex dedicated to membrane rupture protocols. With the pulses being applied at 50 Hz, eliminate current transient generated by the pipette capacitance (i.e., the pipette capacitance transients) by turning the SLOW MAG and FAST MAG τ controls. These dials inject current into the headstage amplifier of the Axopatch 200B, thereby removing their appearance in the recordings. Note that the current required to change the voltage on the pipettes is still provided by the headstage amplifier of the Axopatch 200B.

   Next, apply repeated pulses of gentle suction until membrane ruptures. Rupture of the membrane can be identified when there is a sudden increase of capacitive transients. An alternative method for rupturing the membrane is to use the ZAP function on the Axopatch 200B amplifier. This function applies a large hyperpolarizing voltage (1.3 V_{DC}) to the patch for a controlled duration which is sufficient to electroporate the lipid bilayer, thereby destroying the membrane under the tip of the pipette. We prefer to use a combination: a very brief ZAP application followed immediately by very gentle suction or keeping very gentle suction and applying a very brief ZAP (*see* **Note 12**).

9. Stop applying the 5 mV test pulse at 50 Hz. If the membrane potential recordings are planned, change the patch-clamp mode by setting MODE switch from V-clamp to $I=0$. Using Clampex select the action potential recording protocol that was mentioned above. Once the proper protocol is in place, switch the MODE to I-clamp; run the protocol, record the action potentials from the ES cells, and save the data for later analysis.

10. Then switch the MODE to $I=0$; after hERG tail current recording protocol was loaded properly, switch MODE to V-CLAMP, run membrane test protocol, set holding at −50 mV, and measure cell capacitance (Cm) and access resistant (Ra) (*see* **Note 13**).

11. Cell capacitance transient cancellation:
    Turn the WHOLE CELL CAP. switch to ON. Minimize the capacitance transient by using the WHOLE CELL CAP and SERIES RESISTANCE control simultaneously; turn SERIES RESISTANCE COMP. CORRECTION to 75 % (*see* **Note 14**).

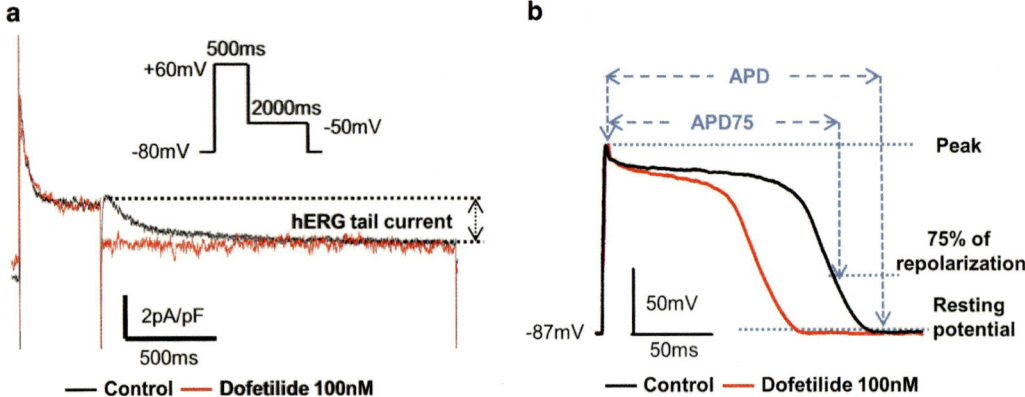

Fig. 3 Examples of representative recordings. (**a**) hERG current and measurement. (**b**) Action potential and measurement

12. Reload hERG tail current recording protocol in Clampex, run the protocol, record the current, and save the data for later analysis.

13. Add dofetilide to external solution (*see* **Note 2**), incubate the cell in dofetilide solution for 5 min, and repeat **steps 9–13** (for recording traces *see* Fig. 3).

3.3 Data Analysis

1. hERG tail current measurement:
Using Clampfit program, measure the current amplitude as shown in Fig. 3a. Current density was calculated by normalized current amplitude to cell capacitance. As shown in Fig. 3a, hERG current is completely blocked by dofetilide at these concentrations.

2. Action potential duration measurement:
Using Clampfit program, measure action potential duration and 75 % repolarization duration as shown in Fig. 3b. Also, we can see that dofetilide prolonged the action potential.

4 Notes

Generally, to carry out patch-clamp experiments efficiently, cells should be viable and of good quality because it is almost impossible to make a seal in dead cells. If necessary, trypan blue is a useful indicator for excluding dead cells especially. Electrical noise should be minimized by effectively grounding all instruments. Also, coating the pipette with Sylgard can reduce pipette capacitive current, and possibly pipette noise, by decreasing the pipette capacitance. Moreover, great patience is needed when experiments are carried on. The following are additional notes for performing electrophysiological experiments.

1. Small particles in solution prevent good seal. It is good to filter the solutions with 2.2 μm microfiber filter.

2. Drugs can be spread onto a cell from 100 μm away through a small tubing (around 100 μm diameter) facing to the cell or by replacing whole external solution in the cell bath with a dofetilide containing external solution.

3. As the ESC-cardiomyocytes are relatively small (around 15 pF) and most currents in them are less than 500 pA, it will be easier to make a good seal with 4–6 MΩ pipette than with the 1–3 MΩ. This should be good enough for most current and action potential recordings.

4. Removing all Ca^{2+} from the digestion solution may prevent over-digestion and increase output of healthy cells from the isolation.

5. When 80 % of cells are separated, it is often a good idea to stop digestion as the remaining cells may have a low viability. The percentage should be adjusted depending on the cell condition when performing patch-clamp experiments. If cell viability is low, even a lower percentage of cells may be left unseparated.

6. It is hard to make giga seal if cell membrane is coated with serum albumin. Culture of cells in serum-free media for 24 h and washout of the culture medium before patch-clamping will help achieve a giga seal.

7. A good AgCl electrode and properly installed agar salt bridge decrease serial resistant and prevent voltage-clamping error.

8. To measure threshold current, in Wave0 page, column A, set first level (mV) at 0 and Delta level (mV) at 2. This means that the amplifier will gradually increase the current output, which is injected to the cell to induce action potential, in the serial sweeps within each run. The threshold current is defined as the smallest stimulation current for inducing an action potential, that is, the value of Delta level (mV) multiplied by the Sweep number of the first action potential.

9. Positive pressure inside the pipette keeps the flow of the small amount of internal solution and prevents pipette being blocked by small particle in the bath solution. Generally, 1.02–1.04-fold of atmosphere pressure is enough. However, it results in local change of external solution around the cell. Therefore, it is important to keep perfusion with external solution.

10. By doing so, you can protect the measuring instruments from being damaged by the static electricity.

11. At around –60 to –50 mV, most channels in ESC-cardiomyocyte are close, I_{K1} current is small, and membrane is in high resistance state. It is a good voltage range for measuring cell capacitance accurately and doing compensation of cell capacitance and serial resistance.

12. Usually, whole-cell configuration is successfully established if seal resistance still shows GΩ range, because ESC-cardiomyocyte membrane resistance is very high at –50 mV in most circumstances.

13. Cell capacitance can also be measured by calculating the area under the capacitance transient. And serial resistance can be calculated with function $\tau = R_{m} \times C$. Serial resistance includes pipette resistance and other resistances in the measuring circuit showed in Fig. 2. The time constant (τ) here can be obtained by fitting the decay of the capacitance transient with mono-exponential function.

14. Under current mode, SERIES RESISTANCE COMP. CORRECTION should be set at full counterclockwise, that is, off.

References

1. Li GR, Deng XL (2011) Functional ion channels in stem cells. World J Stem Cells 3: 19–24

2. Neher E, Sakmann B (1976) Single-channel currents recorded from membrane of denervated frog muscle fibres. Nature 260: 799–802

3. Sakmann B, Neher E (1995) Single-channel recording, 2nd edn. Plenum Press, New York

4. Snyders DJ, Chaudhary A (1996) High affinity open channel block by dofetilide of HERG expressed in a human cell line. Mol Pharmacol 49:949–955

Chapter 19

Optogenetic Control of Cardiomyocytes via Viral Delivery

Christina M. Ambrosi and Emilia Entcheva

Abstract

Optogenetics is an emerging technology for the manipulation and control of excitable tissues, such as the brain and heart. As this technique requires the genetic modification of cells in order to inscribe light sensitivity, for cardiac applications, here we describe the process through which neonatal rat ventricular myocytes are virally infected in vitro with channelrhodopsin-2 (ChR2). We also describe in detail the procedure for quantitatively determining the optimal viral dosage, including instructions for patterning gene expression in multicellular cardiomyocyte preparations (cardiac syncytia) to simulate potential in vivo transgene distributions. Finally, we address optical actuation of ChR2-transduced cells and means to measure their functional response to light.

Key words Optogenetics, Adenovirus, Channelrhodopsin, Cardiomyocytes, Optical mapping, Multiplicity of infection

1 Introduction

Optogenetic control of cardiomyocytes is achieved through the genetic insertion of light-sensitive ion channels and/or pumps of microbial origin (opsins) to render cells controllable by light. This technique can be used for contactless manipulation of activity in excitable tissues such as the brain and heart. In neuroscience, optogenetics has become the tool of choice to dissect neural circuits and brain connectivity in health and disease through cell-specific targeting since its first reported application in 2005 [1]. Since then, a myriad of studies have linked specific neural populations to behavior (for review *see* refs. 2–5) and addressed their contribution to such conditions as depression [6], anxiety [7], addiction [8, 9], and sleep disorders [10]. Pioneering novel optical therapies have stemmed from this technique in the treatment of diseases, such as epilepsy [11, 12], Parkinson's [13–15], and retinal degeneration [16]. Only more recently (since 2010) has optogenetics been applied in the heart (for review *see* ref. 17). In vitro cardiac applications have thus far explored the utility of expressing

Milica Radisic and Lauren D. Black III (eds.), *Cardiac Tissue Engineering: Methods and Protocols*, Methods in Molecular Biology, vol. 1181, DOI 10.1007/978-1-4939-1047-2_19, © Springer Science+Business Media New York 2014

channelrhodopsin-2 (ChR2), an excitatory opsin, through viral means in a mouse embryonic stem cell line followed by differentiation to cardiomyocytes [18, 19] as well as through a tandem cell unit approach where nonexcitable cells, expressing ChR2, are coupled to cardiomyocytes, thus inscribing light sensitivity to the cardiac syncytium [20].

The microbial opsins at the core of optogenetics can produce either depolarizing (excitatory) currents or hyperpolarizing (inhibitory) currents in mammalian cells. Excitatory opsins, such as channelrhodopsin (ChR), can provide fast kinetic currents of sufficient amplitude to trigger action potentials, whereas inhibitory opsins, such as halorhodopsin (HR) and archaerhodopsin (AR) can suppress activity via fast onset of hyperpolarization [21]. When an opsin is activated by a photon of the appropriate wavelength light, the chromophore all-*trans*-retinal, which is covalently bound to the channel, is isomerized to 13-*cis*-retinal resulting in the opening of the channel. Depending on the channel, cations (ChR, AR) or anions (HR) flow with or are pumped against the electrochemical gradient across the cell membrane to change transmembrane potential. Mutant opsins are also available which are designed to improve upon light sensitivity, speed, and spectral response [22, 23]. In the case of cardiac optogenetics, the most commonly used mutant excitatory opsin is ChR2(H134R). ChR2(H134R) was one of the first single amino acid mutants generated and results in a 2–3× increase in channel conductance with minimal effect on kinetics [24].

Optogenetics opens the possibility for an all-optical investigation of neural and cardiac electrophysiology when optical mapping (using voltage- and calcium-sensitive dyes) is combined with these new optical actuation tools. Our 2011 study was the first to combine optical stimulation with high-speed, high-resolution optical mapping [20]. The integration of these techniques allowed the quantitative comparison of wave propagation and conduction properties of both optical and electrical stimulation. As will be discussed in this chapter, care must be taken when choosing opsins and optical sensors, such as calcium- and voltage-sensitive dyes, for a combined experiment, as excitation and emission spectra may overlap making data interpretation difficult.

This chapter describes the process for direct virally mediated opsin expression in cardiomyocytes, including methods for spatial gene patterning in cardiac syncytium. We specifically outline step-by-step procedures for in vitro infection of neonatal rat ventricular cardiomyocytes (NRVMs) with an adenovirus containing the transgene for ChR2(H134R) fused to an eYFP reporter protein with a ubiquitous CMV promoter. This technique is also applicable to (and has been tested in) other cardiac cell types (including adult cells) with appropriately designed optogenetic viruses and optimization of the viral infection protocol as described here.

2   Materials

2.1   Isolating and Preparing Cells for Adenoviral Infection

1. NRVM culture media: M199 media (Invitrogen, Grand Island, NY, USA) supplemented with 10 % fetal bovine serum (FBS), 12 μM L-glutamine, 0.02 μg/mL glucose, 0.05 μg/mL penicillin–streptomycin, and 10 mM HEPES. Store at 4 °C. Warm to 37 °C before use.

2. NRVMs prepared according to Jia et al. [20].

2.2   Determining the Optimal Multiplicity of Infection

1. Custom-designed Ad-ChR2(H134R)-eYFP or other virus containing an appropriate light-sensitive opsin. Viral titer is 9×10^{11} viral particles (VP)/mL (see **Note 1**).

2. Phosphate-buffered saline (PBS).

3. NRVM culture media: M199 media (Invitrogen, Grand Island, NY, USA) supplemented with 10 % FBS, 12 μM L-glutamine, 0.02 μg/mL glucose, 0.05 μg/mL penicillin–streptomycin, and 10 mM HEPES. Store at 4 °C. Warm to 37 °C before use.

4. NRVM infection media: M199 media (Invitrogen, Grand Island, NY, USA) supplemented with 2 % FBS (see **Note 2**), 12 μM L-glutamine, 0.02 μg/mL glucose, 0.05 μg/mL penicillin–streptomycin, and 10 mM HEPES. Store at 4 °C. Warm to 37 °C before use.

5. Fibronectin (BD Biosciences, San Jose, CA, USA): Prepare a stock solution (5 mg/mL) by adding 1 mL distilled water to the vial, aliquot, and store at –20 °C. For coating cell culture dishes, dilute stock solution in PBS for a final working concentration of 50 μg/mL. Unused fibronectin can be stored at 4 °C for up to 1 week.

6. Glass-bottomed dishes (14 or 20 mm) (In Vitro Scientific, Sunnyvale, CA, USA).

7. Tyrode's solution: Prepare a 5× concentrated Tyrode's solution by dissolving the following amounts of compounds in 1 L of distilled water: 1.02 g $MgCl_2$, 2.01 g KCl, 39.45 g NaCl, 0.2 g NaH_2PO_4, and 5.96 g HEPES. This concentrated solution can be stored at 4 °C for several months. To make a working Tyrode's solution mix 400 mL distilled water with 0.46 g D-glucose, 750 μL $CaCl_2$ (for 1.5 mM Ca^{2+} in final volume), and 100 mL of 5× Tyrode's stock. The concentrations (in mM) are as follows: NaCl, 135; $MgCl_2$, 1; glucose, 5; HEPES, 5; KCl, 5.4; $CaCl_2$, 1.5; NaH_2PO_4, 0.33. pH should be adjusted to 7.4 with NaOH.

8. Propidium Iodide (Life Technologies, Grand Island, NY, USA): Stain is packaged as a 1 mg/mL solution. For a final working concentration of 2 μg/mL, dilute 2 μL of stock solution in 1 mL of non-phosphate-containing solution, such as Tyrode's.

2.3 Patterned Transgene Expression	1. Sylgard® 184 Silicone Elastomer Kit (Fisher Scientific, Pittsburgh, PA, USA).
2.4 Optogenetic Actuation and Feedback	1. Quest Rhod-4, AM (AAT Bioquest, Sunnyvale, CA, USA): Prepare a stock solution of 0.5 mM which can be stored at –20 °C until experimental use. The solvent contains 20 % Pluronic (F127) and 80 % dimethyl sulfoxide (DMSO). Dilute to 10 µM in room-temperature Tyrode's solution for cardiomyocyte staining.

3 Methods

3.1 Isolating and Preparing Cells for Adenoviral Infection

Cells specific to this protocol (NRVM) are isolated using a previously published technique [20]. In short, the ventricular apex is excised from 2- to 3-day-old Sprague-Dawley rats and enzymatically digested overnight with trypsin at 4 °C and then with collagenase at 37 °C the following morning. Cardiac fibroblasts are removed from the cell suspension by a two-stage pre-plating procedure. At the end of the isolation procedure, cardiomyocytes are counted and resuspended in NRVM culture media (*see* Subheading 2 for recipe) to a concentration of 1.125×10^6 cells/mL.

3.2 Determining the Optimal Multiplicity of Infection

This section describes the viral infection of cardiomyocytes with opsin genes using a suspension approach with proper dosing for maximum efficiency while maintaining cell viability. Carry out all procedures under a Biosafety Level 2 (BSL-2) cabinet unless otherwise specified (*see* **Note 3**).

1. Prepare a cell suspension of NRVM at a concentration of 1.125×10^6 cells/mL in NRVM culture media (containing 10 % FBS, *see* Subheading 2 for recipe).

2. Remove adenovirus from storage at –20 °C and place on ice (*see* **Note 4**).

3. Separate the total volume of cell suspension into three (or more) conicals for multiplicity of infection (MOI) testing (*see* **Note 5**). For this protocol, we will describe MOI testing at 0 (control), 25, and 100.

4. Calculate the amount of viral particles needed for infection at the specified MOI (*see* **Note 6**):

$$\text{Viral particles} \left(\text{VP}\right) = \left(\text{total number of cells}\right) \times \left(\text{MOI}\right)$$

For example, to infect 2.25×10^6 cells (2 mL cell suspension at 1.125×10^6 cells/mL) at MOI 25, 56.25×10^6 VP are required.

5. Calculate and measure out the volume of virus needed for infection:

$$\text{Viral volume}\,(\mu L) = \frac{\text{Viral particles}\,(\text{VP})}{\text{Viral titer}\,(\text{VP}\,/\,\mu L)}$$

For example, to infect a 2 mL suspension of NRVM (2.25×10^6 total cells), a total viral volume of 62.5 μL is required.

6. Mix the viral volume (μL) calculated in **step 5** with infection media (containing 2 % FBS, *see* **Note 2** and Subheading 2 for recipe), so that the volume of virus and infection media maintain the original cell concentration of 1.125×10^6 cells/mL (*see* **Note 7**).

7. Spin down the cells at $720 \times g$ for 4 min.

8. Aspirate the culture media from the conical taking care not to disrupt the cell pellet, and resuspend the cells in the infection media and viral particle mixture from **step 6**.

9. Incubate the conicals at 37 °C, 5 % CO_2, for 2 h with agitation every 15–20 min during this time (*see* **Note 8**).

10. Spin down the cells at $720 \times g$ for 4 min.

11. Aspirate the infection media from the conical, again taking care not to disrupt the cell pellet, and resuspend the cells in fresh NRVM culture media maintaining the original cell concentration of 1.125×10^6 cells/mL.

12. Plate the cells in a monolayer at a density of 350–470 k cells/cm² on fibronectin-coated (50 μg/mL, *see* **Note 9**) glass-bottomed dishes.

13. Image transgene expression based on fluorescence of the reporter molecule, in this case eYFP (*see* Fig. 1).

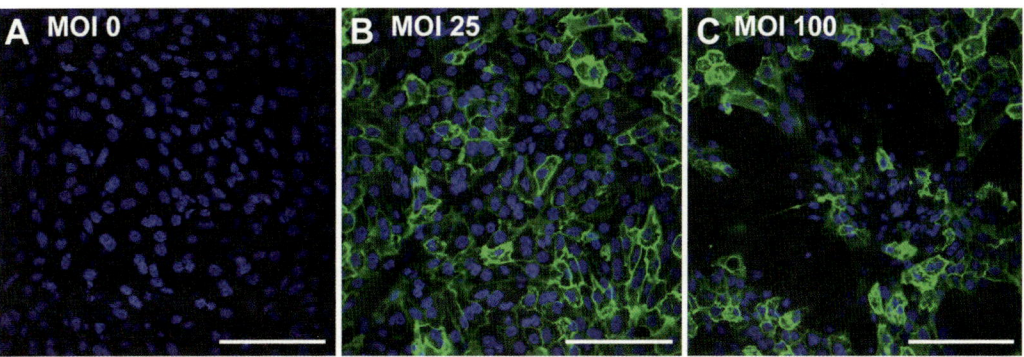

Fig. 1 ChR2 expression in cardiomyocytes: Representative fluorescent images of NRVMs infected with Ad-ChR2(H134R)-eYFP at (**a**) MOI 0, control; (**b**) MOI 25; and (**c**) MOI 100. *Blue stain* is DAPI. Scale bar is 100 μm

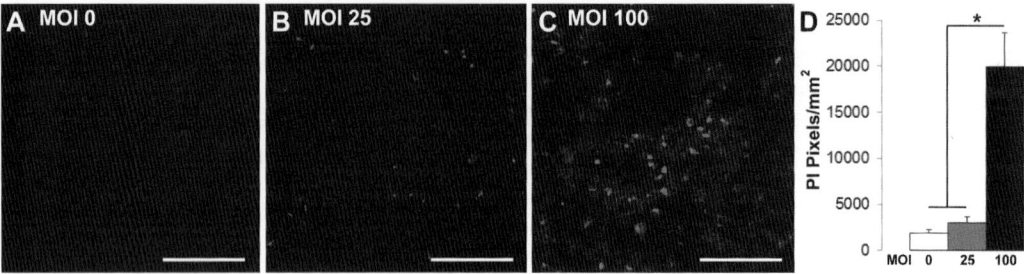

Fig. 2 Cell viability upon viral dosing: Representative fluorescent images of NRVMs infected with Ad-ChR2(H134R)-eYFP and stained with PI at (**a**) MOI 0, control; (**b**) MOI 25; and (**c**) MOI 100. Scale bar is 100 μm. (**d**) Quantification of PI staining presented as average number of pixels with PI stain above threshold normalized to area (mm^2)

Since optimization of the MOI requires maximizing expression efficiency and minimizing cytotoxic effects, propidium iodide (PI) staining can be used to quantify cell death. PI is a membrane-impermeant DNA stain, excluded from viable cells and commonly used to detect dead cells in a given population.

14. Prepare a 2 μg/mL solution of PI diluted in a Tyrode's solution (*see* Subheading 2 for recipe). This fluorescent stain is light sensitive, and therefore the application must be completed in the dark.

15. Remove media and add 0.5 mL of the 2 μg/mL solution to the glass-bottomed dish containing the cells of interest.

16. Incubate the cells with PI for 2 min.

17. Remove the dye solution and wash with fresh Tyrode's solution.

18. Image cell viability based on the fluorescence of PI using appropriate excitation and emission filter sets (*see* Fig. 2). When PI is bound to nucleic acids, the maximum excitation and emission are 535 nm and 617 nm, respectively.

19. Based on the overall transgene (ChR2(H134R)-eYFP) expression and cell viability, determine the optimal MOI to be used in future experiments (*see* **Note 10**). For NRVM infected with our custom-designed Ad-ChR2(H134R)-eYFP, the optimal MOI was found to be 25 based on maximal expression and minimal cell death.

3.3 Patterned Transgene Expression

The ability to infect NRVM cells in suspension, as opposed to in dish infection after cell plating, allows for patterned optogenetic transgene expression, as described in this section. To date, we have patterned islands of ChR2(H134R)-eYFP-expressing cells using silicone elastomer stencils, in addition to graded mixtures of

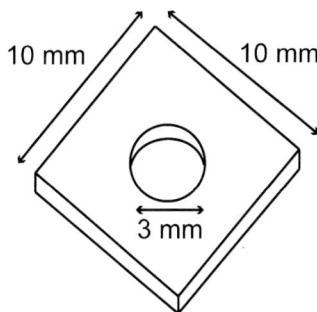

Fig. 3 Patterning stencils: Example schematic of silicone elastomer stencil used to pattern an island of ChR2-expressing NRVMs surrounded by normal, uninfected NRVMs. Not drawn to scale

infected and uninfected cells mimicking potential in vivo transgene expression patterns.

1. Thoroughly mix silicone elastomer in the ratio of 10 parts base to 1 part curing agent (total weight of this mixture is 2.2 g).

2. Pour elastomer mixture into a 60 mm petri dish. Allow the elastomer to spread out and cover the entire bottom of the dish. This volume will yield a 200–250 μm thick layer of silicone.

3. Bake in an oven at 50–60 °C for 2 h. Alternatively, the elastomer can be cured at room temperature for 24 h.

4. In the meantime, coat glass-bottomed dishes with 50 μg/mL of fibronectin (*see* **Note 9**). Cover dishes and leave in the incubator (37 °C, 5 % CO_2) for 1–2 h.

5. When the elastomer is finished setting, cut out stencils, such as that illustrated in Fig. 3, in the shape of the desired pattern. The stencil illustrated here will create an island of light-sensitive cardiomyocytes surrounded by normal, uninfected cardiomyocytes or vice versa.

6. When the glass-bottomed dishes have been coated with fibronectin, remove the solution and carefully place the stencils on the glass-bottomed dish in the desired location (*see* **Note 11**).

7. Plate a droplet of infected cells (350–470 k cells/cm²) in the stencil cutouts and let sit for 30–40 min (*see* **Note 12**). This will be sufficient time to allow the infected NRVM to begin adhering to the dish. The droplet volume is determined in such a way that the open stencil area should be covered at a target density of 350–470 k cells/cm².

8. Carefully, remove the stencil and plate uninfected cells at the same plating density in a monolayer on the glass-bottomed dish.

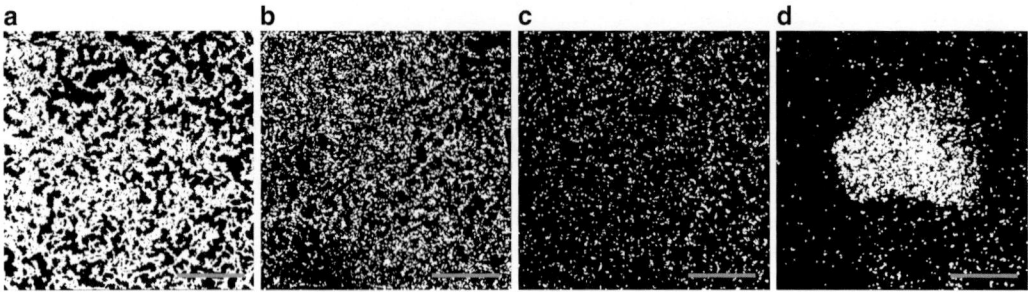

Fig. 4 Spatial patterning of ChR2 expression: Representative binarized distributions of ChR2-expressing NRVMs (eYFP, *white*) in four configurations: (**a**) global expression of ChR2; (**b**) mixture of 35 % ChR2-expressing NRVMs and 65 % non-infected NRVMs; (**c**) mixture of 10 % ChR2-expressing NRVMs and 90 % non-infected NRVMs; (**d**) island of ChR2-expressing NRVMs

9. To plate uniform, graded mixtures of infected and uninfected cells, mix various concentrations, such as 35 % infected cells and 65 % uninfected cells, that have been maintained in suspension during infection, and plate these mixtures at a density of 350–470 k cells/cm².

10. Image transgene expression based on fluorescence of the reporter molecule (*see* Fig. 4 for examples of transgene distributions).

3.4 Optogenetic Actuation and Feedback

This section describes means of functional testing of the cardio-myocyte samples transduced with the opsin of interest. It demonstrates a combined optical actuation and optical sensing. The macroscopic (large field of view) optical mapping described in this section is done using a custom-developed ultrahigh-resolution system which has been previously described in detail [20]. Other charge-coupled diode (CCD)-, photodiode array (PDA)-, and complementary metal-oxide-semiconductor (CMOS)-based optical mapping systems can also be adapted to this protocol (for review *see* ref. 25). Alternatively, the optical sensing can be done with a microscope-integrated photodetector [17]. Here we focus on the choice of optical sensors (i.e., calcium- and voltage-sensitive dyes) and how to trigger optical excitation in NRVM monolayers.

All measurements described here are recorded at room temperature and in the dark so as not to photobleach the fluorescent dyes and/or activate the opsins.

1. Remove a dish with an NRVM monolayer containing cells expressing ChR2(H134R) from the incubator (37 °C, 5 % CO_2).

2. Remove culture media from the dish, and incubate cells with 10 µM Quest Rhod-4AM (*see* Subheading 2 for recipe), a robust calcium-sensitive dye (*see* **Note 13**), diluted in room-temperature Tyrode's solution for 20 min.

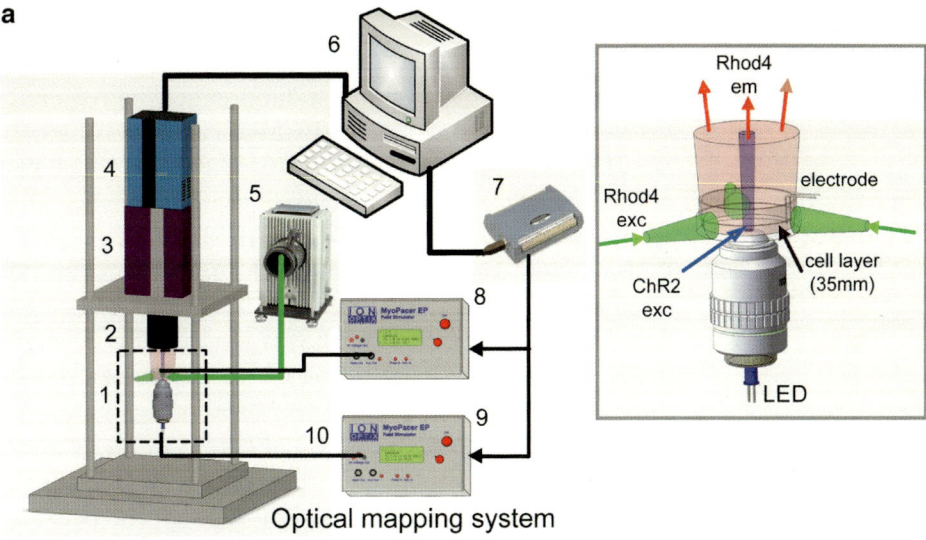

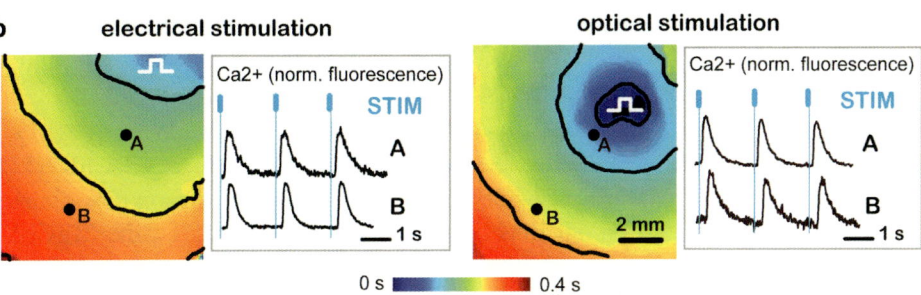

Fig. 5 Combining optical actuation with optical imaging: (**a**) Experimental setup for high-resolution, high-speed optical imaging and optical control of cardiac syncytia. (**b**) Activation maps in cardiac syncytia by electrical and optical pacing at 0.5 Hz. Normalized calcium transients (acquired with Rhod 4 staining) are shown from two locations—A and B

3. Remove dye solution, and wash the cells with a 20-min incubation in fresh room-temperature Tyrode's solution.

4. Place a freshly stained glass-bottomed dish on the optical mapping setup (*see* **Note 14**).

5. Record electrical activation of the monolayer by pacing with custom-built platinum electrodes connected to a Myopacer Cell Stimulator or equivalent (*see* Fig. 5b for example of electrical pacing output and **Note 15**).

6. Record optical activation of the monolayer using light pulses with a wavelength specific to the opsin integrated into the virus (i.e., 470 nm for ChR2(H134R)) (*see* Fig. 5b for example of optical pacing output and **Notes 16–18**). According to a strength–duration curve, the pulse duration can be decreased if higher optical power is used. For example, for our samples, 10-ms optical pulses of $0.01 \ mW/mm^2$ are sufficient to trigger activity in NRVM uniformly expressing ChR2(H134R) (*see* **Note 19**).

4   Notes

1. Depending on the viral titer, a dilution may be required prior to infection. In our case, Ad-ChR2(H134R) (original titer: 9×10^{11} VP/mL) is diluted in 1 mL of PBS for a final working concentration of 9×10^8 VP/mL.

2. High levels of protein in the culture media (with 10 % FBS) can interfere with the binding of the adenovirus (or any viral vector) to their cellular receptors. Therefore, a lower serum infection media (with 2 % FBS) is used during infection as a trade-off with no-serum media which cardiomyocytes do not tolerate well.

3. All work with adenoviruses should be completed under the NIH guidelines for working with recombinant DNA. Infections are performed under a certified BSL-2 hood, and personal protection equipment (gloves, lab coat) are worn at all times while handling infectious materials. In addition, for decontamination, all pipets, dishes, and conicals are rinsed with 10 % bleach and autoclaved.

4. Adenoviruses should not undergo multiple freeze/thaw cycles as they could negatively affect performance. Ad-ChR2(H134R) is stored at −20 °C in a 25 % glycerol solution (2× storage solution: 10 mM Tris pH, 100 mL NaCl, 1 mM $MgCl_2$, 0.1 % BSA, and 50 % glycerol).

5. The amount of virus needed for efficient infection of cells (i.e., successful, widespread expression of the transgene and low cytotoxicity) will differ depending on the cell type. For NRVMs, MOIs ranging from 25 to 100 were tested.

6. VP represent the total amount of particles in a volume irrespective of their function. Transducing units (TU), plaque-forming units (PFU), or infectious units (IFU) represent the number of functional viral particles in a given volume. For most viral preparations the VP:PFU ratio ranges from 20:1 to 50:1.

7. In this protocol, infection is completed with the cells in suspension. Provided the cells (like NRVM) tolerate remaining in suspension for a sufficient period of time, infection in this manner allows for better access of the viral particles to the cell surface (membrane) and (in our hands) results in healthier cells post-infection. NRVMs tolerate this protocol well; however it may not be suitable for all cell types.

8. Make sure to shake hard enough to disrupt the pellet of cells and resuspend them sufficiently in order to fully expose them to the virus for infection.

9. Neonatal cells show preference for fibronectin-coated surfaces, whereas adult cells tend to show preference for laminin-coated surfaces [26].

Table 1
Excitation spectra for opsin, reporter protein, and imaging dyes

	Peak excitation (nm)	Peak emission (nm)
Opsin		
ChR2	470 ± 20	n/a
Reporter protein		
eYFP	515	530
Imaging dyes		
Ca^{2+} (Rhod-4)	525 ± 40	585 ± 40
V_m (di-8-ANEPPS)	525 ± 40	610 ± 75

10. Viral performance may vary between batches. Therefore, a repeat optimization protocol may be necessary.

11. When placing stencils on the fibronectin-coated glass-bottomed dishes, make sure that the dish is dry enough so that the stencil sticks. If not, the cell droplet will spread under the stencil when cells are plated.

12. Ensure that the droplet is of a large enough volume so as not to dry up during this first step of the cell plating procedure. In our experience, a 10 μL droplet is of a sufficient volume and size when placed within a 3 mm diameter stencil.

13. The calcium-sensitive dye—Quest Rhod 4 AM—was chosen in this instance because of the compatibility of its excitation and emission spectra with ChR2 (*see* Table 1 for spectral characteristics). In choosing a dye for optical mapping, it is also critical to account for the reporter gene of the opsin (in this case, eYFP), as it is not ideal to excite it in a way that emits too much light during optical mapping, potentially masking the optical response of the monolayer. As also highlighted in Table 1, since di-8-ANEPPS has the same spectral characteristics as Rhod 4, this voltage-sensitive dye may be used in place of the calcium-sensitive dye in the experiments described in this protocol. It should be noted, however, that di-8-ANEPPS has been reported to have a lower fractional change in fluorescence at the cellular level [25] and, consequently, optics may need to be adjusted in order to collect enough light for an analyzable optical signal.

14. Our custom-developed macroscopic optical mapping system [27], as shown in Fig. 5a, includes a CMOS camera (pco, Germany), capable of recording at 200 frames per second over $1,280 \times 1,024$ pixels; a Gen III fast-response intensifier (Video Scope International, Dulles, VA); light collection optics

(Navitar Platinum lens, 50 mm, f/1.0) and emission filter (610/75 nm); excitation light source (525 nm, Oriel); and an adjustable imaging stage. Excitation light for the calcium- or the voltage-sensitive dye is delivered using tangential illumination (i.e., at a 90° angle with respect to the imaging axis) by a QTH lamp with branching liquid light guides. Excitation light for the opsins is provided through the bottom of the dish (*see* **Notes 16** and **17**).

15. Propagation data, in the cases of both electrical and optical stimulation, are collected using the commercially available software CamWare (pco, Germany). Raw data is then processed and analyzed in custom-developed Matlab software to quantify propagation, conduction velocity, and transient morphology, among other measures. In addition, all data are typically spatially filtered (Bartlett filter; kernel size: 10 pixels) as well as temporally filtered (Savitzky–Golay filter; order: 2, width: 11).

16. To date, we have used two types of illumination systems with which to optically stimulate light-sensitive monolayers: LED and laser based. The choice of illumination system will, in part, depend on the power requirements of the excitable syncytium and the use of any fiber optics downstream of the light source, which can decrease power at its delivery point.

17. As one of the advantages of optogenetics is its high spatiotemporal resolution, it is possible to utilize patterned illumination in the depolarization/hyperpolarization of optogenetically modified excitable syncytium [28, 29]. To date, in our experiments, we have applied single-point (1 mm diameter spot size) illumination, using a fiber optics-coupled high-power blue LED (470 nm, 1.6 A; ThorLabs, Newton, NJ, USA), as well as global (0.79 cm² spot size) illumination, using a fiber optics-coupled diode-pumped solid-state (DPSS) blue laser (470 nm, 120 mW; Shanghai Laser and Optics Century Co., Shanghai, China). We have not directly compared the energies required for point versus global illumination; however, depending on the distribution of optogenetic elements in the syncytium, single-point illumination may not provide sufficient energy for optical excitation.

18. The energy for optical stimulation is measured and reported in terms of irradiance (mW/mm²). Irradiance is measured at the site of light delivery using an optical power meter with an appropriately sized sensor area (0.785 cm²). All reported irradiance measurements are taken at the end of each experiment as slight, unintentional alterations in the position of the sample and/or optical fiber can significantly alter the resultant power measurement. In addition, if optically illuminating from the bottom of the glass-bottomed dish (*see* Fig. 5a), it is also

important to measure the power through a glass-bottomed dish as it could attenuate light. Threshold irradiance levels required to trigger propagating action potentials in test samples are directly related to the expression levels and functionality of the virally introduced opsins.

19. The availability of retinal is necessary and essential for the proper light-sensing abilities of ChR2 and other opsins. As a chromophore, retinal covalently binds to opsins to form photosensitive receptors. Without retinal, no functional opsin channels can be assembled. In our experience with NRVM, we have not had to add exogenous retinal, as NRVMs contain sufficient retinal for opsin function. It has, however, been reported that the addition of exogenous retinal can increase optical responsiveness [30].

Acknowledgements

This work was supported by National Heart, Lung, and Blood Institute Grant R01-HL-111649 (E.E.), an Institutional National Service Research Award T32-DK07521 (C.M.A.), and partially a NYSTEM grant C026716 to the Stony Brook Stem Cell Center.

We would also like to thank Varsha Sitaraman, PhD; Jinzhu Yu, BS; and Kay Chen, BS, for their contributions.

References

1. Boyden ES et al (2005) Millisecond-timescale, genetically targeted optical control of neural activity. Nat Neurosci 8(9):1263–1268

2. Bernstein JG, Boyden ES (2011) Optogenetic tools for analyzing the neural circuits of behavior. Trends Cogn Sci 15(12):592–600

3. Deisseroth K et al (2006) Next-generation optical technologies for illuminating genetically targeted brain circuits. J Neurosci 26(41):10380–10386

4. Fenno L et al (2011) The development and application of optogenetics. Annu Rev Neurosci 34:389–412

5. Zhang F et al (2007) Circuit-breakers: optical technologies for probing neural signals and systems. Nat Rev Neurosci 8(8):577–581

6. Airan RD et al (2007) High-speed imaging reveals neurophysiological links to behavior in an animal model of depression. Science 317(5839):819–823

7. Kim SY et al (2013) Diverging neural pathways assemble a behavioural state from separable features in anxiety. Nature 496(7444):219–223

8. Britt JP, Bonci A (2013) Optogenetic interrogations of the neural circuits underlying addiction. Curr Opin Neurobiol 23:1–7

9. Lobo MK et al (2010) Cell type-specific loss of BDNF signaling mimics optogenetic control of cocaine reward. Science 330(6002):385–390

10. Carter ME et al (2009) Sleep homeostasis modulates hypocretin-mediated sleep-to-wake transitions. J Neurosci 29(35):10939–10949

11. Krook-Magnuson E et al (2013) On-demand optogenetic control of spontaneous seizures in temporal lobe epilepsy. Nat Commun 4:1376

12. Tonnesen J et al (2009) Optogenetic control of epileptiform activity. Proc Natl Acad Sci U S A 106(29):12162–12167

13. Aravanis AM et al (2007) An optical neural interface: in vivo control of rodent motor cortex with integrated fiber optic and optogenetic technology. J Neural Eng 4(3):S143–S156

14. Kravitz AV et al (2010) Regulation of parkinsonian motor behaviours by optogenetic control of basal ganglia circuitry. Nature 466(7306):622–626

15. Tonnesen J et al (2011) Functional integration of grafted neural stem cell-derived dopaminergic neurons monitored by optogenetics in an in vitro Parkinson model. PLoS One 6(3):e17560

16. G N et al (2013) Channelrhodopsins: visual regeneration and neural activation by a light switch. N Biotechnol 30(5):461–474

17. Entcheva E (2013) Cardiac optogenetics. Am J Physiol Heart Circ Physiol 304(9): H1179–H1191

18. Bruegmann T et al (2010) Optogenetic control of heart muscle in vitro and in vivo. Nat Methods 7(11):897–900

19. Abilez OJ et al (2011) Multiscale computational models for optogenetic control of cardiac function. Biophys J 101(6):1326–1334

20. Jia Z et al (2011) Stimulating cardiac muscle by light: cardiac optogenetics by cell delivery. Circ Arrhythm Electrophysiol 4(5):753–760

21. Kleinlogel S et al (2011) A gene-fusion strategy for stoichiometric and co-localized expression of light-gated membrane proteins. Nat Methods 8(12):1083–1088

22. Mattis J et al (2012) Principles for applying optogenetic tools derived from direct comparative analysis of microbial opsins. Nat Methods 9(2):159–172

23. Yizhar O et al (2011) Optogenetics in neural systems. Neuron 71(1):9–34

24. Lin JY (2012) Optogenetic excitation of neurons with channelrhodopsins: light instrumentation, expression systems, and channelrhodopsin variants. Prog Brain Res 196:29–47

25. Herron TJ et al (2012) Optical imaging of voltage and calcium in cardiac cells & tissues. Circ Res 110(4):609–623

26. Lam ML et al (2002) The 21-day postnatal rat ventricular cardiac muscle cell in culture as an experimental model to study adult cardiomyocyte gene expression. Mol Cell Biochem 229(1–2):51–62

27. Entcheva E, Bien H (2006) Macroscopic optical mapping of excitation in cardiac cell networks with ultra-high spatiotemporal resolution. Prog Biophys Mol Biol 92(2):232–257

28. Arrenberg AB et al (2010) Optogenetic control of cardiac function. Science 330(6006): 971–974

29. Papagiakoumou E et al (2010) Scanless two-photon excitation of channelrhodopsin-2. Nat Methods 7(10):848–854

30. Ullrich S et al (2013) Degradation of channelopsin-2 in the absence of retinal and degradation resistance in certain mutants. Biol Chem 394(2):271–280

Chapter 20

Methods for Assessing the Electromechanical Integration of Human Pluripotent Stem Cell-Derived Cardiomyocyte Grafts

Wei-Zhong Zhu, Dominic Filice, Nathan J. Palpant, and Michael A. Laflamme

Abstract

Cardiomyocytes derived from human pluripotent stem cells show tremendous promise for the replacement of myocardium and contractile function lost to infarction. However, until recently, no methods were available to directly determine whether these stem cell-derived grafts actually couple with host myocardium and fire synchronously following transplantation in either intact or injured hearts. To resolve this uncertainty, our group has developed techniques for the intravital imaging of hearts engrafted with stem cell-derived cardiomyocytes that have been modified to express the genetically encoded protein calcium sensor, GCaMP. When combined with the simultaneously recorded electrocardiogram, this protocol allows one to make quantitative assessments as to the presence and extent of host–graft electrical coupling as well as the timing and pattern of graft activation. As described here, this system has been employed to investigate the electromechanical integration of human embryonic stem cell-derived cardiomyocytes in a guinea pig model of cardiac injury, but analogous approaches should be applicable to other human graft cell types and animal models.

Key words Human pluripotent stem cells, Zinc finger nuclease, GCaMP3, Cardiac repair, Intravital imaging, Electromechanical coupling

1 Introduction

In recent years, there has been tremendous interest in stem cell- and tissue engineering-based approaches to replace the muscle that is lost following myocardial infarction. For example, our group and others have shown that human embryonic stem cell-derived cardiomyocytes (hESC-CMs) can partially remuscularize the infarct scar and improve contractile function in rodent models of myocardial infarction [1–4]. However, to contribute functionally meaningful new force-generating units and avoid creating electrically excitable but imperfectly coupled tissue that might promote arrhythmias, graft cells must undergo appropriate

Milica Radisic and Lauren D. Black III (eds.), *Cardiac Tissue Engineering: Methods and Protocols*, Methods in Molecular Biology, vol. 1181, DOI 10.1007/978-1-4939-1047-2_20, © Springer Science+Business Media New York 2014

electromechanical integration and contract synchronously with host myocardium during systole.

A number of strategies have been employed to assess the electromechanical integration of various graft cell types following transplantation into either intact or injured hearts. Historically, the most common approach has been to transplant graft cells tagged with a static fluorescent label, such as green fluorescent protein (GFP) or DiO. With this approach, the engrafted heart is later harvested and stained with a spectrally distinct fluorescent voltage- or calcium-sensitive dye that reports cardiomyocyte activation [5–10]. One then looks for evidence of activation (e.g., optical action potentials or calcium transients) arising from the tagged graft tissue. However, a major challenge with this approach is the need to reliably distinguish between graft and host, since both will have been labeled with the dye. Graft implants commonly overlie surviving host myocardium, so distinguishing between the two becomes particularly problematic when imaging the heart from the epicardial surface. Illustrating this problem, our group recently used an independent technique to identify hearts with *uncoupled* grafts, and we found that deep surviving host tissue labeled with the voltage dye RH237 emitted optical action potentials [3]. Unlike the graft, these optical action potentials were perfectly synchronized with the host ECG, and we would have otherwise erroneously interpreted them as graft derived. It may therefore be cause for concern that nearly all studies using this sort of an approach have reported apparently "seamless" graft integration; that is, both graft and host tissues showed similar optical action potential amplitudes, patterns of graft activation, and conduction velocities.

One partial solution to this problem is to use higher resolution techniques such as two-photon microscopy to distinguish between host and graft cardiomyocytes at the single-cell level [8], but the latter does not allow one to assess graft behavior at the macroscopic level or acquire critical tissue-level parameters (e.g., graft conduction velocity). An arguably better solution—and the one that forms the basis for the protocol described here—is to use a graft-autonomous reporter in which the activation signal arises *solely* from the graft. The genetically encoded fluorescent calcium sensor GCaMP serves as a convenient reporter for this purpose, since cardiomyocytes exhibit cytosolic calcium transients with each depolarization and contractile cycle. A large family of increasingly bright GCaMP variants has now been created, each of which incorporates moieties including circularly permuted GFP (cpGFP), calmodulin (CaM), and a CaM-interacting M13 domain from myosin light chain kinase [8, 11–13]. Cardiomyocytes expressing GCaMP show robust fluorescent transients in vitro with each contractile cycle, and they form intracardiac grafts that emit fluorescence transients that can then be correlated with the ECG of recipient hearts [3, 14].

Below, we describe how to use zinc finger nuclease (ZFN)-mediated transgenesis [15, 16] to generate a stable hESC line expressing a GCaMP variant called GCaMP3 as well as how to generate large quantities of enriched GCaMP3$^+$ hESC-CMs. We then provide detailed methods for creating a guinea pig model of cardiac injury, transplanting GCaMP3$^+$ hESC-CMs in this model, and ultimately the intravital imaging of GCaMP3$^+$ hESC-CM grafts. Our lab recently used these same methods to demonstrate that, while hESC-CMs exhibit reliable 1:1 host–graft coupling in uninjured hearts, their electromechanical integration is imperfect following transplantation in injured hearts (presumably due to intervening scar tissue) [3]. While this protocol is written for this specific cell type and animal model, the same general approach should be useful in assessing the electromechanical integration of other graft cell types as well as engineered tissue constructs. In addition to overcoming the aforementioned concerns about factitious graft-derived signals, the ZFN-mediated targeting strategy used to insert the GCaMP3 expression cassette can be directly applied to any human cell type.

2   Materials

2.1   Generation of the GCaMP3 Reporter hESC Line

1. H7/WA07 hESC line (Wicell Research Institute, Madison, WI) (see **Note 1**).
2. AAVS1 ZFN plasmid (see **Notes 2** and **3**).
3. pZDonor-GCaMP3 plasmid (see **Notes 2** and **3**).
4. Electroporator.
5. Human stem cell Nucleofector kit 1 (Lonza).
6. LY294002 stock solution: 5 mg LY294002 is dissolved in 1.6 mL DMSO to make 10 mM stock solution. Aliquot, and store at −20 °C.
7. 50 mg/mL G418.
8. Accutase solution.
9. Mouse embryonic fibroblast conditioned medium (MEF-CM) and Matrigel-coated plates (please see ref. 17, 18).
10. Human basic fibroblast growth factor (hbFGF): Dissolve into PBS (with 0.2 % BSA) to make 10 μg/mL stock solution. Store aliquots at −20 °C.

2.2   Generation of GCaMP3$^+$ hESC-Derived Cardiomyocytes

1. Versene.
2. RPMI-B27 medium: 98 % (v/v) RPMI 1640, 2 % B27 serum supplement minus insulin, and 2 mM L-glutamine.
3. Activin A: Make 10 μg/mL stock solution in PBS (with 0.2 % BSA). Store aliquots at −20 °C.

4. BMP-4: Make 1 μg/mL stock solution in PBS (with 0.2 % BSA). Store aliquots at −20 °C.

5. Y27632 dihydrochloride (ROCK inhibitor): Dissolve in sterile water to make 10 mM stock solution. Store aliquots at −20 °C.

2.3 Cryopreservation of GCaMP3+ hESC-CMs

1. Trypsin–EDTA 0.05 % solution and defined trypsin inhibitor.

2. Insulin-like growth factor-1 (IGF-1).

3. Cyclosporine A (CsA).

4. DNase I: Dissolved to 400 U/μL with PBS. Sterile filter, and store at −20 °C.

5. Cryostor CS10 (BioLife Solutions).

6. Controlled-rate freezer.

2.4 Preparation of GCaMP3+ hESC-CMs for Intracardiac Transplantation

1. Matrigel-based pro-survival cocktail (PSC): Thaw 1 mL undiluted Matrigel on ice. Mix CsA (200 nM), ZVAD-FMK (10 μM), BCL-XL (50 nM), IGF-1 (0.1 μg/mL), and pinacidil (50 μM) into Matrigel to make PSC.

2. RPMI based PSC: Add all the preceding factors into RPMI medium instead of Matrigel.

3. Insulin syringes with 27G needle.

2.5 Cardiac Cryoinjury and Transplantation of GCaMP3+ hESC-CMs into Guinea Pig Hearts

1. Male guinea pigs (650–750 g).

2. A general surgical instrument set, including needle holder, tweezers, forceps, scissors, eye scissors, retractor, blades, and scalpel handle.

3. Sutures: 4-0 Vicryl coated, 4-0 silk, 5-0 nylon.

4. 8 mm diameter aluminum cryoprobe, liquid nitrogen, and thermal container.

5. Pediatric laryngoscope and 14G shielded intravenous catheter.

6. Heating pad.

7. Skin cleaning agents: 70 % ethanol, chlorhexidine gluconate, Betadine.

8. Lubrifresh opthalmic ointment.

9. Rodent ventilator and anesthesia vaporizer with isoflurane.

10. Anesthetics: Ketamine (50 mg/kg), xylazine (2 mg/kg), isoflurane (1.5 %).

11. Analgesics: 0.025 % (v/v) Bupivacaine, buprenorphine (0.05 mg/kg).

12. Immunosuppressive drugs: CsA and methylprednisolone.

13. Antibiotic: 0.3 mL/kg Baytril® (2.27 % enrofloxacin injectable solution).

2.6 Intravital Imaging of GCaMP3+ hESC-CM Grafts in Guinea Pig Hearts

1. Epifluorescence stereomicroscope.

2. Fluorescence illumination system.

3. High-speed CCD camera (e.g. Andor iXon).

4. Data acquisition system.

5. Bioamplifier.

6. Langendorff heart system.

7. Langendorff perfusion solution (mM): 25.0 $NaHCO_3$, 1.2 $MgSO_4$, 4.7 KCl, 118.0 NaCl, 1.2 KH_2PO_4, 11.0 glucose, 1.0 sodium pyruvate, 1.8 $CaCl_2$. Oxygenate with a submerged bubbler with 95 % O_2/5 % CO_2 gas for about 10 min before adding $CaCl_2$ to stabilize pH, then 0.2 μm filtered.

8. Blebbistatin (10 μM): 10 mM Stock solution in DMSO and store at –20 °C. Dilute in Langendorff perfusion solution (above) immediately before use.

2.7 Analysis of GCaMP3 Fluorescence Signals

1. Matlab software (Mathworks) and a PC.

3 Methods

3.1 Generation of the GCaMP3 Reporter hESC Line (See Notes 1–4)

Transgenic hESCs stably expressing the GCaMP3 fluorescent reporter are generated by ZFN-mediated insertion of the GCaMP3 expression cassette into the AAVS1 locus using methods adapted from Hockemeyer et al. [15] (see also Fig. 1). In brief, this protocol involves transient transfection with two plasmids (the AAVS1 ZFN and pZDonor-GCaMP3 vectors), both of which are available from our group upon request or can be generated de novo as described in **Note 2**. Prior to transfection, undifferentiated wild-type hESCs are cultured under feeder-free conditions, using Matrigel-coated plates and daily feeding with MEF-CM plus bFGF (4 ng/mL). Methods for the feeder-free growth of hESCs have been detailed elsewhere by our group and others [17, 18].

1. For at least 4 h prior to transfection, treat hESCs with MEF-CM supplemented with bFGF (4 ng/mL) and LY294002 (10 μmol/L).

2. Prepare plasmid DNA cocktail containing 5 μg of AAVS1 ZFN and 20 μg of pZDonor GCaMP3 donor vectors. Immediately prior to transfection, mix the DNA cocktail with Amaxa Human Stem Cell Nucleofector Solution 1 (see **Notes 3** and **5**).

3. Rinse hESCs with PBS and then treat with Accutase for approximately 5 min. Remove the Accutase, rinse with PBS,

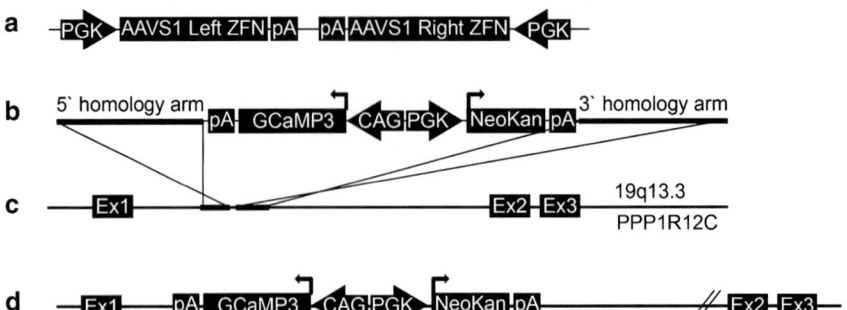

Fig. 1 Zinc finger nuclease (ZFN)-mediated targeting of the GCaMP3 expression cassette to the AAVS1 locus. (**a**) The AAVS1 ZFN plasmid includes two expression cassettes, in which separate human PGK promoters drive the expression of the AAVS1 site-specific ZFN pair. (**b**) The pZDonor-GCaMP3 plasmid includes GCaMP3 and neomycin selection cassettes that are flanked by AAVS1 site homology arms (each ~800 bp). GCaMP3 expression is driven by the CAG promoter, while the neomycin resistance gene is driven by the human PGK promoter. (**c**) The AAVS locus is located between exons 1 and 2 of the PPR1R12C gene on chromosome 19. (**d**) Co-transfection of AAVS1 ZFN and ZDonor-GCaMP3 plasmids in human cells results in targeting of the latter vector to the AAVS1 locus by homologous recombination (*see* Subheading 3.1 for details). *pA* bovine growth hormone polyadenylation signal, *Ex* exon, *NeoKan* neomycin/kanamycin resistance

and then triturate hESCs into a single-cell suspension in MEF-CM.

4. Count hESCs using a hemacytometer. Aliquot 4×10^5 cells into a microcentrifuge tube and spin down at $300 \times g$ for 5 min. Resuspend the pellet in Nucleofector Solution 1 with the DNA cocktail, and place it in a nucleofector cuvette. Proceed with nucleofection as per vendor instructions using program setting B-16.

5. After transfection, gently resuspend hESCs in MEF-CM and then replate onto Matrigel-coated 10 cm plates in 10 mL volume of MEF-CM supplemented with bFGF (4 ng/mL) and LY294002 (10 μmol/L). Thereafter, re-feed the cells with daily exchanges of this same medium.

6. Begin antibiotic selection of successfully targeted hESCs 4 days after nucleofection, a time point by which typical hESC colonies should have reformed. For this, treat cells with MEF-CM supplemented with bFGF (4 ng/mL) and G418 (75 μg/mL), but no LY294002.

7. Continue antibiotic selection by daily re-feeding of cells with MEF-CM supplemented with bFGF (4 ng/mL) and 75 μg/mL G418 for a total of 4 days. Thereafter, reduce the G418 to a maintenance concentration of 20 μg/mL.

8. After obtaining stable GCaMP3+ hESC cultures, verify their karyotype and confirm proper targeting of the AAVS1 locus by Southern blotting (*see* **Note 6**).

3.2 Generation of GCaMP3⁺ hESC-Derived Cardiomyocytes

GCaMP3+ hESCs are differentiated into cardiomyocytes using methods we have previously reported for wild-type cultures [17, 18]. Please *see* Fig. 2a for an overview of the cardiac differentiation protocol. Prior to the onset of differentiation, transgenic GCaMP3+ hESCs are maintained in the undifferentiated state by culture on Matrigel-coated plates in the presence of MEF-CM plus bFGF (4 ng/mL). Importantly, because there should be no residual feeder cells present, cultures should be maintained under feeder-free conditions for at least 2–3 passages prior to attempting cardiac differentiation. When the undifferentiated hESC colonies occupy approximately two-thirds of the surface area, they are typically ready for cardiac differentiation.

1. Treat GCaMP3⁺ hESCs with 0.2 mL/cm² Versene at 37 °C until all of the cells are rounded up but not yet detached. This typically requires 3–7 min, but cultures should be checked at least every 1–2 min during this step.

2. Gently aspirate the Versene, and replace it with 0.1 mL/cm² MEF-CM supplemented with bFGF (4 ng/mL) and the

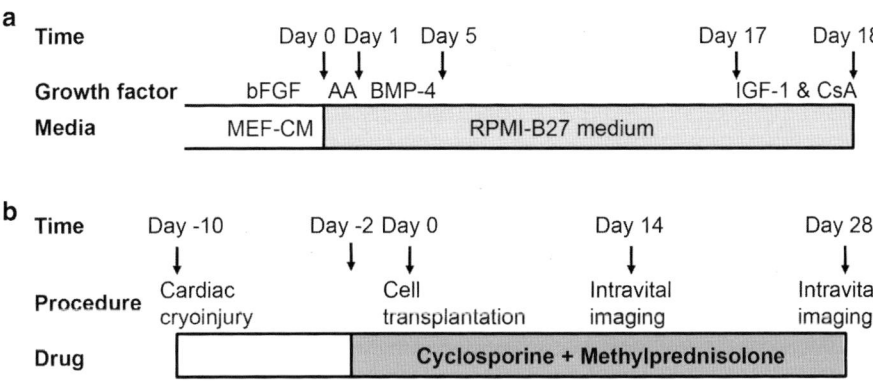

Fig. 2 Timelines for the cardiac differentiation of GCaMP+ hESCs and the intravital imaging of GCaMP3+ hESC-CM grafts in injured hearts. (**a**) GCaMP3+ hESC-CMs are generated using our directed cardiac differentiation protocol, previously described for wild-type hESCs [17] (*see* Subheading 3.2 for details). In brief, undifferentiated GCaMP3+ hESCs are expanded under standard feeder-free conditions using mouse embryonic fibroblast-conditioned medium (MEF-CM) supplemented with human basic fibroblast growth factor (bFGF). Undifferentiated cultures are then replated to form a dense monolayer in MEF-CM plus bFGF. After a compact monolayer is formed, hESCs are switched to RPMI-B27 medium and serially pulsed with activin A (AA) on day 0 and bone morphogenetic protein-4 (BMP-4) from days 1 to 5. Thereafter, the differentiating cultures are grown in RPMI-B27 medium without any exogenous factors. One day before harvest, cells are heat-shocked and treated with insulin-like growth factor-1 (IGF-1) and cyclosporine (CsA) to improve graft survival. Cells are typically harvested on ~day 18 for cryopreservation or immediate transplantation. (**b**) To evaluate the electromechanical integration of GCaMP3+ hESC-CM grafts in a subacute model of cardiac injury, guinea pigs undergo a thoracotomy and direct cryoinjury of the left ventricle. 10 days later, a repeat thoracotomy is performed, and GCaMP3+ hESC-CMs are injected into the injured heart. Intravital imaging is typically performed either 14 or 28 days post-transplantation. To avoid graft rejection, animals are treated from day -2 to day 28 post-transplantation with methylprednisolone and CsA

ROCK inhibitor Y-27632 dihydrochloride (10 µmol/L). Dislodge the cells by pipetting the aforementioned medium over them, collect the cells into a 50 ml tube, and gently triturate them into a single-cell suspension (*see* **Note 7**).

3. Plate hESCs at $1.8–2.5 \times 10^5$ cells/cm^2 onto Matrigel-coated plates. Thereafter, feed the cultures daily with 0.3 mL/cm^2 MEF-CM plus bFGF (4 ng/mL).

4. When the resultant hESC monolayer becomes fully confluent with cells that are tightly packed but still have discernible borders, the cultures are ready for the switch to differentiating conditions At this point, replace the medium with 0.25 mL/cm^2 RPMI-B27 medium supplemented with 100 ng/mL activin A (*see* **Note 8**).

5. 24 h after treatment with activin A, gently replace the latter medium with 0.5 mL/cm^2 RPMI-B27 medium supplemented with 10 ng/mL BMP-4. Culture the cells for an additional 4 days without medium changes.

6. Four days after treatment with BMP-4, gently replace the latter medium with 0.5 mL/cm^2 RPMI-B27 medium without exogenous growth factors.

7. Feed cells every other day with 0.5 mL/cm^2 RPMI-B27 medium without exogenous growth factors. Spontaneous contractile activity typically commences around 9–12 days after the initial induction with activin A. One can expect total cell yields of $3–5 \times 10^5$ cells/cm^2 and ~60–90 % cardiomyocytes (based on the flow cytometry for cardiac troponin T or α-actinin) [3].

3.3 Cryopreservation of GCaMP3+ hESC-CMs

For logistical reasons and to minimize batch-to-batch variation in cell yields and purity, it is generally advisable to cryopreserve and "bank" hESC-CMs prior to use in intracardiac cell transplantation. Cells can be harvested for cryopreservation anytime 16–25 days after the initial induction with activin A. Particularly if cryopreservation is to be employed, we strongly recommend heat-shocking and treating the cells with IGF-1 and CsA prior to freezing them, as these interventions have been previously shown to significantly enhance graft survival [2, 19].

1. 24 h prior to harvesting for cryopreservation, heat-shock cultures with 30-min exposure to RPMI-B27 medium prewarmed to 42 °C. After 30 min, re-feed with 37 °C RPMI-B27 medium supplemented with IGF-1 (0.1 µg/ml) and CsA (0.25 µg/ml), and return the cultures to the incubator.

2. On the day of harvest, turn on the controlled-rate freezer and pre-chill it to 0 °C. Prepare 0.05 % trypsin–EDTA supplemented with 200 U/mL DNAse I and pre-warm to 37 °C. Chill Cryostor CS-10 solution on ice.

3. To begin harvesting GCaMP3+ hESC-CMs, briefly rinse the cultures twice with PBS supplemented with 200 U/mL DNase I. Next, incubate the cells with 0.2 mL/cm² of 0.05 % trypsin–EDTA plus 200 U/mL DNase I at 37 °C for 5 min. Cells should then readily detach from the surface with brisk tapping of the flask. If necessary, cells can be further dislodged by gentle agitation and/or flowing the trypsin solution over the cell surface.

4. Stop the enzymatic reaction by adding an equivalent volume of RPMI-B27 medium supplemented with 10 % FBS. Transfer the cells to a 50 ml tube and gently triturate to obtain single-cell suspension. Centrifuge the cells down at $300 \times g$ for 5 min to remove the trypsin and then resuspend in RPMI-B27 medium. This is a convenient time point to take small aliquots for cell counting (e.g., by hemacytometer) and/or determination of cardiomyocyte purity (e.g., by flow cytometry).

5. Centrifuge the cells down again at $300 \times g$ for 5 min, and then resuspend them in chilled Cryostor CS-10 solution at 1 ml per 1×10^8 total cells. Aliquot into pre-labeled 2 ml cryovials on ice.

6. Transfer the cryovials to a controlled-rate freezer already at 0 °C. Set controlled-rate freezer to cool at 1 °C per minute until it reaches −80 °C. Transfer the vials to a −80 °C freezer and leave overnight.

7. The next day, transfer the vials to a liquid nitrogen tank for storage until transplantation. hESC-CMs have been successfully transplanted after >1 year in cryopreservation.

3.4 Preparation of GCaMP3+ hESC-CMs for Intracardiac Transplantation

The exact procedure for preparing GCaMP3+ hESC-CMs for transplantation will depend on whether one is using "live" cultures or cryopreserved aliquots (prepared as described in Subheading 3.3).

3.4.1 Preparation of "Live" (i.e., Non-cryopreserved) Cells for Implantation

1. Heat-shock and detach cells as described in **steps 1–4** of Subheading 3.3.

2. Before commencing cell harvesting, prepare the Matrigel-based (*see* **Note 9**) and RPMI based PSC. Pretreatment and delivery in PSC have been previously shown to enhance hESC-CM graft survival [2, 19].

3. Resuspend the cell pellet in the RPMI based PSC, and then transfer the cell suspension to a single Eppendorf tube.

4. Centrifuge the latter at $300 \times g$ for 5 min, and then resuspend the cells in the Matrigel-based PSC. Add Matrigel-based PSC to a final volume of up to 5×10^5 cells per µl. (Our general practice is to prepare aliquots of 1×10^8 cells in 200 µl, each of which is injected into a single guinea pig heart.) Gently draw

the cell suspension into a precooled insulin syringe with 27G needle. Keep the syringe at 4 °C for use within 1 h.

3.4.2 Thawing and Preparation of Cryopreserved Cells for Implantation

1. Prepare PSC as described in **step 2** of Subheading 3.4.1. Preheat RPMI-B27 medium supplemented with 200 U/mL DNase I to 37 °C.

2. Transfer cryovial(s) containing the cryopreserved GCaMP3+ hESC-CMs to a bucket of dry ice until ready for use.

3. Thaw the vial in a 37 °C water bath with gentle agitation until the pellet is almost completely melted (this generally requires <1.5 min).

4. Add 1 mL of preheated RPMI-B-27 medium with 200 U/mL DNase I to each cryovial, and mix by gentle shaking. Slowly transfer the resultant cell suspension in a dropwise fashion to a 50 mL tube with 25 mL of 37 °C RPMI-B27 medium. Add RPMI-B27 medium to a final volume of 50 mL.

5. Wash away the cryopreservative by centrifugation at $300 \times g$ for 5 min and resuspend the cell pellet in RPMI-B27 medium with 200 U/ml DNase I. Proceed with **steps 3–4** of Subheading 3.4.1 to prepare syringes of cells.

3.5 Cardiac Cryoinjury and Transplantation of GCaMP3+ hESC-CMs into Guinea Pig Hearts

All animal procedures should be performed in compliance with institutional, local, and federal regulations. Please *see* Fig. 2b for the typical sequence of events involved in cardiac cryoinjury and cell transplantation. Because the guinea pig is relatively challenging to intubate, we recommend that anesthesia be induced with injectable ketamine/xylazine, followed by maintenance with inhalational isoflurane. Note that exposure to isoflurane anesthesia for long periods of time (>2 h) or at high concentrations (≥2 %) can adversely suppress cardiac and respiratory functions in the guinea pig. It is also essential that the animal's body temperature be maintained at 37.0 ± 1.0 °C during all procedures using a heating pad regulated by a rectal thermistor probe (*see* **Note 9**).

3.5.1 Induction of Cardiac Cryoinjury

1. Induce anesthesia with an intraperitoneal injection of ketamine (50 mg/kg)/xylazine (2 mg/kg).

2. After confirming anesthesia by toe-pinch, carefully shave the animal's chest and left axillary region. Clean the skin site by serial wiping with ethanol (70 %) and chlorhexidine gluconate. Coat the animal's eyes with Lubrifresh ointment to prevent dehydration.

3. Intubate the animal using a pediatric laryngoscope to visualize the vocal cords. A tracheal tube can be formed using a 14G shielded intravenous catheter bent into a curve. One can confirm proper intubation by placing a small amount of water in the catheter ahead of the procedure. Thus, if the endotracheal

tube is properly positioned, this water should be expelled with the animal's first breath. After confirming proper intubation, connect the catheter to a mechanical ventilator set at a rate of 60 breaths per minute and a tidal volume of 3 mL/kg. At this point, the animal can be switched to inhalational anesthesia with 1.5 % isoflurane in oxygen.

4. Place the animal on the heating pad, and position it to expose the surgical site. Sterilize the site by gentle wiping with Betadine three times.

5. Using a surgical blade, make a 3–4 cm long skin incision from the left axilla to the area of the xiphoid process using blade and scissors. Use blunt forceps to gently separate the connective tissues between the skin and underlying muscle. Separate the first layer of muscle with blunt forceps, and then cut the second layer of muscle with scissors to expose the chest wall.

6. After it has been adequately exposed, open the chest wall using blunt forceps. Take care to avoid injury to the adjacent viscera. Separate the second and third ribs with a retractor.

7. After the beating heart is clearly visible, gently remove the pericardium with blunt forceps.

8. Induce cardiac cryoinjury by firmly contacting the anterior left ventricle with the aluminum cryoprobe that has been pre-cooled in liquid nitrogen. The cryoprobe should be applied four times for 30-s duration each, with re-cooling for at least 30 s in liquid nitrogen between each application.

9. After inspecting the heart and lungs for bleeding or other unexpected injuries, carefully close the chest wall and deep tissues using 4-0 silk sutures. Suture the skin wound closed using nylon 5-0 sutures.

10. For analgesia, infiltrate the wound site during closure with 0.025 % (v/v) topical bupivacaine. Inject buprenorphine (0.05 mg/kg subcutaneous in a 1 mL volume) every 8–12 h for up to 48 h or as directed by veterinary staff (*see* **Note 10**).

11. To avoid perioperative dehydration, inject 4 mL of sterile saline subcutaneously (divided between at least two injection sites).

3.5.2 Cell Transplantation

To ensure survival of the human xenografts in the guinea pig hearts, we recommend a regimen combining cyclosporine (15 mg/kg/day subcutaneous for 7 days, later reduced to 7.5 mg/kg/day) and methylprednisolone (2 mg/kg/day intraperitoneal). To ensure adequate levels of immunosuppression at the time of cell transplantation, administration of the cyclosporine and methylprednisolone should commence 2 days prior to transplantation. We have also found that a one-time perioperative administration of

Baytril (enrofloxacin) greatly reduces the incidence of infections associated with this second round of survival surgery in chronically immunosuppressed animals.

1. Prepare syringes loaded with GCaMP3+ hESC-CMs as described above in Subheading 3.4. Bend the needle tip to facilitate injection into the myocardium.

2. Immediately before surgery, inject the animal with 0.3 mL/kg intraperitoneal Baytril.

3. Repeat **steps 1–7** of Subheading 3.5.1 for anesthesia, intubation, and repeat thoracotomy.

   Note that exposure of the chest wall and heart can sometimes be more complicated due to scarring and adhesion related to the earlier procedure.

4. After the heart is exposed, inject cells into three locations (i.e., the center of the scar and each of the flanking border zones) with each site receiving approximately one-third of the total injection. Proper cell injection is indicated by blanching and the appearance of a white halo surrounding the needle entry site.

5. Close the chest wall and deep tissues using 4-0 vicryl sutures. Administer analgesia (i.e., bupivacaine and buprenorphine), and rehydrate as detailed in **steps 10–11** of Subheading 3.5.1.

3.6 Intravital Imaging of GCaMP3+ hESC-CM Grafts in Guinea Pig Hearts

Our laboratory has successfully imaged normal and cryoinjured guinea pig hearts with GCaMP3+ hESC-CM grafts at both 2 and 4 weeks post-transplantation [3]. In this work, we have employed two different experimental preparations, each of which provides slightly different information and has its own advantages and disadvantages. The first approach employs an open chest preparation, performed as a terminal procedure in an already anesthetized animal. This prep has the advantage that all (or nearly all) of the neurohormonal signaling of the intact animal will be in place, but image acquisition is somewhat limited by motion artifact. The second approach involves an ex vivo preparation in which the heart is harvested and mounted on a Langendorff apparatus. While the latter preparation obviously cannot model all of the signaling and other factors that would apply in the intact animal, it does allow the use of mechanical uncouplers (e.g., blebbistatin [20]) to remove motion artifacts. It also allows one to image hearts with GCaMP3+ grafts after treatment with drugs and/or other fluorescent indicators (e.g., voltage-sensitive dyes). Below, we describe the methods for both experimental preparations separately.

3.6.1 Imaging Using an Open Chest Preparation

1. Follow **steps 1–4** of Subheading 3.5.1 to prepare the animal for surgery as well for anesthesia, intubation, and mechanical ventilation.

2. Place the anesthetized animal in a supine position on a heated surface, and outfit it for standard surface ECG recordings.

3. Perform a wide thoracotomy by incising the full width of the epicardium and reflecting chest wall away anteriorly.

4. Use an appropriate optical system to visualize the GCaMP3 grafts. In brief, our own system employs a high-speed EM-CCD camera that is mounted on an epifluorescence stereomicroscope (*see* Fig. 3a). GCaMP3 is excited at 470 ± 20 nm, and emitted light is collected. Image acquisition is typically performed at 80–140 frames per second, and signals from the CCD camera and the surface ECG are fed through a computer for digital storage and off-line analysis as described below (*see* **Note 11**).

3.6.2 Imaging Using an Ex Vivo Langendorff Preparation

1. Prepare Langendorff perfusion buffer as described above. (A total of ~4 L is usually sufficient for a typical imaging experiment.) Set aside two 1 L bottles, and add blebbistatin to it for a final concentration of 10 μM.

2. Connect 1 L reservoirs containing the control and blebbistatin-supplemented perfusate to a non-recirculating Langendorff apparatus, and bubble each solution with 95 % O_2–5 % CO_2 gas mix using submerged bubblers. The Langendorff apparatus should be configured to deliver retrograde coronary perfusion at ~70 mmHg constant pressure, and it should warm the perfusate to 37 °C prior to flow into the heart, e.g., by flow through a glass heat exchange coil fed by a warm circulating water bath.

3. Euthanize the animal, and excise the engrafted heart for mounting onto the Langendorff apparatus. To harvest the heart, perform a wide thoracotomy, quickly identify the aorta, and cut across the aorta and other great vessels to dissect the heart from the surrounding tissues. Cannulate the aorta without introducing air bubbles, connect the heart to the Langendorff system, and perfuse the heart with the blebbistatin-free solution.

4. Monitor the electrical activity of the isolated heart by placing negative and positive ECG leads at the base of the right ventricle and left ventricular apex, respectively. These leads are then connected to an appropriate bioamplifier system for recording and off-line storage.

5. Once the preparation is electrically stable, switch to the blebbistatin-containing perfusate. Motion arrest typically occurs within 10 min.

6. After mechanical arrest, maneuver the heart into focus on an appropriate imaging system (*see* Fig. 3 and Subheading 3.6.1). Record the epicardial GCaMP3 signals as described above (*see* **Note 11**).

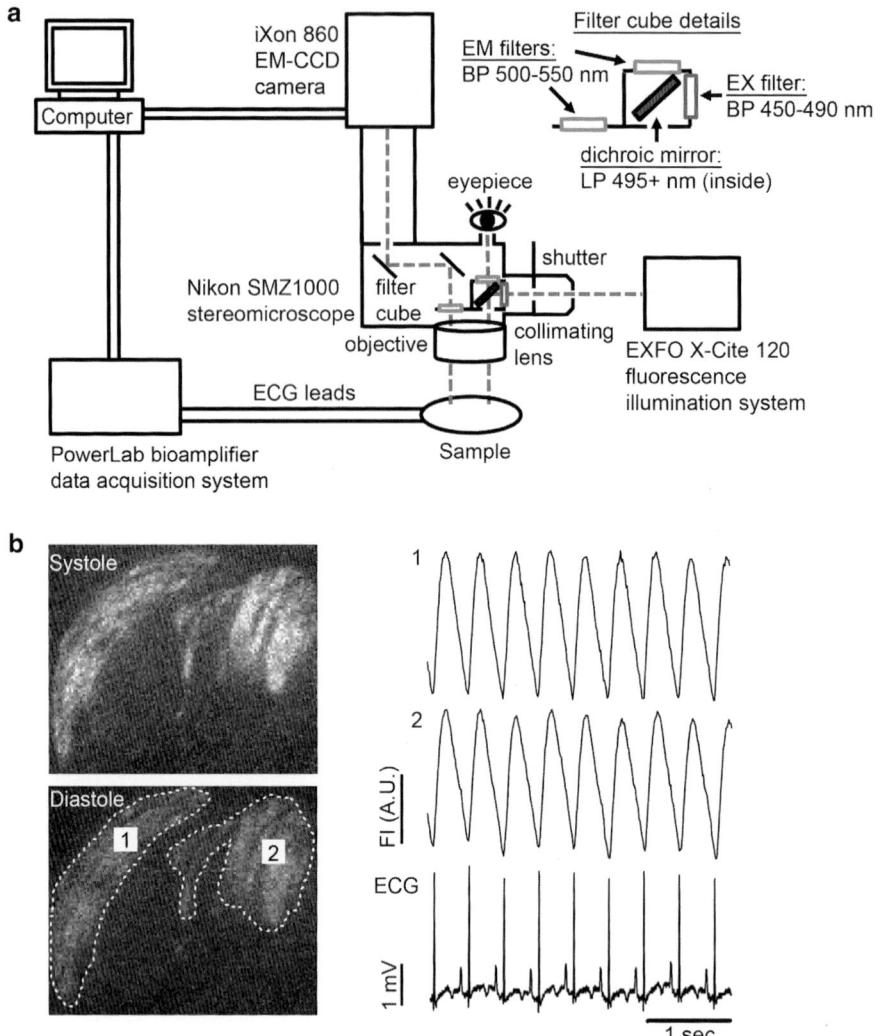

Fig. 3 Schematic diagram of our GCaMP3 imaging system and representative recordings. (**a**) Our intravital imaging system includes a high-speed, high-sensitivity CCD camera (Andor iXon 860 EM-CCD) mounted on an epifluorescence stereomicroscope (Nikon SMZ1000). GCaMP3 is excited with a 450–490 band-pass filter, and its emission passes through a 500–550 nm band-pass filter before detection. The host electrocardiogram (ECG) is simultaneously recorded using a bioamplifier and data acquisition board (ADInstruments PowerLab system). Fluorescence and ECG signals are digitized and fed to a computer for storage and off-line analysis. (**b**) *Left*: Representative still images of GCaMP3+ hESC-CM grafts during systole and diastole. Two graft regions of interest (ROIs) have been *circled* by the *dotted line*. *Right*: Representative GCaMP3 fluorescence traces for the aforementioned ROIs as well as the simultaneously acquired ECG. Note that, in this case, fluorescence transients in both graft ROIs occur in a 1:1 relationship with the host ECG, indicating reliable graft–host coupling

3.7 Analysis of GCaMP3 Intravital Imaging Data

Our lab has created a number of custom Matlab scripts for analyzing the experiments described in the preceding section, and these tools are available upon request. Among other functions, they allow one to examine the relative intensity of GCaMP3 fluorescence transients, the relationship between the activation of GCaMP3+ graft(s) and host myocardium, and, in some circumstances, even the timing and pattern of graft activation. (*See* Fig. 3b for representative images of GCaMP3+ hESC-CM grafts during systole and diastole as well as representative fluorescence intensity traces.) The methods below assume that the reader will apply these tools, although we do describe alternative approaches for those who prefer to work outside of the Matlab software environment (*see* **Note 11**).

Arguably the most common reason for intravital imaging of GCaMP3+ grafts is to determine the temporal relationship between host and graft activation. For this, one obviously needs to precisely correlate the graft-derived GCaMP3 fluorescence transients and the host ECG. In our case, we have found that one convenient way of accomplishing this is to use the shutter fire signal emitted by the Andor EM-CCD to "time-sync" the imaging and ECG data recordings. ADInstruments' PowerLab multichannel data acquisition system can simultaneously receive both this TTL signal from the camera shutter and the digitized ECG. This information can be used later to align the GCaMP3 and ECG traces, either in a spreadsheet or using our custom Matlab scripts.

1. In either custom Matlab scripts or in the camera controller software package (in our case Andor Solis), play the recorded data and identify graft regions of interest (ROIs).

2. Calculate the mean pixel intensity of each ROI, and plot this against the time-synced ECG. Assess the extent of graft coupling by visually identifying which ROIs activate in synchrony with the QRS complex of the ECG. We recommend recording at least ten contractile cycles to ensure appropriate image smoothing and statistical analyses.

3. Our custom Matlab scripts can also be used to assess the activation times for graft regions that are reliable coupled to the host ECG. For this, apply available spatial and temporal filters to the imaging data and background subtract the imaging data. Next, identify the peaks and troughs in the mean pixel intensity traces for each coupled ROI as well as the corresponding R wave of the host ECG. Select a time window that spans several graft activations, and then determine (on a per pixel basis) the time difference between the beginning of the rise in the GCaMP3 signal and the corresponding R wave of the host ECG. These time intervals can then be stored in a separate matrix for each cycle of activation. Average these matrices to generate an ensemble-averaged map of activation times. Our scripts greatly automate this analysis and generate two-dimensional color maps to help visualize the pattern and timing of graft activation (*see* **Note 12**).

4   Notes

1. While this protocol is written for the H7/WA07 hESC line, we have had good success using an equivalent ZFN-mediated targeting strategy to introduce the GCaMP3 expression cassette into multiple human pluripotent stem cell lines, including RUES2 hESCs and IMR90 human induced pluripotent stem cells. Greater transfection efficiency may be possible by using line-optimized electroporation settings and solutions in the nucleofection system.

2. Two plasmids are required for ZFN-mediated insertion of the GCaMP3 expression cassette into the AAVS1 locus: the AAVS1 ZFN and pZDonor-GCaMP3 vectors (*see* Fig. 1a, b). Both are readily available from our group upon request. Alternatively, these plasmids can be re-created using previously described methods [3]. To create the AAVS1 ZFN vector, we had the left and right arms of the AAVS1-specific ZFN de novo synthesized using sequences reported by Hockemeyer et al. [15]. These arms were then cloned into a pUC57 backbone plasmid to create a polycistronic vector in which the expression of each ZFN arm is driven by a separate human PGK promoter. This ZFN pair causes a double-strand break in the intron between exon 1 and 2 of the AAVS1 locus (19q13.3, *see* Fig. 1c). The donor vector was generated by modifying the pZDonor plasmid (Sigma Aldrich, #Z3027), which includes a multi-cloning site and ~800 bp homology arms flanking the AAVS1 ZFN cut site. Two expression cassettes were inserted between these homology arms: one in which the CAG promoter drives expression of GCaMP3 (Addgene, #22692) and another in which the PGK promoter drives neomycin resistance. Please *see* Fig. 1d for a schematic of the correctly targeted transgene including both expression cassettes.

3. Plasmid DNA should be prepared using a "midi-prep" kit (Qiagen), rather than phenol/chloroform extraction. The DNA should then be concentrated to a volume of <10 μL in water. Salts used in other DNA buffers (e.g., TE buffer) can interfere with transfection efficiency.

4. As an alternative to ZFN-mediated insertion of the GCaMP3 expression cassette, both TALEN- and CRISPR-mediated targeting of the human AAVS1 locus has been described [21, 22], and suitable vector systems are now available via the Addgene plasmid repository. These systems should also be compatible with the pZDonor-GCaMP3 vector described above.

5. Immediately before electroporation, prepare the nucleofection solution as per kit instructions. In brief, the nucleofection solution is a 4.5:1 mix of solution 1 to supplement solution (corresponding to 81.8 μL:18.2 μL for a typical 100 μL reaction).

6. The following primer sets can be used to generate Southern blot probes useful for confirming proper targeting of the GCaMP3 expression cassette to the AAVS1 locus: genomic probe (forward): GGAGGTGGTGCGCTTCTTGG, genomic probe (reverse): CGCATCCCCTCCCAGAAAGAC, neomycin cassette forward: GAAGAACTCGTCAAGAAGGCG, and neomycin cassette reverse: GAAGAACTCGTCAAGAAGGCG. DNA were cut with NdeI and NheI overnight then isolated by ethanol precipitation and loaded into a 1 % agarose gel. DNA transfer and membrane preparation was performed using Biorad Zeta-probe GT instruction manual section 2.1 and probe labeling was accomplished using Amersham Megaprime DNA labeling system (GE Healthcare RPN1604/5 kit). Southern blotting with the genomic probe produces a wild-type/nontargeted band at 3,968 bp, while the properly targeted allele shifts to 5,604 bp. In properly targeted cells, the neomycin probe will label a 5,604 bp band, whereas off-targeted integration events will be seen at other positions.

7. This seeding density works well for the cardiac differentiation of GCaMP3+ hESCs generated from the H7/WA07 hESC line. However, the optimal seeding density varies slightly from line to line, so it should be determined empirically when testing other lines. The goal should be to obtain a uniformly distributed monolayer of hESCs that occupies 80–100 % of the growth area 1 day post-plating.

8. The timing of this step is critical. The properly timed application of activin A should be based on the appearance of the culture (i.e., the formation of a compact monolayer without gaps), rather than a pre-specified interval of time. This will vary slightly from line to line, depending on its plate efficiency, growth kinetics, etc. We have also found that the efficiency of this cardiac differentiation protocol can be improved for some (but not all) lines by adding a variation of the Matrigel "overlay" step proposed by Kamp and colleagues [23]. The latter is performed during the activin A induction step. For this, thaw growth factor-reduced Matrigel on ice for at least 2 h. Chill RPMI-B27 medium on ice for 10 min. Mix the two reagents thoroughly to prepare a 1:60 dilution of Matrigel in medium. Warm the mixture to 37 °C, add activin A (100 ng/mL), and then gently overlay this onto cells at a volume of 0.25 mL/cm². The remainder of the protocol is unchanged from that described above in Subheading 3.2 (i.e., continue on with **steps 5–7**).

9. The guinea pig has a large tongue, narrow oral cavity, and other anatomic features that make endotracheal intubation in this species particularly challenging. While beyond the scope of the present protocol, other authors have provided invaluable advice both on obtaining and confirming successful intubation in this model [24, 25]. One other issue with the guinea pig is

that this species has excellent collaterals, so experimental myo-cardial infarction cannot be induced by coronary ligation as in other rodents. However, cardiac cryoinjury is a reasonable alternative and results in a highly reproducible area of injury and degree of dysfunction.

10. Like other rodents, guinea pigs usually recover from thoracot-omy and cardiac cryoinjury procedures remarkably well. We allow animals to recover in a 37 °C warmed environment in which they can be closely monitored, and we typically find that they resume their normal activities within 3–4 h post-procedure. Animals that recover well are then returned to their cages with some hydrating gel and food placed directly onto the bedding, so that they do not become dehydrated due to incision discomfort when reaching up to the cage insert with their water bottle and food pellets. Veterinary staff should be consulted if there are con-cerning signs such as weight loss, altered posture, unusual vocal-izations, listless behavior, and respiratory distress. Our experience suggests that a <25 % mortality rate can be expected after two rounds of survival surgery (i.e., cardiac cryoinjury and cell trans-plantation), with most deaths occurring shortly after cryoinjury.

11. Although our group has found that custom Matlab scripts pro-vide more powerful analysis tools, there are simpler alternatives for those uncomfortable with work in the Matlab software environment. Most relevant EM-CCD systems are controlled by acquisition software that usually include limited analysis functions. For example, our imaging system includes an Andor iXon EM-CCD, which is controlled by the Andor Solis soft-ware package. Solis allows one to directly export the mean pixel intensity values for rectangular ROIs into a spreadsheet such as Microsoft Excel. Alternatively, if ROI shapes other than rect-angles are desired, Solis data can be exported as a stack of *.tiff files into ImageJ. There, the freehand selection tool can be used to define a custom ROI and determine the mean pixel intensity. In either case, this pixel intensity data can then be manually aligned with the digital ECG recordings in a spreadsheet using the available camera-derived signals (e.g., shutter fire pulse) as a "time stamp." This approach is tedious, but it works.

12. Ensemble averaging of activation time matrices can only be per-formed if the signal is periodic and steady. If the signal signature changes with time, then this method cannot be reliably applied.

Acknowledgements

This work was supported by funding from the National Institutes of Health (grants R01-HL064387, P01-HL094374, U01-HL100405, and R01-HL117991). The authors would also like to thank Mr. Benjamin Van Biber for helpful comments on this manuscript.

References

1. Caspi O, Huber I, Kehat I et al (2007) Transplantation of human embryonic stem cell-derived cardiomyocytes improves myocardial performance in infarcted rat hearts. J Am Coll Cardiol 50:1884–1893

2. Laflamme MA, Chen KY, Naumova AV et al (2007) Cardiomyocytes derived from human embryonic stem cells in pro-survival factors enhance function of infarcted rat hearts. Nat Biotechnol 25:1015–1024

3. Shiba Y, Fernandes S, Zhu WZ et al (2012) Human ES-cell-derived cardiomyocytes electrically couple and suppress arrhythmias in injured hearts. Nature 489:322–325

4. van Laake LW, Passier R, Monshouwer-Kloots J et al (2007) Human embryonic stem cell-derived cardiomyocytes survive and mature in the mouse heart and transiently improve function after myocardial infarction. Stem Cell Res 1:9–24

5. Gepstein L, Ding C, Rahmutula D et al (2010) In vivo assessment of the electrophysiological integration and arrhythmogenic risk of myocardial cell transplantation strategies. Stem Cells 28:2151–2161

6. Lai PF, Panama BK, Masse S et al (2013) Mesenchymal stem cell transplantation mitigates electrophysiological remodeling in a rat model of myocardial infarction. J Cardiovasc Electrophysiol 24:813–821

7. Rota M, Kajstura J, Hosoda T et al (2007) Bone marrow cells adopt the cardiomyogenic fate in vivo. Proc Natl Acad Sci U S A 104:17783–17788

8. Rubart M, Soonpaa MH, Nakajima H et al (2004) Spontaneous and evoked intracellular calcium transients in donor-derived myocytes following intracardiac myoblast transplantation. J Clin Invest 114:775–783

9. Xue T, Cho HC, Akar FG et al (2005) Functional integration of electrically active cardiac derivatives from genetically engineered human embryonic stem cells with quiescent recipient ventricular cardiomyocytes: insights into the development of cell-based pacemakers. Circulation 111:11–20

10. Costa AR, Panda NC, Yong S et al (2012) Optical mapping of cryoinjured rat myocardium grafted with mesenchymal stem cells. Am J Physiol Heart Circ Physiol 302:H270–H277

11. Chen TW, Wardill TJ, Sun Y et al (2013) Ultrasensitive fluorescent proteins for imaging neuronal activity. Nature 499:295–300

12. Tallini YN, Ohkura M, Choi BR et al (2006) Imaging cellular signals in the heart in vivo: cardiac expression of the high-signal $Ca2+$ indicator GCaMP2. Proc Natl Acad Sci U S A 103:4753–4758

13. Tian L, Hires SA, Mao T et al (2009) Imaging neural activity in worms, flies and mice with improved GCaMP calcium indicators. Nat Methods 6:875–881

14. Roell W, Lewalter T, Sasse P et al (2007) Engraftment of connexin 43-expressing cells prevents post-infarct arrhythmia. Nature 450:819–824

15. Hockemeyer D, Soldner F, Beard C et al (2009) Efficient targeting of expressed and silent genes in human ESCs and iPSCs using zinc-finger nucleases. Nat Biotechnol 27:851–857

16. Zou J, Maeder ML, Mali P et al (2009) Gene targeting of a disease-related gene in human induced pluripotent stem and embryonic stem cells. Cell Stem Cell 5:97–110

17. Zhu WZ, Van Biber B, Laflamme MA (2011) Methods for the derivation and use of cardiomyocytes from human pluripotent stem cells. Methods Mol Biol 767:419–431

18. Xu C, Inokuma MS, Denham J et al (2001) Feeder-free growth of undifferentiated human embryonic stem cells. Nat Biotechnol 19:971–974

19. Laflamme MA, Gold J, Xu C et al (2005) Formation of human myocardium in the rat heart from human embryonic stem cells. Am J Pathol 167:663–671

20. Dou Y, Arlock P, Arner A (2007) Blebbistatin specifically inhibits actin-myosin interaction in mouse cardiac muscle. Am J Physiol Cell Physiol 293:C1148–C1153

21. Holkers M, Maggio I, Liu J et al (2013) Differential integrity of TALE nuclease genes following adenoviral and lentiviral vector gene transfer into human cells. Nucleic Acids Res 41:e63

22. Mali P, Yang L, Esvelt KM et al (2013) RNA-guided human genome engineering via Cas9. Science 339:823–826

23. Zhang J, Klos M, Wilson GF et al (2012) Extracellular matrix promotes highly efficient cardiac differentiation of human pluripotent stem cells: the matrix sandwich method. Circ Res 111:1125–1136

24. Blouin A, Cormier Y (1987) Endotracheal intubation in guinea pigs by direct laryngoscopy. Lab Anim Sci 37:244–245

25. Nambiar MP, Gordon RK, Moran TS et al (2007) A simple method for accurate endotracheal placement of an intubation tube in Guinea pigs to assess lung injury following chemical exposure. Toxicol Mech Methods 17:385–392

Chapter 21

Quantifying Electrical Interactions Between Cardiomyocytes and Other Cells in Micropatterned Cell Pairs

Hung Nguyen, Nima Badie, Luke McSpadden, Dawn Pedrotty, and Nenad Bursac

Abstract

Micropatterning is a powerful technique to control cell shape and position on a culture substrate. In this chapter, we describe the method to reproducibly create large numbers of micropatterned heterotypic cell pairs with defined size, shape, and length of cell–cell contact. These cell pairs can be utilized in patch clamp recordings to quantify electrical interactions between cardiomyocytes and non-cardiomyocytes.

Key words Micropatterning, Cardiac cell therapy, Electrical coupling, Stem cells, Patch clamp

1 Introduction

Systematic studies of structural and functional interactions between cardiomyocytes and other cells are critical for understanding cardiac development, function, and diseases (e.g., infarction, arrhythmias) as well as for improving the safety and efficacy of cardiac cell therapies. Specifically, a large variety of different donor cells including skeletal myoblasts [1], bone marrow-derived mesenchymal stem cells (MSCs) [2], heart-derived stem cells [3], pluripotent stem cell-derived cardiomyocytes [4], and genetically engineered somatic cells [5, 6] are currently being evaluated for their utility in cell-based cardiac repair. Direct studies of in situ interactions between these cells and host cardiomyocytes, although critical, are hampered by our inability to access and identify sites of host–donor cell contact and by high variability among different hearts and treatment outcomes. Traditional cell co-cultures have been used to aid in general understanding of in situ host–donor cell interactions; however, heterocellular studies at a single-cell level have been hampered by the inability to control individual cell size, shape, number, and type of interacting cells. To overcome these challenges, we

Milica Radisic and Lauren D. Black III (eds.), *Cardiac Tissue Engineering: Methods and Protocols*, Methods in Molecular Biology, vol. 1181, DOI 10.1007/978-1-4939-1047-2_21, © Springer Science+Business Media New York 2014

have utilized micropatterning techniques to develop and optimize methods for the generation of large numbers of individual heterotypic cell pairs with controllable size, shape, and cell–cell contact length [7, 8]. Dual whole-cell voltage clamp recordings [9] are used to quantify the electrical coupling between the two cells in the pair, while whole-cell current clamp is employed to assess the resulting changes in electrical properties of cardiomyocytes [8].

2   Materials

2.1   Micropatterning Components

Reagents

1. Negative photoresist (SU8-2, Microchem Inc.).
2. Silane solution [(tridecafluoro-1,1,2,2-tetrahydro octyl)-1-trichlorosilane].
3. Propylene glycol methyl ether acetate (PGMEA): CAUTION: Wear protective goggles, gloves, and clothing, and ensure proper ventilation.
4. Isopropyl alcohol.
5. Poly-dimethylsiloxane (PDMS) silicone elastomer base and curing agent (Sylgard 184, Dow Corning).
6. Ethanol, 70 % in DI H_2O.
7. Fibronectin (Human), powder.
8. Dulbecco's phosphate-buffered saline, $1\times$ (DPBS).
9. Triton X-100.
10. Paraformaldehyde, 16 % in DI H_2O (PFA): CAUTION: Wear protective goggles, gloves, and clothing, and ensure proper ventilation.

Equipment

1. 4″ Silicon wafer.
2. 22 mm Glass cover slips.
3. Tissue culture 12-well plate.
4. Polystyrene tissue culture Petri dish, 100 mm diameter.
5. Crystallizing dishes, 100 mm diameter, 50 mm tall.
6. Programmable spin coater.
7. Digital hot plate.
8. Mask aligner with 350 W mercury arc lamp UV light source.
9. Vacuum desiccator and vacuum pump.
10. Sonicator bath.
11. UV decontamination system.
12. N_2 tank and filtered N_2 gun.

13. Wafer grip tweezers.

14. Fine forceps, straight.

15. Standard forceps, curved.

16. Syringe needle (21G 1½″) with tip bent 90° using needle-nose pliers.

17. Oven at 80 °C.

18. Incubator at 37 °C, 5 % CO_2.

19. Water bath at 37 °C.

20. Metallurgical microscope and standard 10× objective.

21. Miscellaneous: Transfer pipettes, razor blades, Pasteur pipettes (vacuum linked).

Software

1. Metamorph image acquisition software package (Molecular Devices).

2. AutoCAD software package (Autodesk).

3. MATLAB software package (The Mathworks).

2.2   Cell Sources and Media

1. For a list of potential cell types, their respective culture media, and the source from which they can be obtained please *see* Table 1.

2.3   Patch Clamp Components

1. Custom-made Faraday cage for electromagnetic interference reduction.

2. Anti-vibration N_2 air table.

3. Inverted fluorescence microscope.

4. Multiclamp 700B amplifier (Axon Instruments, Inc.).

5. NIDAQ-MX computer interface (National Instruments).

6. WINWCP software package (John Dempster, University of Strathclyde) or any other patch clamp analysis software.

7. Patch glass pipette 1.5 mm O.D. × 1.16 mm I.D.

8. Micropipette puller.

9. MF-200 microforge (World Precision Instruments, Inc.) and stereomicroscope (with 40× objective) for fire-polishing glass pipette.

10. Patch pipette is filled with pipette solution containing (in mM) 140 KCl, 10 NaCl, 1 $CaCl_2$, 10 EGTA, 10 HEPES, 2 $MgCl_2$, and 5 MgATP (adjusted to pH 7.2 with KOH and 270–280 mOsm with glucose) [8].

11. MicroFil syringe needle for filling micropipettes 34 gauge/67 mm long (World Precision Instruments, Inc.). The needle is attached to a 10 mm syringe via a 22 μm syringe filter.

Table 1
A listing of the potential cell types that can be used in cell-pair patterning experiments, their respective culture media, and the sources for each cell type

Cell type	Media	Cell sources
Neonatal rat ventricular myocytes (NRVMs)	*Seeding media*: DMEM/F-12 media supplemented with 10 % calf serum and 10 % horse serum [7] *Maintenance media*: M199 media supplemented with 2 % horse serum, HEPES (10 mM), MEM nonessential amino acids, GlutaMAX (2 mM, Invitrogen), vitamin B_{12} (2 μg/ml), and penicillin-G (20 U/ml) [8]	Dissociated from the ventricles of 2-day-old Sprague-Dawley rats by trypsin and collagenase and then enriched using two 1-h differential preplating steps to remove faster-adhering nonmyocytes [10]
Neonatal rat ventricular fibroblasts (NRVFs)	*Fibroblast growth media*: M199 medium supplemented with 10 % FBS and HEPES (10 mM), nonessential amino acid solution (×1), L-glutamine (2 mM), dextrose (3.5 mg/ml), vitamin B12 (4 μg/ml), and penicillin G (100 U/ml) [11]	Recovered from NRVM isolation during the preplating steps and, upon reaching confluence, passaged at 1:2 ratio and used at passage 1 [7]
Adult rat mesenchymal stem cells (MSCs)	*MSC growth media*: α-MEM medium supplemented with 20 % fetal bovine serum (FBS), 50 U/ml penicillin, and 50 μg/ml streptomycin [12]	Isolated from the bone marrow of 3-month-old male Sprague-Dawley rats using previously published methods [13]
Adult rat skeletal myoblasts (SKMs)	*SKM media*: DMEM supplemented with 20 % FBS and 50 μg/ml gentamicin [7]	Isolated from the hindlimb soleus muscle of female Sprague-Dawley rats as previously described [14]
Mouse C2C12 myoblasts	*Myoblast growth media*: DMEM supplemented with 10 % FBS, 10 % calf serum, 2.8 ml chick embryo extract, and 50 μg/ml gentamicin [11]	American Type Culture Collection (ATCC): Cells need to be kept below 70 % confluence to prevent myotube formation
Mouse embryonic stem cell-derived cardiomyocytes (mESC-CMs) or cardiovascular progenitor cells (mESC-CVPs)	*Maintenance media*: N2B27 media [15] with 0 or 2 % FBS	Derived from D3 mESCs (ATCC CRL-11632) using previously published differentiation protocols [16, 15]
Wild-type or genetically engineered human embryonic kidney 293 cells (HEK293)	*HEK media*: DMEM low glucose supplemented with 10 % FBS, 50 U/ml penicillin, and 50 μg/ml streptomycin [11]	Wild-type HEK293 cells were purchased from ATCC (CRL-1573). Cells were stably transfected and selected for monoclonal lines with Cx43 [11], Kir2.1 + Cx43 [8], or $Na_v1.5$ + Kir2.1 + Cx43 [17]

12. Perfusion chamber: Cell culture dish 35 mm outer diameter with glass bottom (23.5 mm in diameter). Perfusion chamber is filled with Tyrode's solution containing (in mM) 135 NaCl, 5.4 KCl, 1.8 $CaCl_2$, 5 HEPES, 5 glucose, 1 $MgCl_2$, and 0.33 sodium phosphate [8].

13. Temperature controller with heating stage.

14. Micromanipulator with controller.

15. Straight microelectrode holder with Ag/AgCl wire, 1 mm pin, and 2 mm pressure port, for 1.5 mm O.D. glass. Frequent chloriding of the electrode is done by immersion in household bleach for about 15 min.

16. Pressure system: Pressure port on microelectrode holder is connected to a custom-made U-tube manometer via silicone tubing. A three-way stopcock is used to switch between a constant positive pressure of 10 cmH_2O and manual suction by mouth or syringe.

17. Bath electrode is made of Ag/AgCl pellet. Frequent chloriding of the electrode is done by immersion in household bleach for about 15 min.

3 Methods

3.1 Cell Pair Micropatterning

3.1.1 Photomask Design

Draw micropatterns using custom AutoLISP software in AutoCAD (Autodesk, San Rafael, CA), and then print high-resolution photomasks (chrome on soda lime) [18]. The patterns consist of pairs of rectangular islands, with one island held constant at 30×30 μm^2 while the other varied from 20×10 μm^2 to 20×36 μm^2 in 2 μm increments (Fig. 1). The contact length between the islands is held constant at 20 μm. The patterns should be arranged such that approximately 5,000 island pairs fit within a single 22 mm diameter cover slip [8]. Other custom patterns can be generated using AutoLISP and AutoCAD software.

3.1.2 High-Resolution Photolithography

All photolithography steps should be performed in a chemical fume hood inside a clean room, wearing proper lab coats, shoe covers, bonnets, gloves, and UV goggles.

1. Use wafer tweezers to transfer a silicon wafer to a 200 °C hot plate for 5 min to dehydrate the surface. Transfer wafer to a Petri dish and let cool to room temperature.

2. Clean dust from the wafer with the N_2 gun and UV-decontaminate for 8 min.

3. Using wafer tweezers, transfer and center the wafer onto the spin coater and apply vacuum.

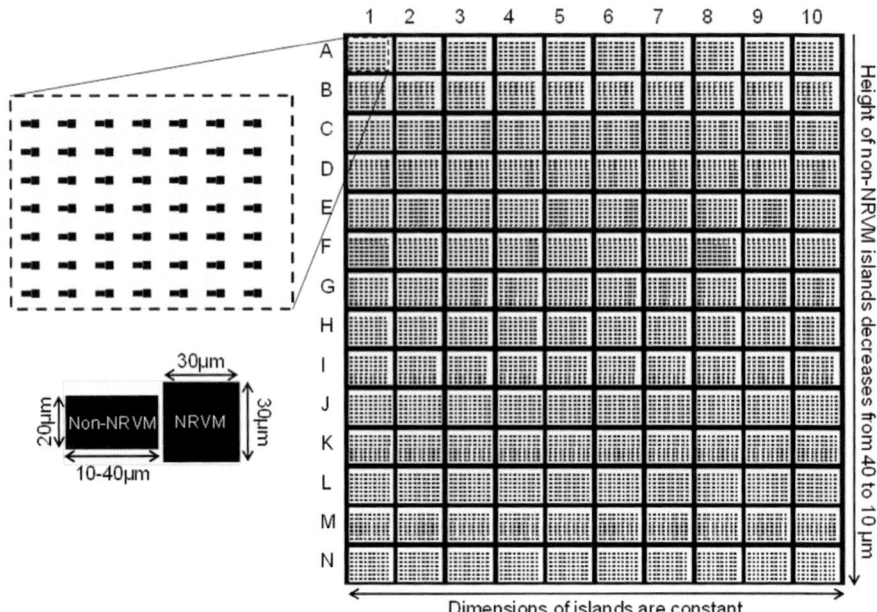

Fig. 1 Photomask design of the micropatterned cell pairs

4. Transfer approximately 5 mL of Su8-2 photoresist onto the wafer with a transfer pipette.

5. To obtain 5 μm tall features, spin-coat at 1,000 rpm for 1 min.

6. Release the vacuum, and, wearing sterile gloves, transfer the wafer by hand to a 65 °C hot plate and cover with a crystallizing dish, avoiding contact with the photoresist-coated surface of the wafer (*see* **Note 1**).

7. Heat at 65 °C for 1 min and then at 95 °C for 3 min.

8. Turn off hot plate and let cool to room temperature. Using wafer tweezers, return wafer to Petri dish. PAUSE POINT: Once baked, photoresist-coated wafers can be stored in dark, ambient conditions.

9. Position the wafer and glass photomask properly in the mask aligner, and apply chamber vacuum for "hard contact."

10. Expose to 350 W collimated UV light for 20 s.

11. Using wafer tweezers, transfer wafer back to the hot plate (now at room temperature). Ramp up to 65 °C, bake for 1 min, ramp up to 95 °C, and bake for an additional 3 min. Finally, ramp down to 25 °C at 8 °C/h (*see* **Note 2**).

12. Using wafer tweezers, transfer wafer to a crystallizing dish containing PGMEA for 2 min, then remove, and rinse with isopropyl alcohol over an empty crystallizing dish (*see* **Note 3**).

13. Dry wafer with N_2 gun, and observe under the metallurgic microscope to ensure uniform patterned lines and line spac-

ings, the absence of dust or other impurities, and no feature detachment. The finished wafer should be void of any defects.

14. Leave the template wafer overnight (12 h) at room temperature inside a vacuum desiccator containing a micro slide with one drop of silane solution. The evaporating silane solution will coat the wafer, creating a "non-stick" surface. Store silanized wafer in a Petri dish at room temperature. PAUSE POINT.

PDMS stamp molding

1. Mix approximately 20 g PDMS base and 2.0 g curing agent (10:1 by mass) thoroughly in a weigh boat for 1 min using a transfer pipette. Degas in vacuum desiccator for 20 min or until all air bubbles are extracted.

2. Gently pour the PDMS solution onto the template wafer in the 100 mm diameter Petri dish, filling the dish to a height of approximately 5 mm.

3. Cure the PDMS-covered wafer in an oven at 80 °C for 2 h. Remove and let cool to room temperature.

4. Tap Petri dish containing PDMS-covered wafer upside down until PDMS mold (and attached wafer) release from dish (*see* **Note 4**). Gently peel PDMS mold from wafer (avoid cracking the wafer), and place PDMS mold, feature side up, on a hard surface. Cut around the square border of the desired pattern with a clean razor blade. PAUSE POINT.

PDMS coating of glass cover slips

1. Mix PDMS base and curing agent (10:1 by mass) thoroughly in a weigh boat for 1 min using a transfer pipette. Degas in vacuum desiccator for 20 min or until all air bubbles are extracted.

2. Center a 22 mm diameter glass cover slip on the spin coater chuck, and apply vacuum. Using a transfer pipette, transfer approximately 250 μL of PDMS onto the cover slip, and spin-coat at 4,000 rpm for 20 s. Transfer cover slip by hand to a covered Petri dish, avoiding contact with the PDMS-coated surface (*see* **Note 5**).

3. Cure the PDMS-coated cover slips in an oven at 80 °C for 2 h. Remove and let cool to room temperature. Store cover slips covered at room temperature for microcontact printing.

3.1.3 Microcontact Printing

1. Sonicate PDMS stamp in 70 % ethanol for 45 min.

2. Remove stamp with standard curved forceps, dry with N_2 gun in a sterile hood, and place, pattern side up, onto a Petri dish.

3. Coat stamp with 200 μL of 15 μg/ml fibronectin solution, spreading with the side of a sterile pipette tip, if necessary, to

ensure even distribution. Coat at room temperature for at least 1 h.

4. Near the end of the fibronectin coating of the PDMS stamp, UV-decontaminate the surface of the PDMS-coated cover slip in a Petri dish for 8 min with lid off.

5. Immediately after UV decontamination of the PDMS-coated cover slips, use standard curved forceps to remove stamp from the Petri dish, then dry stamp again with N_2 gun in a sterile hood, and return to the Petri dish, pattern side up.

6. Using fine forceps, gently place each cover slip PDMS side down onto a dried, fibronectin-coated stamp.

7. Very lightly, trace one tip of the forceps on the back of the cover slip along the features of the underlying stamp to ensure contact. Other than the weight of the forceps, no applied force should be necessary. Wait for 30 min to allow protein transfer (see **Note 6**).

8. Gently slide one tip of the fine forceps between the edge of the cover slip and the stamp. Slowly pry the cover slip from the stamp by running the tip of the forceps along the border of the stamp (see **Note 7**).

9. Transfer the cover slip into the well of a 12-well plate and coat with Pluronic F-127 (0.2 %, Invitrogen) for 10 min to block cell adhesion in unwanted areas.

10. Wash cover slip once with PBS and leave in 2 ml PBS in tissue culture hood at room temperature until ready for cell seeding.

Cell Seeding

1. Aspirate PBS. Seed NRVMs at a density of 10^4 cells/cm^2 in cardiac seeding media. Incubate at 37 °C 5 % CO_2 incubator overnight (day 0).

2. The morning after seeding, aspirate cardiac seeding media and wash with PBS. Gently agitate the plate to remove dead and unattached cells before aspirating PBS and replacing with seeding media (day 1).

3. About 48 h after seeding, aspirate seeding media and replace with cardiac maintenance media (day 2). Replace with fresh maintenance media on day 4.

4. On day 5, aspirate maintenance media before seeding non-NRVM cells to form heterotypic cell pairs. Seed non-NRVM cells at a density of 2×10^3 cells/cm^2 (see **Note 8**). Incubate cells at 37 °C 5 % CO_2 incubator.

5. On day 6, cell pairs are ready for patch clamp recordings.

3.2 Patch Clamp

In this section, we describe the patch clamp procedures using our setup and in accordance with previously described patch clamp methodology [19]. When different equipment is used, it is advisable for the user to follow the manufacturer's instructions.

Software and equipment setup

1. Open the N_2 tank connected to the anti-vibration table.

2. Open the perfusion system to fill the perfusion chamber with Tyrode's solution (bath solution). After the chamber is filled to about 3 mm, adjust the perfusion output such that the solution level is maintained at a steady level. Make sure that the bath (ground) electrode, temperature probe, perfusion line, and suction line are all immersed in the bath solution.

3. Turn on the temperature controller, and set the bath temperature to 37 °C.

4. Turn on the micromanipulator controller.

5. Turn on Multiclamp 700B amplifier. Open Multiclamp software, and make sure that voltage clamp mode (VC) is selected.

6. Open WinWCP software. Create a new data file. Open the "Pipette seal test/Signal monitor" window, and apply a constant test pulse of 5 mV.

Patch pipette preparation

1. Install the glass pipette inside the pipette puller, and run the program. Program parameters depend on the type and size of pipette, position and shape of the heating filament, and intended use of the pipette. Follow the manufacturer's instructions to optimize the program for the desired pipette tip size and shape.

2. Use a Bunsen burner to heat and bend the pipette in the middle at about a 35° angle. Depending on the relative position of the micromanipulator to the horizontal plane, this angle should be adjusted such that after clamped to the manipulator, the pipette is oriented as vertically as possible.

3. Using the microforge setup, fire-polish the pipette tip to about 1–2 μm in inner diameter.

4. Backfill the pipette with pipette solution (*see* **Note 9**).

5. Mount the pipette onto the electrode holder. Make sure that the pipette solution makes contact with the chlorided part of the silver electrode.

Whole-cell patch clamp

The current response to a 5 mV test pulse (applied with WinWCP software) is useful to help recognize the different steps of the patch clamp procedure (Fig. 2).

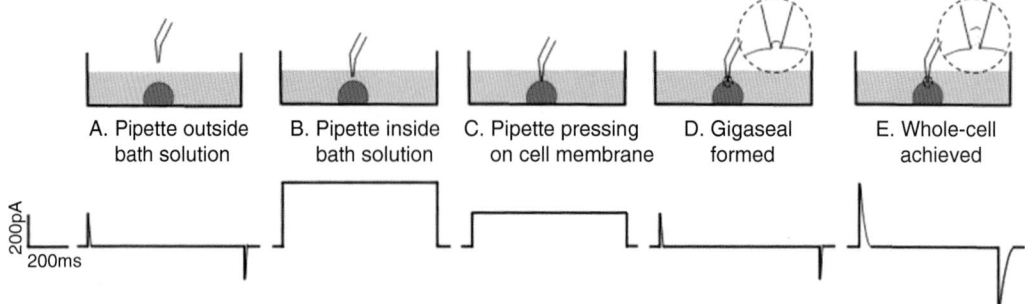

Fig. 2 Current response to a 5 mV test pulse during each step of the whole-cell patch clamp procedure

1. With the manipulator at its coarsest mode, drive the pipette tip to the objective view, and then lower it down to the bath solution (*see* **Note 10**). After the pipette reaches the bath solution, the current response shown in the WINWCP software will change to a square shape (Fig. 2b). The displayed pipette resistance should be 1–2 MΩ (*see* **Note 11**).

2. Continue adjusting the position of the pipette tip so that it locates right above the cell to be patched (*see* **Note 12**).

3. Lower the pipette slowly and gradually until initial contact is made with the cell, which is shown as a slight increase in the pipette resistance (Fig. 2c). Keep pressing down on the cell until the resistance is increased by about 1.5–2 times.

4. Close the positive pressure applied on the pipette solution. The removal of the positive pressure should induce some initial suction on the cell membrane, shown as a further increase in pipette resistance.

5. Apply a slight, constant negative pressure (preferably by mouth suction) on the cell membrane until gigaseal is formed, indicated by an increase in pipette resistance to at least 1 GΩ (Fig. 2d). A good seal is usually formed instantly, within less than 5 s of suction.

6. Apply further negative pressure to rupture the membrane patch between the pipette tip and the cell to achieve whole-cell patch clamp configuration (Fig. 2e). This is indicated by a relatively broad capacitive transient in the current response on the WINWCP window (*see* **Note 13**).

Action potential recordings

1. After whole-cell configuration is achieved, switch to current clamp mode with zero applied current (*I*=0 mode in Multiclamp 700B window). Record the displayed membrane potential. If the cardiomyocyte is at rest and not contracting spontaneously, this is its resting membrane potential.

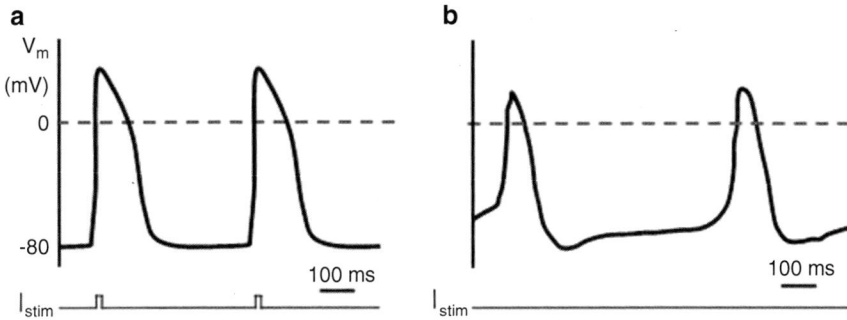

Fig. 3 Action potential traces in NRVMs triggered by a stimulus current (I_{stim}) (**a**) or in a spontaneously contracting cell (**b**)

2. To record action potentials, switch to current clamp mode (IC in Multiclamp 700B) and use the "Record" function of WinWCP. In cells without spontaneous pacemaking activity, action potentials can be elicited by applying a 10-ms current pulse at 1.1× threshold amplitude (Fig. 3a). In spontaneously contracting cells, no applied current is necessary (Fig. 3b).

Dual whole-cell voltage clamp recordings

1. With the cardiomyocyte kept in the stable whole-cell configuration, switch the manipulator to control the left-side manipulator.

2. Repeat the same procedure to get the second pipette to form gigaseal with the non-cardiac cell in the cell pair, and then break into whole-cell configuration.

3. In VC modes, hold the NRVM at 0 mV while applying a 4-s voltage pulse of various potentials from −40 to 40 mV on the non-cardiac cell.

4. For each voltage pulse, determine the gap-junctional conductance between the two cells by dividing the measured junctional current by the applied voltage difference after correcting for series resistance and membrane resistance as previously described [20].

4   Notes

1. Dust can adhere to wet photoresist, ultimately disrupting the micropattern. Make sure that gloves and surrounding region are clean. Transfer and cover wafer quickly.

2. Rapid cooling introduces stress to the photoresist–silicon bond, resulting in feature fragility. Gradual, ramped cooling is essential.

3. If white residue is seen during isopropyl alcohol rinse, return wafer to PGMEA for an addition minute and re-rinse with isopropyl alcohol. Repeat as necessary.

4. If the wafer and attached PDMS mold fail to detach from the polystyrene Petri dish, use clippers to make several cuts along the circumference of the Petri dish wall and carefully break off wall segments until the contents of the dish can easily be removed.

5. Repeat for each cover slip. Dust can adhere to the wet PDMS and interfere with micropatterned protein transfer. Make sure that gloves and surrounding region are clean. Transfer and cover cover slips quickly.

6. The pattern should gradually appear as seal is formed between the cover slip and the PDMS stamp. Applied force other than the weight of the forceps could result in loss of the pattern due to sagging, which is also visible. In such case, reduce protein transfer time (no less than 15 min) to prevent further sagging.

7. Excessive force could result in breaking the cover slip or forcing undesired stamp contact and protein transfer. An alternative way is to bend the PDMS stamp gently to "pop off" the cover slip.

8. Use a syringe to hold the NRVM-stamped cover slip in place.

9. Make sure that the needle is inserted in the pipette as close to the pipette tip as possible before injecting the solution with the syringe to prevent large bubbles. To get rid of bubbles at the tip, gently flick the pipette with your finger near the tip end until the bubbles float up.

10. A slight positive pressure of around 10 cm of H_2O should be applied on the pipette solution before lowering the pipette to the bath solution. This will help protect the pipette tip from debris at the air–solution interface.

11. If the resistance is too large (>5 MΩ), it usually means that either the pipette tip is too small or there is an air bubble at the pipette tip. Replace with new pipette in such case.

12. To locate the pipette in the field of view, it is easier to switch to a 10× objective. Move the pipette around in the x–y plane until its slightly dark, moving shadow can be recognized. Once the rough position of the pipette is identified, gradually lower the pipette while move up the focal plane of the microscope until the pipette tip comes into focus as a small circle in a dark background. Then, lower the focal plane and the pipette iteratively, making sure that the focal plane always lies between the cover slip and the pipette tip to prevent the pipette from accidentally hitting the cover slip. As the pipette tip gets close enough to

the cell such that both are recognizable (but not in focus) in the same field of view, switch to a 40× objective, and switch micromanipulator to a finer mode.

13. There are various methods that can be used to break the membrane patch to get to whole-cell mode. However, for NRVMs, applying gradual negative pressure with a 3 mm syringe seems to give the highest success rate.

Acknowledgements

We thank Dr. Robert Kirkton for reading of the manuscript. This work was supported by the National Heart, Lung, and Blood Institute of the NIH under Award Numbers R01-HL-104326 and R21-HL-106203. The content is solely the responsibility of the authors and does not necessarily represent the official views of the NIH.

References

1. Menasche P, Hagege AA, Vilquin JT et al (2003) Autologous skeletal myoblast transplantation for severe postinfarction left ventricular dysfunction. J Am Coll Cardiol 41(7):1078–1083

2. Assmus B, Schachinger V, Teupe C et al (2002) Transplantation of progenitor cells and regeneration enhancement in acute myocardial infarction (TOPCARE-AMI). Circulation 106(24):3009–3017

3. Chugh AR, Beache GM, Loughran JH et al (2012) Administration of cardiac stem cells in patients with ischemic cardiomyopathy: the SCIPIO trial: surgical aspects and interim analysis of myocardial function and viability by magnetic resonance. Circulation 126(11 Suppl 1):S54–S64. doi:10.1161/circula-tionaha.112.092627

4. Laflamme MA, Chen KY, Naumova AV et al (2007) Cardiomyocytes derived from human embryonic stem cells in pro-survival factors enhance function of infarcted rat hearts. Nat Biotechnol 25(9):1015–1024. doi:10.1038/nbt1327

5. Cho HC, Marban E (2010) Biological therapies for cardiac arrhythmias: can genes and cells replace drugs and devices? Circ Res 106(4):674–685. doi:10.1161/CIRCRESAHA.109.212936

6. Gepstein L (2010) Cell and gene therapy strategies for the treatment of postmyocardial infarction ventricular arrhythmias. Ann N Y Acad Sci 1188:32–38. doi:10.1111/j.1749-6632.2009.05080.x

7. Pedrotty DM, Klinger RY, Badie N et al (2008) Structural coupling of cardiomyocytes and noncardiomyocytes: quantitative comparisons using a novel micropatterned cell pair assay. Am J Physiol Heart Circ Physiol 295(1):H390–H400. doi:10.1152/ajpheart.91531.2007

8. McSpadden LC, Nguyen H, Bursac N (2012) Size and ionic currents of unexcitable cells coupled to cardiomyocytes distinctly modulate cardiac action potential shape and pacemaking activity in micropatterned cell pairs. Circ Arrhythm Electrophysiol 5(4):821–830. doi:10.1161/CIRCEP.111.969329

9. del Corsso C, Srinivas M, Urban-Maldonado M et al (2006) Transfection of mammalian cells with connexins and measurement of voltage sensitivity of their gap junctions. Nat Protoc 1(4):1799–1809. doi:10.1038/nprot.2006.266

10. Bursac N, Parker KK, Iravanian S et al (2002) Cardiomyocyte cultures with controlled macroscopic anisotropy: a model for functional electrophysiological studies of cardiac muscle. Circ Res 91(12):e45–e54

11. Klinger R, Bursac N (2008) Cardiac cell therapy in vitro: reproducible assays for comparing the efficacy of different donor cells. IEEE Eng Med Biol Mag 27(1):72–80. doi:10.1109/MEMB.2007.913849

12. Pedrotty DM, Klinger RY, Kirkton RD et al (2009) Cardiac fibroblast paracrine factors alter impulse conduction and ion channel expression of neonatal rat cardiomyocytes. Cardiovasc Res 83(4):688–697. doi:10.1093/cvr/cvp164

13. Mangi AA, Noiseux N, Kong D et al (2003) Mesenchymal stem cells modified with Akt prevent remodeling and restore performance of infarcted hearts. Nat Med 9(9):1195–1201. doi:10.1038/nm912

14. Taylor DA, Silvestry SC, Bishop SP et al (1997) Delivery of primary autologous skeletal myoblasts into rabbit heart by coronary infusion: a potential approach to myocardial repair. Proc Assoc Am Physicians 109(3):245–253

15. Liau B, Christoforou N, Leong KW et al (2011) Pluripotent stem cell-derived cardiac tissue patch with advanced structure and function. Biomaterials 32(35):9180–9187. doi:10.1016/j.biomaterials.2011.08.050

16. Christoforou N, Miller RA, Hill CM et al (2008) Mouse ES cell-derived cardiac precursor cells are multipotent and facilitate identification of novel cardiac genes. J Clin Invest 118(3):894–903. doi:10.1172/JCI33942

17. Kirkton RD, Bursac N (2011) Engineering biosynthetic excitable tissues from unexcitable cells for electrophysiological and cell therapy studies. Nat Commun 2:300. doi:10.1038/ncomms1302

18. Badie N, Satterwhite L, Bursac N (2009) A method to replicate the microstructure of heart tissue in vitro using DTMRI-based cell micropatterning. Ann Biomed Eng 37(12):2510–2521. doi:10.1007/s10439-009-9815-x

19. Hamill OP, Marty A, Neher E et al (1981) Improved patch-clamp techniques for high-resolution current recording from cells and cell-free membrane patches. Pflugers Arch 391(2):85–100

20. Veenstra RD, Brink PR (1992) Patch-clamp analysis of gap junctional currents. In: Stevenson B, Paul DL, Gallin W (eds) Cell-cell interactions: a practical approach. Oxford University Press, Oxford

INDEX

Milica Radisic and Lauren D. Black III (eds.), *Cardiac Tissue Engineering: Methods and Protocols*, Methods in Molecular Biology, vol. 1181, DOI 10.1007/978-1-4939-1047-2, © Springer Science+Business Media New York 2014

Printed by Printforce, the Netherlands